A Building Craft Foundation

The NVQ Construction series titles are:

WOOD OCCUPATIONS by Peter Brett
(covers all wood occupations at Level 1)

A BUILDING CRAFT FOUNDATION (2nd edition) by Peter Brett
(covers the common core units at Levels 1 and 2)

SITE CARPENTRY AND JOINERY (2nd edition) by Peter Brett
(covers the Site Carpentry units at Level 2)

BENCH JOINERY by Peter Brett
(covers the Bench Joinery units at Level 2)

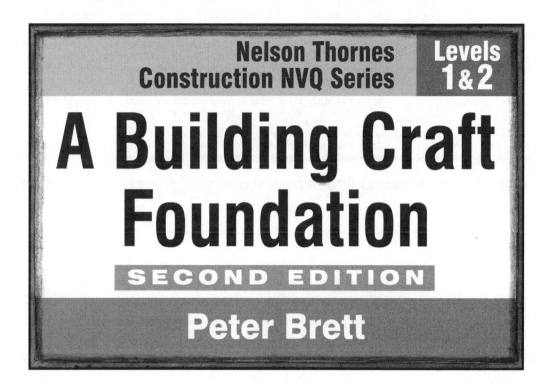

Nelson Thornes Construction NVQ Series

Levels 1&2

A Building Craft Foundation

SECOND EDITION

Peter Brett

First published in 1991 by:
Stanley Thornes (Publishers) Ltd
Second edition published in 2002 by:

Nelson Thornes Ltd
Delta Place
27 Bath Road
CHELTENHAM
GL53 7TH
United Kingdom

02 03 04 05 06 / 10 9 8 7 6 5 4 3 2 1

A catalogue record for this book is available from the British Library

ISBN 0 7487 6531 X

Page make-up by Florence Production Ltd, Stoodleigh, Devon

Printed in Great Britain by Scotprint

Contents

Acknowledgements vii

National Vocational Qualifications – NVQs ix

Introduction xi

How to use this book xiii

1. The Construction Industry **1**

Introduction to the construction industry 1
Careers 3
The building team 7
Principles of construction 15

2. Health and Safety **35**

Accidents 35
Health and safety controls 39
The Health and Safety at Work Act (HASAWA) 40
Management of H&S legislation 47
The regulations 49
Regulation of hazardous substances 62
General safety 67

3. Communications **81**

Scale 81
Drawings, symbols and abbreviations 86
Construction activity documents 94
Messages 109
Personal communications 119

4. Numerical Skills **121**

The number system 121
Types of numbers 126
Basic rules for numbers 129
Units of measurement 142
Applied mathematics 152
Statistics 157
Powers and roots of numbers 161
Angles and lines 163
Shapes and solids 167
Formulae 173
Measuring and costing materials 183

5. Scaffolding **203**

Using scaffolding 203
Types of tube scaffolding 206
Scaffold components 210
Tying in to a building 211
Other working platforms 216

6. Materials **225**

On-site provision for storage of materials 225
Storage of bulk durable building materials 231
Storage of hazardous products 237
Storage of fragile or perishable materials 239
Storage of miscellaneous materials 243

Acknowledgements

The author wishes to acknowledge the following:

The official forms included in this package are reproduced with the permission of the Controller of The Stationery Office who reserve Crown Copyright.

The £1 and 10 pence coin on page 151 are reproduced courtesy of the Royal Mint; the £5 note on the same page is reproduced courtesy of the Bank of England.

Word-square searches were kindly produced by James Brett.

This 'Building Craft Foundation' is dedicated to Matthew, Christopher and Rebecca.

National Vocational Qualifications – NVQs

The work of a skilled person in the construction industry can be divided into various tasks: build a brick wall; fix plasterboard; prepare and paint surfaces; assemble a door; hang a door, etc. These tasks along with many others are grouped into 'units of competence'. You can consider these units of competence as a menu to select from, according to your own or employers' skill requirements.

Traditional barriers to gaining a qualification such as age, length of training, mode of training, how and where skills are acquired have been removed. Individuals may acquire units of competence, in any order as and when and where they want. Units of competence are accredited individually and may be transferred to any appropriate NVQ award.

Credits for units of competence, which can be accumulated over any period of time, may be built into a full NVQ award at three levels.

NVQ Level 1 Introduction to industry, a 'foundation' common core plus occupational base skills, e.g. Wood occupations, Trowel occupations and Decorative occupations, etc.

NVQ Level 2 A set number of units of competence in a recognisable work role, e.g. Carpentry and joinery, Sitework, Benchwork and brickwork, etc.

NVQ Level 3 A more complex set of units of competence again in a recognisable work role including some work of a supervisory nature.

The Qualification and Curriculum Authority QCA is the accrediting body for NVQ qualifications in the construction industry.

The Construction Industry Training Board CITB establish standards for the units of competence and the qualification structure for the industry.

Collecting evidence

You will need to collect evidence of your satisfactory performance in each element of a unit of competence.

This evidence can be either:

- **Work-based**. This will be evidence from your employers and supervisors, etc. that confirms you have demonstrated the full range of practical skills required for a unit. This should be supported by drawings, photographs and other associated documentation used/produced as part of the activity.

- **Simulation**. Where work-based evidence is not available or appropriate, simulated activities may be undertaken in a training or assessment environment. Again, supporting documentation will be required as with work-based evidence, so that the total provides sufficient evidence to infer that you can repeat the skills competently in a work-based environment.
- **A combination** of **work-based and simulation evidence**. Again with supporting documentation to infer competence.

Introduction

This book you are about to start is one of the Construction NVQ Series, and is aimed at those working, intending to work or undergoing training in the Construction Industry. It covers the NVQ Construction mandatory core units at levels 1 & 2 for **trainees from all crafts**, and is intended as an introduction to the industry and its 'house keeping practices'. Once successfully completed it forms part of an NVQ occupational award.

All units should be undertaken in their entirety by the new entrant. This will provide a foundation on which other occupational-specific Units of Competence can be built. As an alternative, persons with prior achievement may choose to undertake individual foundation units as refreshers, to support other level Units of Competence according to their needs.

The following six mandatory cores at NVQ Levels 1 & 2 are common to all construction craft options:

Level 1

Unit No. 01	Load and unload materials
Unit No. 02	Contribute to erecting and dismantling working platforms
Unit No. 03	Contribute to maintaining work relationships

Level 2

Unit No. 07	Store resources ready for work
Unit No. 08	Erect and dismantle working platforms
Unit No. 09	Contribute to efficient working practices

All of the mandatory core units are fully covered in this book.

How to use this book

This is a self-study package designed to be supported by:

- tutor reinforcement and guidance
- group discussion
- films, slides and videos
- text books
- practical learning tasks.

You should read/work through each section of a unit, one at a time as required. Discuss its content with your group, tutor, or friends wherever possible. Attempt to answer the *Questions for you* in that section. Progressively read through all the sections, discussing them and answering the questions as you go. At the same time you should be either working on the matching practical learning task/assessment set by your college/training centre, or alternatively be carrying out the practical competence and recording its successful completion in the workplace.

This process is intended to aid learning and enable you to evaluate your understanding of the particular section and to check your progress through the units and entire package. Where you are unable to answer a question, further reading and discussion of the section is required.

Throughout this learning package 'Harry' the general foreman and his thoughts will prompt you to undertake an activity or task.

JUST FOLLOW MY THOUGHTS TO COMPLETE THIS LEARNING PACKAGE SUCCESSFULLY

The *Questions for you* in this package are either multiple choice or short answer questions.

Multiple-choice questions consist of a statement or question followed by four possible answers. Only *one* answer is correct, the others are distractors. Your response is recorded by filling in the line under the appropriate letter.

Example

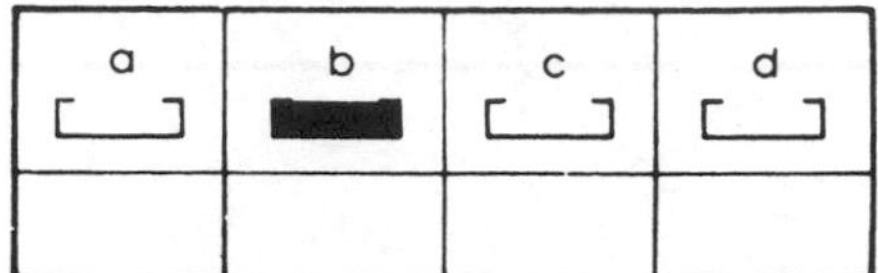

This indicates that you have selected (b) as the answer.
If after consideration you want to change your mind, fill in the box under your first answer and then fill in the line under the new letter.

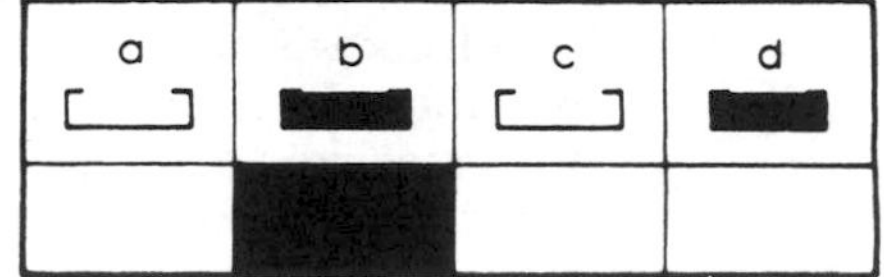

This changes the answer from (b) to (d).

Short-answer questions consist of a task to which a short written answer is required. The length will vary depending on the 'doing' word in the task, *Name* or *List* normally require one or two words for each item, *State, Define, Describe* or *Explain* will require a short sentence or two. In addition, sketches can be added to any written answer to aid clarification.

Example

Name the architect's on-site representative.

Typical answer: The clerk of works.

Example

Define the term 'the building team'.

Typical answer: The team of professionals who work together to produce the required building or structure. Consists of the following parties: client, architect, quantity surveyor, specialist engineers, clerk of works, local authority health and safety inspector, building contractor, sub-contractors and material suppliers.

Example

Make a sketch to show the difference in size between a brick and a block.

Typical answer:

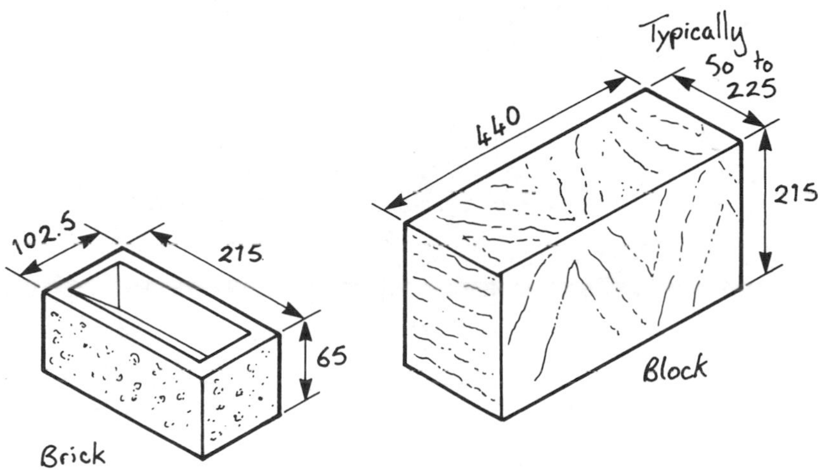

In addition this package also contains *Learning tasks*. Follow the instructions given with each exercise. They are intended to reinforce the work undertaken in this package. They give you the opportunity to use your newly acquired awareness and skills.

In common with NVQ knowledge and understanding assessments, the learning exercises in this package may be attempted orally. You can simply tell someone your answer, point to a diagram, indicate a part in a learning pack or textbook, or make sketches, etc.

1 The construction industry

Introduction to the construction industry

The 'construction industry' in its widest sense covers the following **four** main areas of work. They all work to the same purpose, that is the provision of shelter and other services to the population as a whole.

Building

This is the construction, maintenance and adaptation of buildings ranging from office blocks, industrial complexes and shopping centres to schools, hospitals, recreation centres and homes. Included in this area are the specialist builders who concentrate on the provision of one skill, e.g. glazing, cladding, tiling and roofing, etc.

Civil engineering

This is the construction and maintenance of public works such as roads, railways, bridges, airports, docks, sewers, etc.

Mechanical engineering

This is the installation, commissioning and maintenance of lifts, escalators and heating, ventilation, refrigeration, sprinkler and plumbing systems, etc.

Electrical engineering

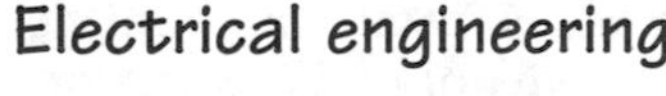

This is the installation, commissioning and maintenance of various electrical and electronic devices.

Range of work

Contained within these four main work areas are a variety of specific job functions and careers giving employment to 1.9 million people. About 65 per cent are directly employed and the remaining 35 per cent being self-employed. Approximately 1 in 5 of Britain's self-employed workers are engaged in construction activities.

The value of the work carried out in the UK annually is £65 billion, accounting for about 8 per cent of the whole country's gross domestic product (GDP). At the time of writing £7.5 billion of this is spent by government departments and agencies, with plans to increase this up to £19 billion per annum over the next three years. Examples of this work can be seen almost everywhere. They range from basic housing and maintenance through to multi-million pound developments such as airports, motorways, bridges, tunnels, business and retail parks (factories, offices and shops) and new town developments.

About 60 per cent of the annual total can be attributed to new work and 40 per cent to maintenance, refurbishment and renovation work. Of this about 37 per cent is spent on housing, with the remaining 63 per cent being spent on other works.

New: a building that has just been built.

Maintenance: the **repairs** undertaken to a building and its services in order to keep it at an acceptable standard so that it may fulfill its function.

Refurbishment: to bring an existing building up to standard, or make it suitable for a new use by **renovation**.

Restoration: to bring an existing building back to its original condition.

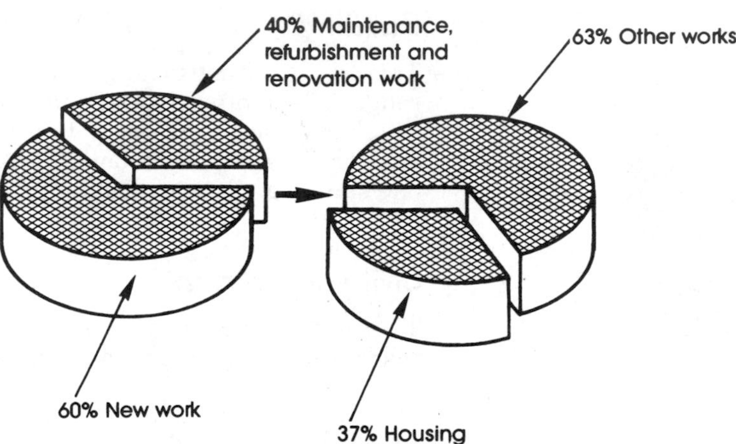

Although a major employer, the industry is possibly one of the most fragmented. Firms vary widely in size, from the small-scale local builder employing perhaps two or three people, to the international companies employing thousands. Of the total workforce, the private sector contractors (privately owned firms who undertake building work in order to make a profit) account for about 88 per cent of those employed. The remaining 12 per cent are employed by the building departments of public sector authorities (council building departments for housing, education and hospitals, etc, often termed direct or public works departments), a large proportion of whose work is concerned with repairs, maintenance and restoration.

Statistics show there are around 75 000 contracting firms who employ two or more people, and whilst there are a number of very large firms the vast majority of firms in the industry (92 per cent) employ fewer than 25 people.

Firms are classified by size into three groups according to the number of employees employed:

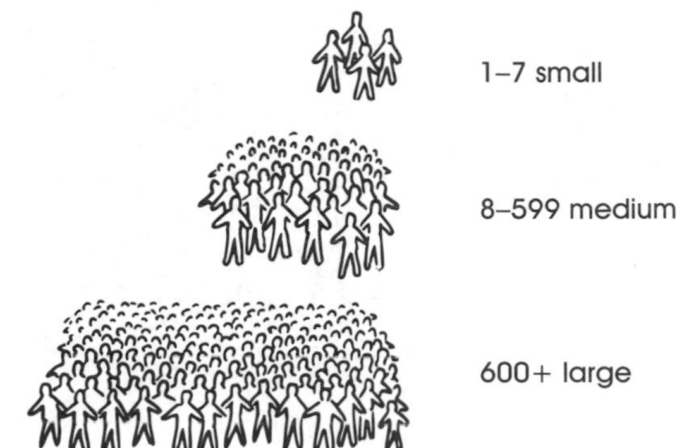

Careers

The construction industry offers employment in four distinct career areas: professional, technician, building crafts, building operatives.

Professional

These graduate-entry positions include the following:

Architect – designs and supervises the construction of buildings.

Engineer – can be either a civil engineer (concerned with roads and railways, etc.), structural engineer (concerned with the structural aspects of a building's design) or a service engineer who plans building-service systems.

Surveyor – can be either a land surveyor (who determines positions for buildings, roads and bridges, etc.), a building surveyor (who is concerned with the administration of maintenance and adaptation works as well as new buildings), a quantity surveyor (who measures and describes building works using information contained on architects' drawings; in addition they also prepare valuations of works in progress).

Technician

This is the link level in the industry between the professional and craft areas. The main job functions of technicians are as follows:

Architectural technician – involved with the interpretation and presentation of the architect's design information, into a form suitable for use by the builder.

Building technician – involved with the estimating, purchasing, site surveying, site management and documentation of building works.

Building surveying technician – may specialize in building maintenance, building control or structural surveys, etc.

Quantity surveying technician – calculates costs and payments for building works.

Building crafts

The building crafts involve the skilled operatives who work with specific materials and actually undertake the physical tasks of constructing a building. The main examples are as follows:

Bricklayer – works with bricks and mortar to construct all types of walling, also concerned with maintenance and adaptation of existing works.

Carpenter and/or joiner – works with timber, other allied materials, metal/plastic items and ironmongery. They make, fix and repair all timber components in buildings. Carpenters work on building sites, whereas joiners work mainly in a workshop at the bench.

Electrician – works with metals, plastics, wire and cables, and installs and maintains electrical systems.

Formworker – works with timber, metal and plastic, etc. to produce a structure that supports and shapes wet concrete until it has become self-supporting.

Painter and decorator – works with paint, paper, fabrics and fillers, to decorate or re-decorate new and existing works; they sometimes glaze windows and carry out sign writing.

Plasterer – works with plaster, cement mixes, plasterboard and expanded metal, to finish walls, ceilings and floors; also makes and fixes plaster decorations.

Plumber – works with metals, plastics and ceramics; installs tanks, baths, toilets, sinks, basins, rainwater goods, boilers, radiators, and gas appliances; also cuts and fixes sheet-metal roof covering and flashing and sometimes glazing; also maintains existing works.

Roof slater and tiler – works with felt, timber, metals, mortar and a wide variety of slates and tiles; covers new or existing pitched roofs with slates or tiles; also maintains existing works.

Shopfitter – works with timber, metal, glass and plastics, etc.; makes and installs shop fronts and interiors, also for banks, hotels, offices and restaurants.

Stonemason – works with stone and mortar; 'bankers' cut and smooth stone while 'fixers' erect prepared stones.

Woodworking machinist – operates a wide range of woodworking machines; prepares timber for the production of timber-building components.

Building operatives – Two main types are employed on site:

The *general building operative* uses various items of plant, e.g. hand tools, power tools, compressors and concreting equipment, etc.; mixes concrete, mortar and plaster; lays drainage, kerb stones and concrete, etc.; off-loads materials and transports around site; also generally assists the work of craft operatives.

The *specialist building operative* carries out specialist building operations, e.g. ceiling fixer, dry liner, glazier, mastic asphalter, built-up felt roofer, plant mechanic, roof sheeter and cladder, scaffolder, wall and floor tiler, etc.

TRY AND ANSWER THESE

Questions for you

1. Name **THREE** of the main areas of work which make up the building industry.

____

____

____

2. State the difference between private and public sector building work.

____

____

____

____

3. Name **THREE** examples of building work.

____

____

____

TRY AND
ANSWER
THESE

4. Name **THREE** examples of civil engineering work.

5. State the difference between maintenance, refurbishment and restoration.

6. State the main purpose of the construction industry.

7. State the classification of a contractor with 100 employees.

WELL, HOW
DID YOU
DO?
WORK
THROUGH THE SECTION
AGAIN IF YOU HAD
ANY PROBLEMS

The building team

The construction of a building is a complex process, which requires a team of professionals working together to produce the desired results. This team of professionals, which is collectively known as the building team, is a combination of the following parties:

Client
Architect
Quantity surveyor
Specialist engineers
Clerk of works
Local authority
Health and safety inspector
Building contractors
Sub-contractors
Suppliers

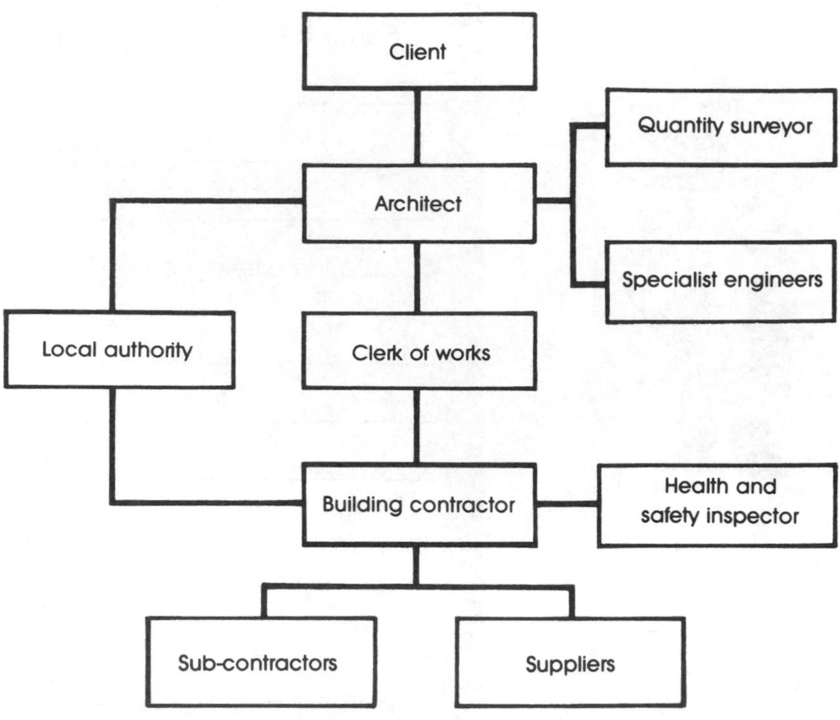

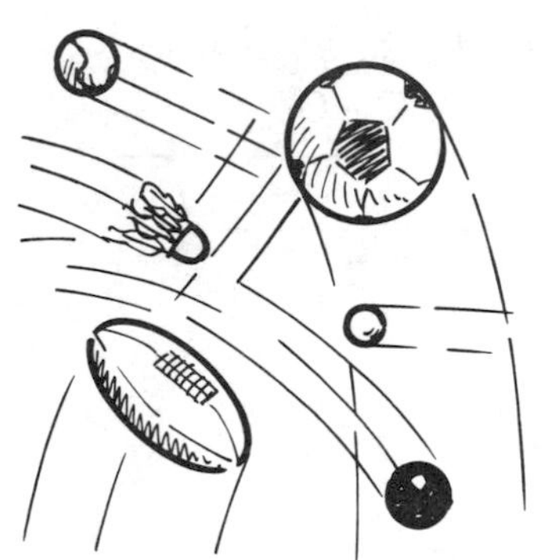

Client

This is the person or persons who have an actual need for building work e.g. the construction of a new house, office block, factory or extensions repairs and alterations to existing buildings. The client is the most important member of the building team, without them the work would simply not exist. He/she is responsible for the overall financing of the work and in effect employs either directly or indirectly the entire team. The type of client can vary from a single person to a very large organisation, for example:

Private individual
Association
Partnership
Public company
Local authority
Nationalised industry
Statutory undertaking
Government department.

Architect

The architect is the client's agent and is considered to be the leader of the building team. The role of an architect is to interpret the client's requirements, translate them into a building form and generally supervise all aspects of the work until it is completed. All architects must be registered with the Architects' Registration Council, the majority of them also being members of the Royal Institute of British Architects, using the designatory letters RIBA.

Quantity surveyor

In effect, the quantity surveyor or QS as they are often termed, is the client's building economic consultant or accountant. This specialist surveyor advises during the design stage as to how the building may be constructed within the client's budget, and measures the quantity of labour and materials necessary to complete the building work from drawings and other information prepared and supplied by the architect. These quantities are incorporated into a document known as the **bill of quantities** which is used by the building contractors when pricing the building work. During the contract, the quantity surveyor will measure and prepare valuations of the work carried out to date to enable interim payments to be made to the building contractor. At the end of the building contract they will prepare the final account for presentation to the client. In addition, the quantity surveyor will advise the architect on the cost of additional works or variations.

Specialist engineers

These are engaged as part of the design team to assist the architect in the design of the building within their specialist fields, e.g. civil engineer, structural engineer, service engineer.

They will prepare drawings and calculations to enable specialist contractors to quote for these areas of work. In addition, during the contract the specialist engineers will make regular inspections to ensure the installation is carried out in accordance with the design.

Clerk of works

The clerk of works is appointed by the architect/client to act as their on-site representative. On large contracts they will be resident on-site whilst on smaller ones they will only visit periodically. The clerk of works or COW is an 'inspector of works' and as such will ensure that the contractor carries out the work in accordance with the drawings and other contract documents. This includes inspecting both the standard of workmanship and the quality of materials. The COW will make regular reports back to the architect, keep a diary in case of disputes, make a daily record of the weather, and personnel employed on-site and any stoppage. They also agree general matters directly with the building contractor although these must be confirmed by the architect to be valid.

The local authority

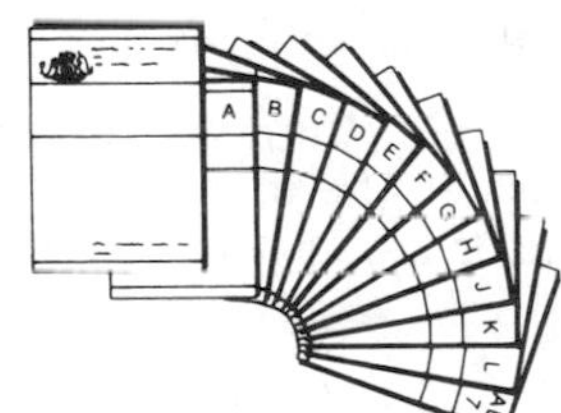

The local authority normally has the responsibility of ensuring that proposed building works conform to the requirements of relevant planning and building legislation. For this purpose, they employ planning officers and building control officers to approve and inspect building work. In some areas they are called building inspectors or district surveyors (DS).

Health and safety inspector

The health and safety inspector (also known as the factory inspector) is employed by the Health and Safety Executive. It is the inspector's duty to ensure that the government legislation concerning health and safety is fully implemented by the building contractor.

Building contractor

The building contractor enters into a contract with the client to carry out, in accordance with the contract documents, certain building works. Each contractor will develop their own method and procedures for tendering and carrying out building work which in turn, together with the size of the contract, will determine the personnel required.

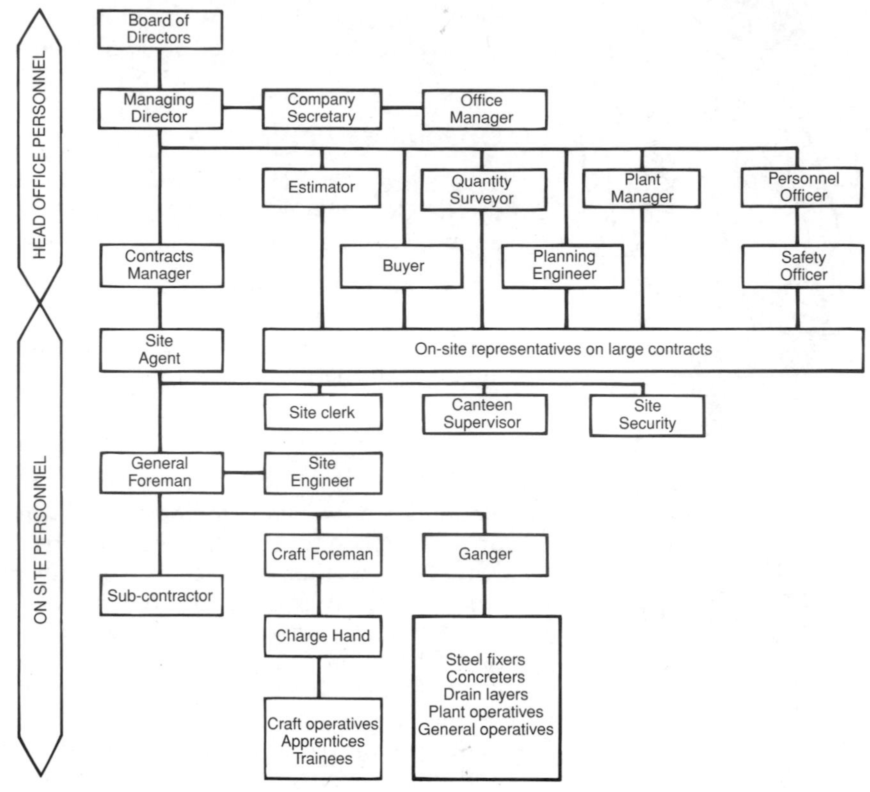

A typical contractor's organisation structure

Estimator – arrives at an overall cost for carrying out a building contract. In order to arrive at the overall cost they will break down each item contained in the bill of quantities into its constituent parts (labour, materials and plant) and apply a rate to each, representing the amount it will cost the contractor to complete the item. Added to the total cost of all items will be a percentage for overheads (head/site office costs, site management/administration salaries) and profit.

Buyer – responsible for the purchase of materials; they will obtain quotations, negotiate the best possible terms, order the materials and ensure that they arrive on-site at the required time, in the required quantity and quality.

Building contractor's quantity surveyor – the building contractor's building economist; they will measure and evaluate the building work carried out each month including the work of any subcontractors. An interim valuation is prepared by them on the basis of these measurements and passed on to the client for payment. They are also responsible for preparing interim costings to see whether or not the contract is within budget; finally they will prepare and agree the final accounts on completion of the contract.

Planning engineer – responsible for the pre-contract planning of the building project. It is their role to plan the work in such a way as to ensure the most efficient/economical use of labour, materials, plant and equipment. Within their specialist field of work planning engineers are often supported by a work study engineer (to examine various building operations to increase productivity) and a bonus surveyor (to operate an incentive scheme which is also aimed at increasing productivity by awarding operatives additional money for work completed over a basic target).

Plant manager – responsible for all items of mechanical plant (machines and power tools) used by the building contractor. At the request of the contracts manager/site agent they will supply from stock, purchase or hire, the most suitable plant item to carry out a specific task. The plant manager is also responsible for the maintenance of plant items and the training of operatives who use them.

Safety officer – responsible to senior management for all aspects of health and safety. They advise on all health and safety matters, carry out safety inspections, keep safety records, investigate accidents and arrange staff safety training.

Note: Each of the head-office personnel previously mentioned is the leader of a specialist service section and depending on the size of the firm will employ one or more technicians for assistance. On very large contracts they may also have a representative resident on-site.

Contracts manager – the supervisor/coordinator of the site's management team, on a number of contracts. The contracts manager has an overall responsibility for planning, management and building operations. They will liaise between the head office staff and the site agents on the contracts for which they are responsible.

Site agent – also known as the site manager or project manager, the site agent is the building contractor's resident on-site representative and leader of the site work force. They are directly responsible to the contracts manager for the day-to-day planning, management and building operations.

General foreman – works under the site agent and is responsible for co-ordinating the work of the craft foreman, ganger and subcontractors. They will also advise the site agent on constructional problems, liaise with the clerk of works and may also be responsible for the day-to-day employing and dismissing of operatives. On smaller contracts which may not require a site agent the general foreman will have total responsibility for the site.

Site engineer – sometimes called the surveyor, works alongside the general foreman. They are responsible for ensuring that the building is the correct size and in the right place. They will set out and check the line, level and vertical (plumb) of the building during its construction.

Craft foreman – works under the general foreman to organise and supervise the work of a specific craft, e.g. foreman bricklayer and foreman carpenter.

Ganger – like the craft foreman, the ganger also works under the general foreman but is responsible for the organisation and supervision of the general building operatives.

Chargehand – on large contracts employing a large number of craft operatives in each craft (normally bricklayers and carpenters), chargehands are often appointed to assist the craft foreman and supervise a subsection of the work. For example, a foreman carpenter may have chargehands to supervise the carcassing team (floor joists and roofs); the first fixing team (flooring, frames and studwork); the second fixing team (doors, skirting, architraves and joinery fitments). Chargehands are often known as working foremen because, in addition to supervising their small team, they also carry out the skilled physical work of their craft.

Operative – the person who carries out the actual physical building work. Operatives can be divided into two main groups:

Craft operatives are the skilled craftsmen who perform specialist tasks with a range of materials, e.g. bricklayer, carpenter, electrician, painter, plasterer and plumber.

Building operatives are further subdivided into general building operatives, who mix concrete, lay drains, off-load material and assist craft operatives; and specialist building operatives, e.g. ceiling fixer, glazier, plant mechanic and scaffolder.

Site clerk – responsible for all site administrative duties and the control of materials on site. They will record the arrival and departure of all site personnel, prepare wage sheets for head office, record the delivery and transfer of plant items, record and check delivery of materials and note their ultimate distribution (assisted by a storekeeper).

The sub-contractor

The building contractor may call upon a specialist firm to carry out a specific part of the building work; for this they will enter into a subcontract, hence the term sub-contractor. Building contractors generally sub-contract work such as structural steelwork, formwork, mechanical services, electrical installations, plastering, tiling and often painting. However, at certain times they may also sub-contract the major crafts of bricklaying and carpentry. Sub-contractors may be labour-only (where they contract to fit the building contractors material), or they may contract to supply and fix their own material. Architects can name or nominate a specific sub-contractor in the contract documents and this sub-contract must be used. They are then known as nominated sub-contractors.

The suppliers

Building materials, equipment and plant are supplied by a wide range of merchants, manufacturers and hirers. The building contractor will negotiate with these to supply their goods in the required quantity and quality, at the agreed price, and finally in accordance with the building contractor's delivery requirements. Architects may nominate specific suppliers who must be used and are therefore termed nominated suppliers.

Questions for you

8. Name and describe the role of **FOUR** members of the building team.

9. List the persons who would be employed in the design of a building.

10. Name the person who is responsible for checking the standard of work on behalf of the client on site.

11. Name the person who is responsible for protection against injury on site.

12. Name **THREE** craft operatives and give an example of the work **EACH** carries out.

TRY AND ANSWER THESE

13. Explain the term 'general operative'.

14. Explain the term 'sub-contractor'.

15. A specification for building works is normally prepared for a client by the:
(a) building control officer
(b) clerk of works
(c) estimator
(d) quantity surveyor

a	b	c	d

16. Name the job title of the person who controls a number of building contracts.

17. List ten occupations of people employed by a building contractor.

WELL, HOW DID YOU DO?
WORK THROUGH THE SECTION AGAIN IF YOU HAD ANY PROBLEMS

Principles of construction

Types of building

A structure is defined as an organised combination of connected parts (elements) which are constructed to perform a required function, e.g. a bridge. A building takes this idea a step further and is used to define structures that enclose space using an external envelope.

The external envelope is simply the walls or covering material which provide the desired internal conditions for the building's occupants with regard to: security, safety, privacy, warmth, light and ventilation, etc.

Buildings are classified into three main categories according to their height.

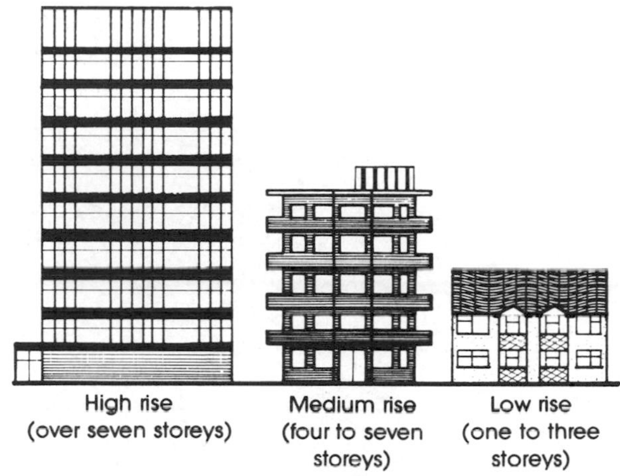

These categories are further subdivided into a wide variety of basic shapes, styles and groupings, for example:

Detached – a building that is unconnected with adjacent ones.

Semi-detached – a building which is joined to one adjacent building but is detached from any other. It will share one dividing or party wall.

Terraced – a row of three or more adjoining buildings, the inner ones of which will share two party walls.

Another method of categorising buildings, which is used in statutory regulations, groups buildings according to their purpose:

Detached

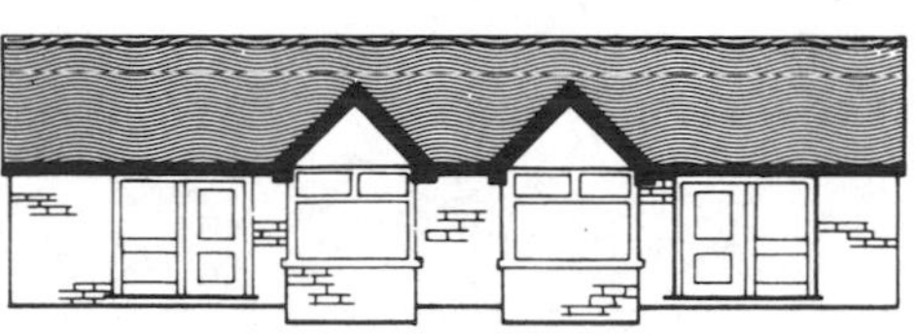

Semi-detached

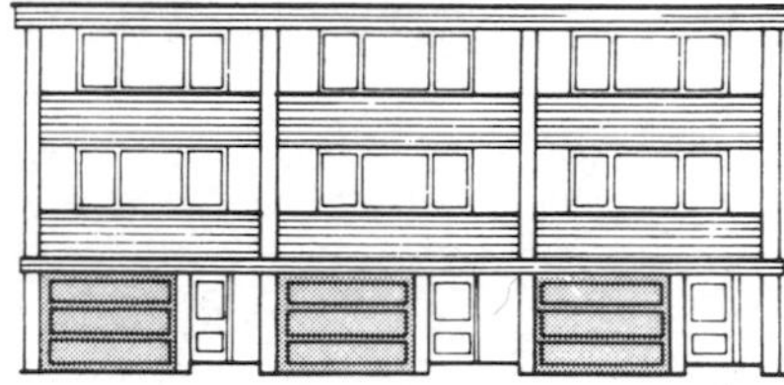

Terraced

Dwellings and premises: Purpose groups

Main category	*Purpose group*	*Intended use*
Residential	*Dwelling house* (not a flat or maisonette)	Private dwelling house
	Flat (including a maisonette)	Self-contained dwelling not being a house
	Institutional	Hospitals, schools and homes used as living accommodation for persons suffering from disabilities owing to illness, old age, physical or mental disorders and those under five years old, where these persons sleep on the premises
	Other residential	Residential accommodation not included in previous groups, e.g. hotels, boarding houses, and hostels, etc.
Non-residential	*Assembly*	Public building or assembly building where people meet for social, recreational, or business activities (not office, shop or industrial)
	Office	All premises used for administration procedures, e.g. clerical, drawing, publishing and banking, etc.
	Shop (retail outlet)	All premises used for the retail sale of goods or services, including restaurants, public houses, cafes, hairdressers and hire or repair outlets
	Industrial	All premises defined as a factory in Section 175 of the Factories Act (1961), not including slaughter houses etc.
	Other non-residential	All places used for the deposit or storage of goods, the parking of vehicles and other premises not covered in the previous non-residential groups

Structural form

Solid structures – also known as mass wall construction. Walls are constructed of either brickwork, blockwork or concrete. They form a stable box-like structure, but are normally limited to low-rise, short-span buildings.

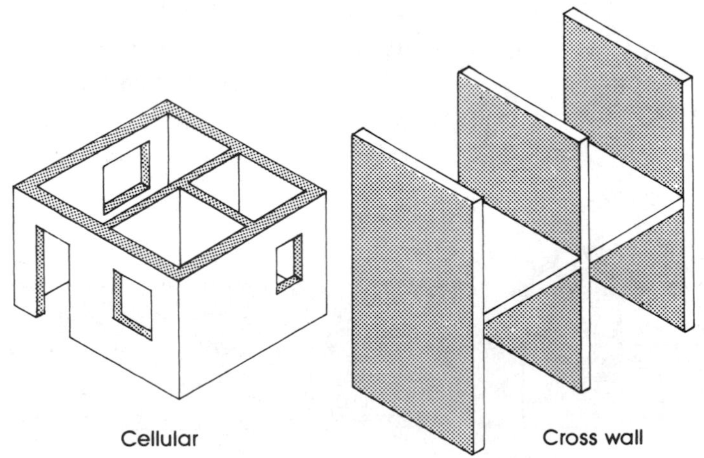

Framed structures – also termed skeleton construction, this consists of an interconnected framework of members having a supporting function. The protecting external envelope is provided by either external cladding or infill walls. Frames made from steel, concrete or timber are often pre-made in a factory as separate units, which are simply and speedily erected on site. Framed construction is suitable for a wide range of buildings and civil engineering structures from low to high rise.

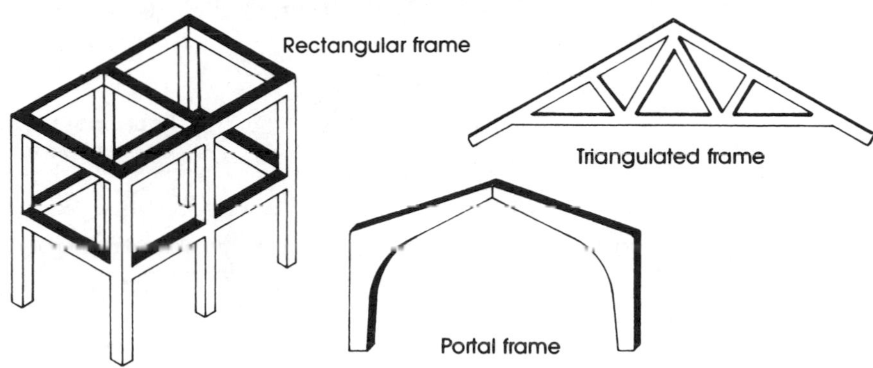

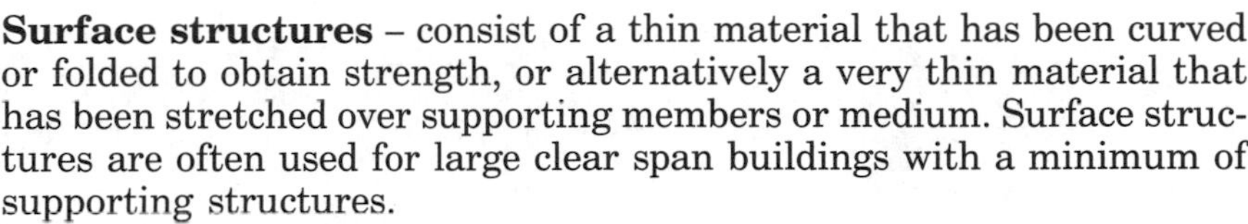

Surface structures – consist of a thin material that has been curved or folded to obtain strength, or alternatively a very thin material that has been stretched over supporting members or medium. Surface structures are often used for large clear span buildings with a minimum of supporting structures.

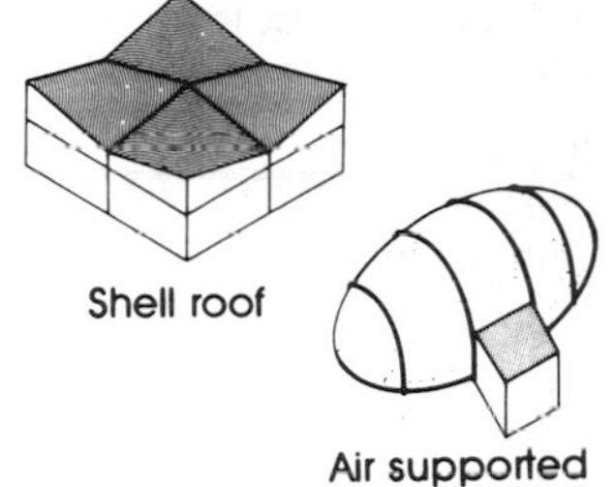

Surface structures

Structural parts

All structures consist of two main parts: that below ground and that above ground.

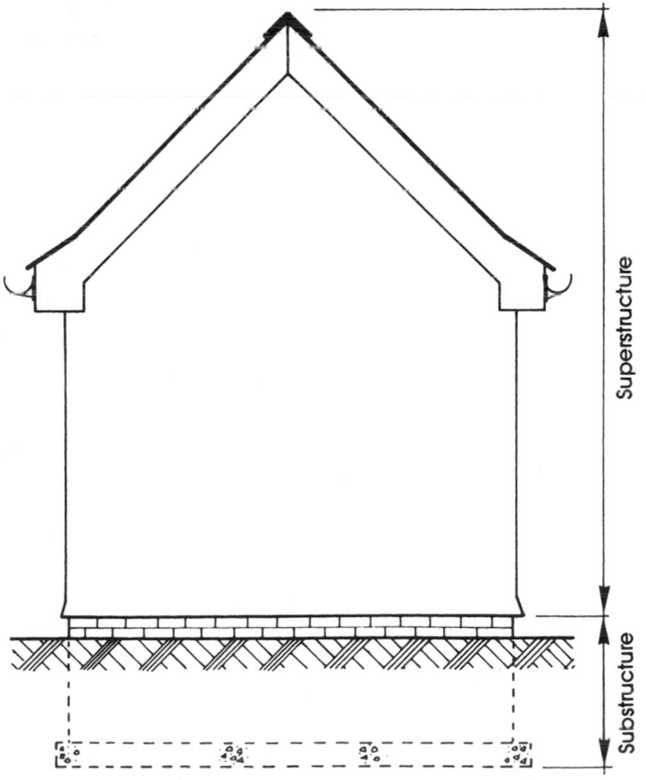

Substructure – all of the structure below ground and that up to and including the ground floor slab and damp-proof course. Its purpose is to receive the loads from the main building superstructure and its contents and transfer them safely down to a suitable load-bearing layer of ground.

Superstructure – all of the structure above the substructure both internally and externally. Its purpose is to enclose and divide space, and transfer loads safely on to the substructure.

STUDY THESE DIAGRAMS
LOADS

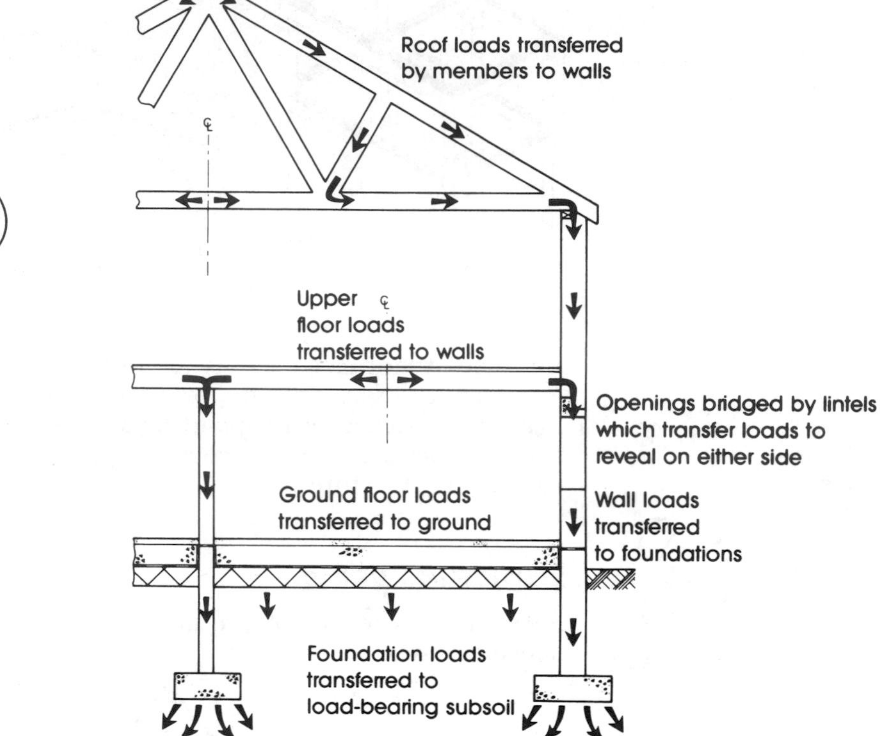
Roof loads transferred by members to walls
Upper floor loads transferred to walls
Openings bridged by lintels which transfer loads to reveal on either side
Ground floor loads transferred to ground
Wall loads transferred to foundations
Foundation loads transferred to load-bearing subsoil

Elements

A constructional part of the sub- or superstructure of a building having its own functional requirements. These include foundations, walls, floors, roofs, stairs and structural framework.

Primary elements – the main supporting, enclosing or protection elements of a building. Also those that divide space and provide floor-to-floor access.

Foundations are a primary element that transfer the loads of a structure safely on to the ground.

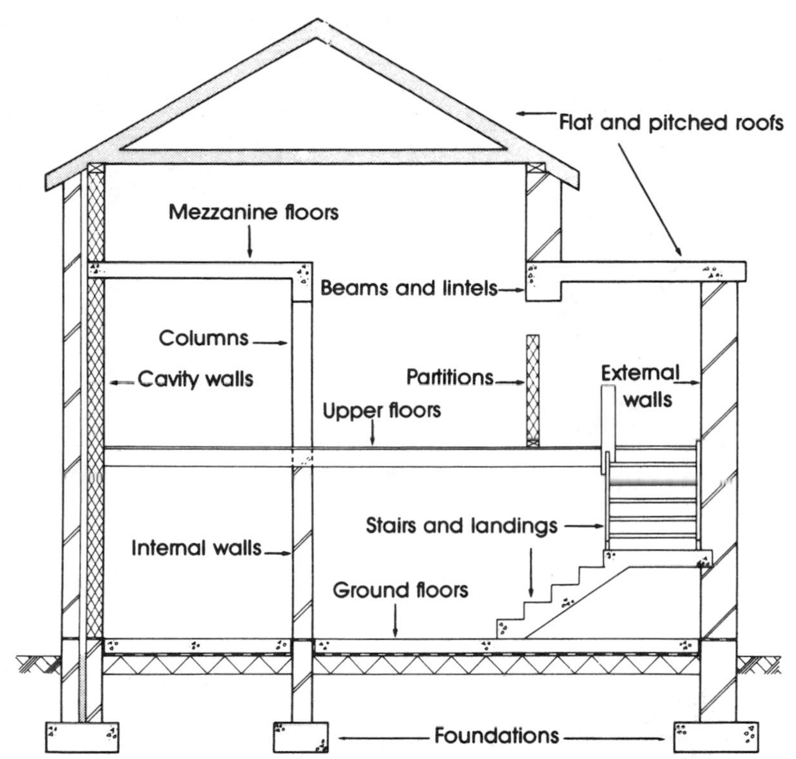

Primary elements

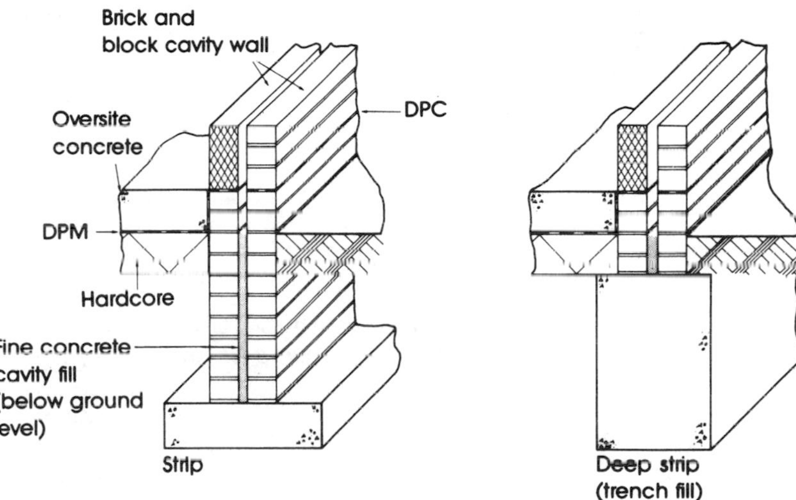

Strip foundations

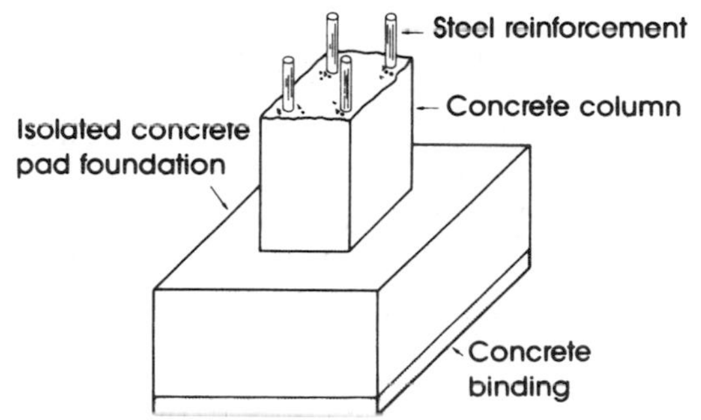

Pad foundation

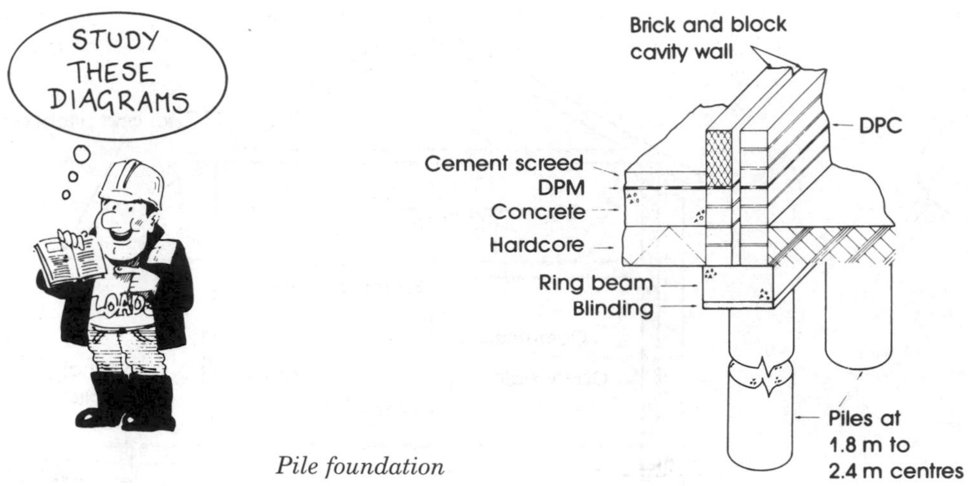

Pile foundation

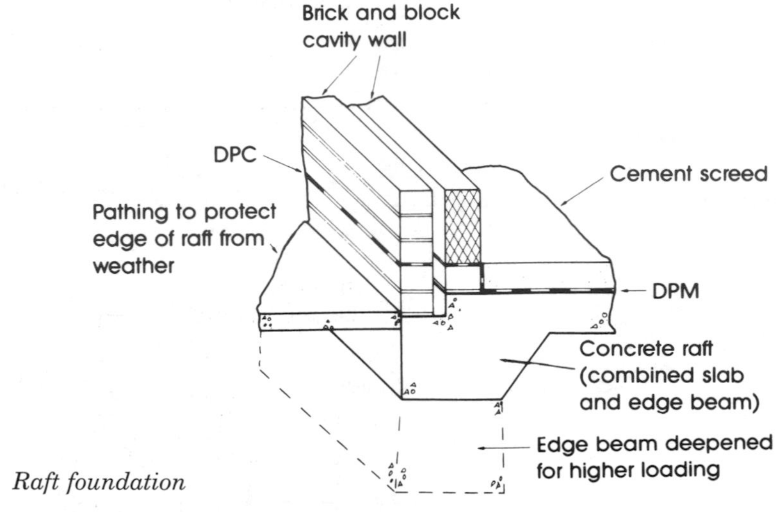

Raft foundation

Walls are the vertical and dividing elements of a building. They may be load-bearing or simply to divide space (non-load-bearing).

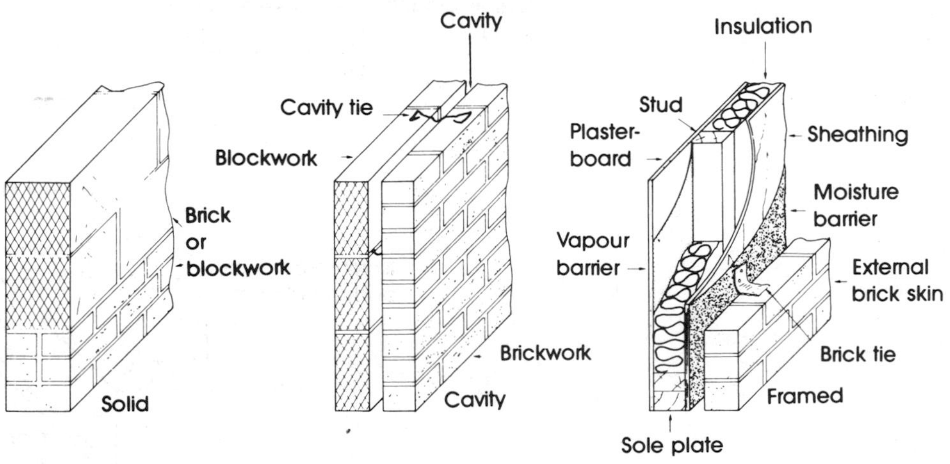

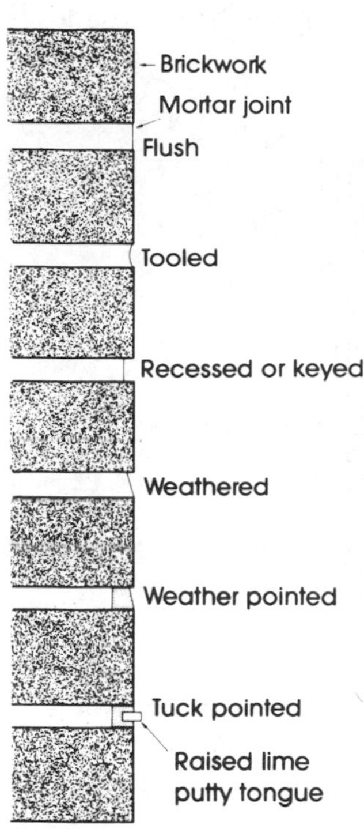

Wall jointing and pointing

Openings in walls for doors and windows are spanned by steel or concrete lintels. These bridge the opening and transfer loads to the reveal on either side.

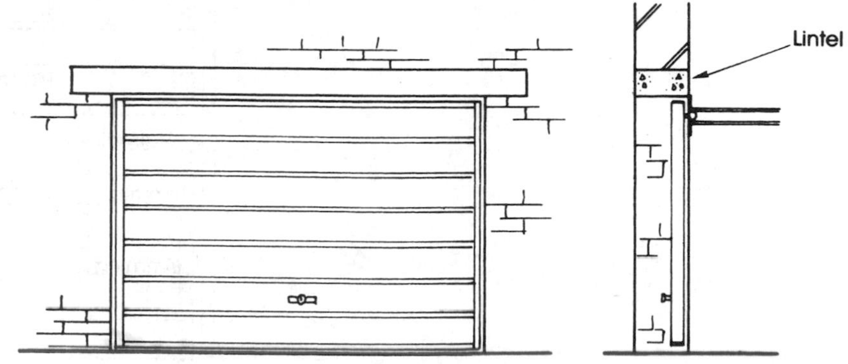

Floors are the ground or upper levels in a building which provide an acceptable surface for walking, living and working.

Ground floors

Upper floors

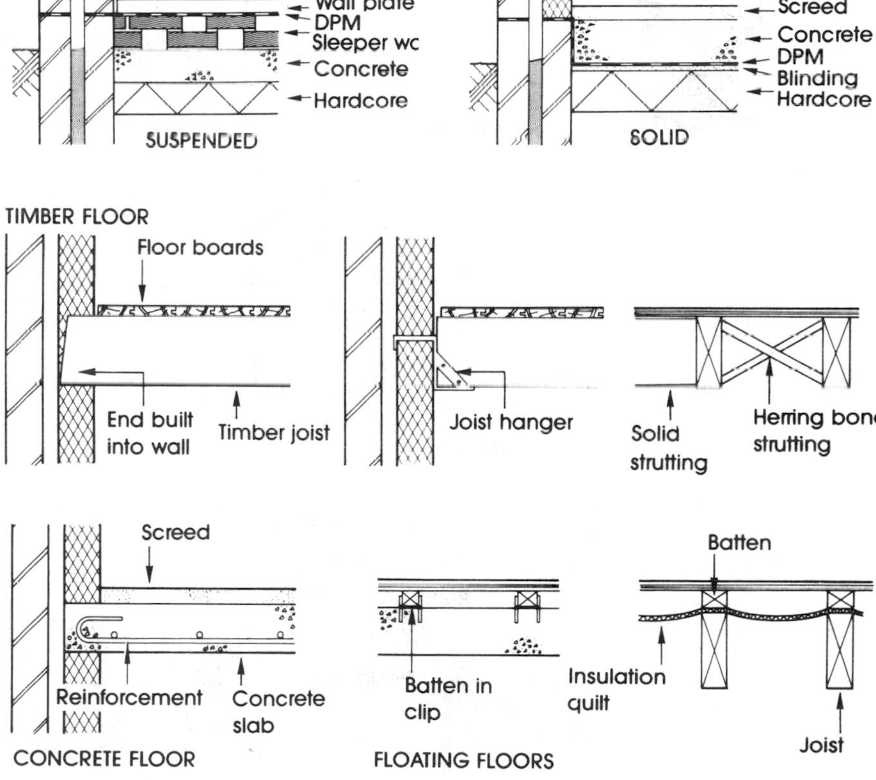

Where openings for stairs etc. occur in timber upper floors, the joists have to be framed or trimmed

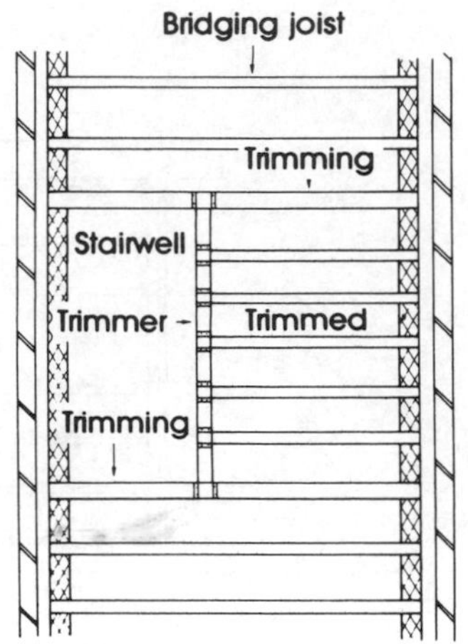

Roofs are the uppermost part of a building that span the walls to provide weathering and insulation. Mainly flat or pitched but variously named according to shape.

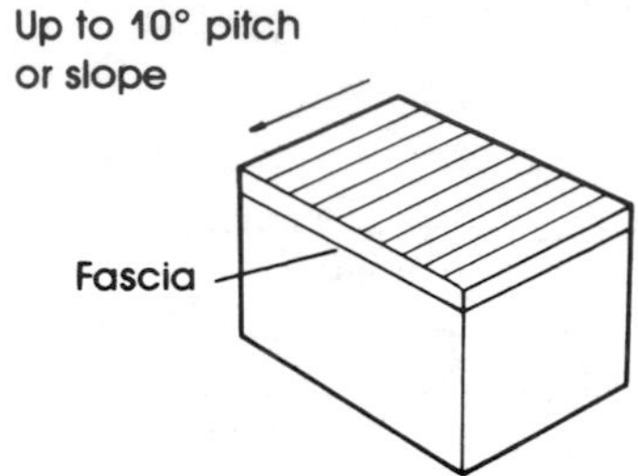

Flat roof

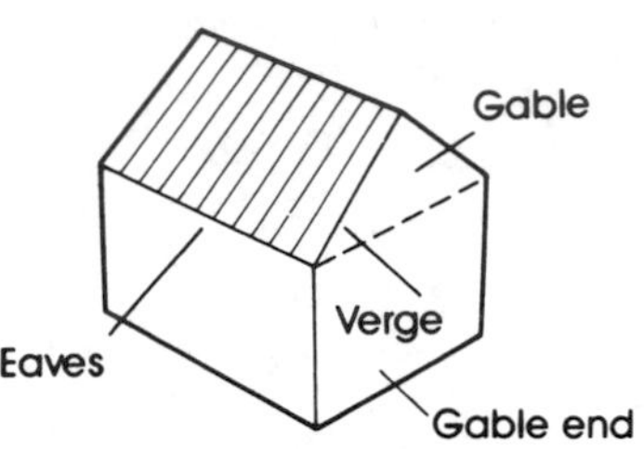

Gable roof

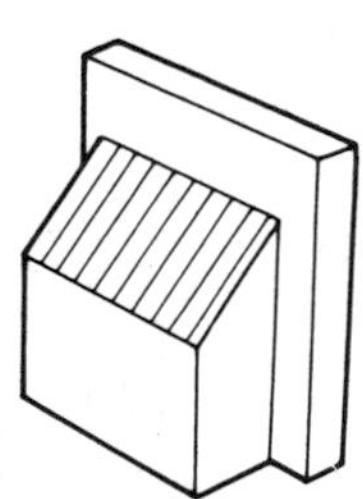

Lean-to roof

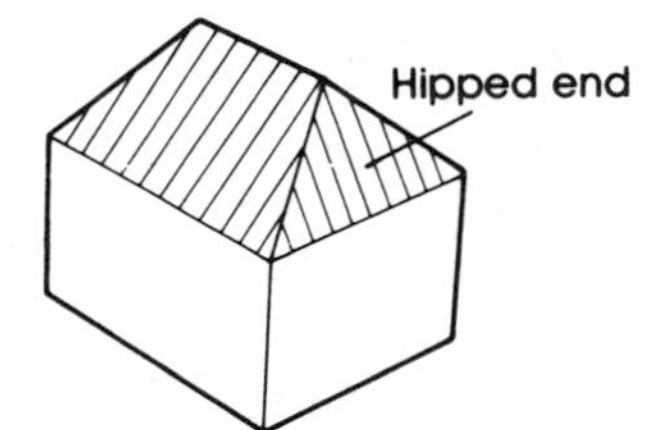

Hipped roof

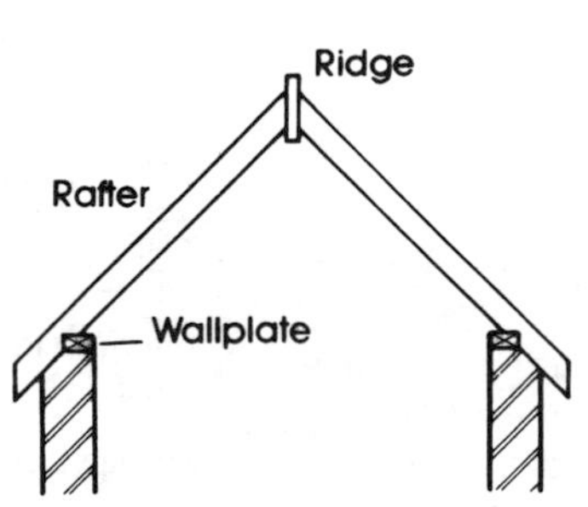

Single roof

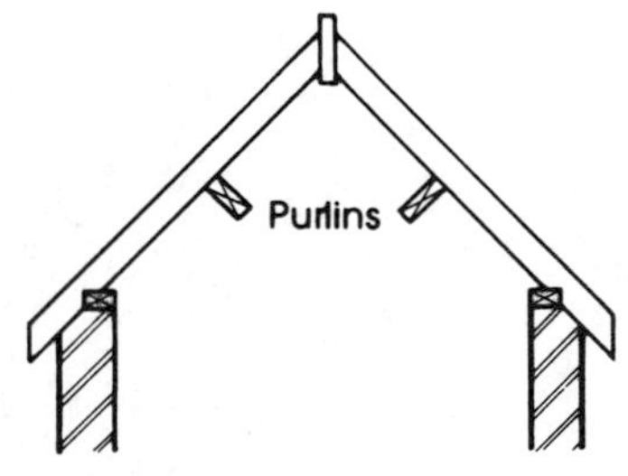

Double roof

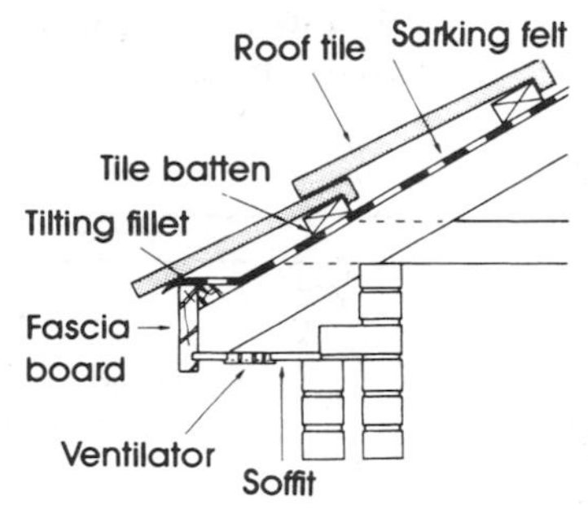

Eaves

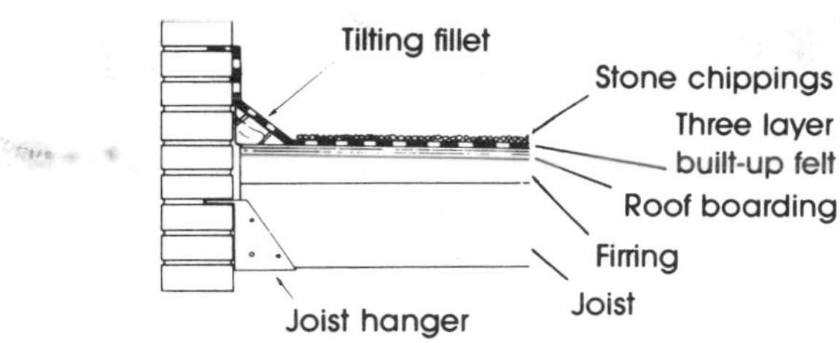

Flat roof at abutment

Stairs are the series of steps (combination of tread and riser) that form a stairway.

Stairway – a series of steps (including any balustrade and handrail) giving floor-to-floor access. Each continuous set of steps in between floors or landings is termed a flight of stairs. Landings may be introduced between floor levels, to break up long flights and give resting points for users. Alternatively landings may be used to change the direction of the stair.

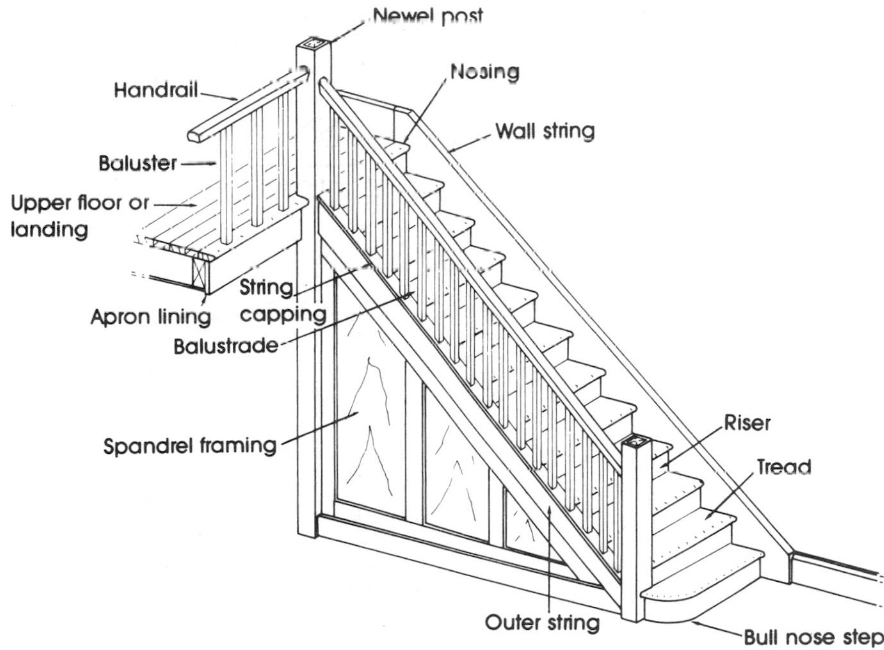

Secondary elements – the non-essential elements of a structure, having mainly a completion role around openings in primary elements.

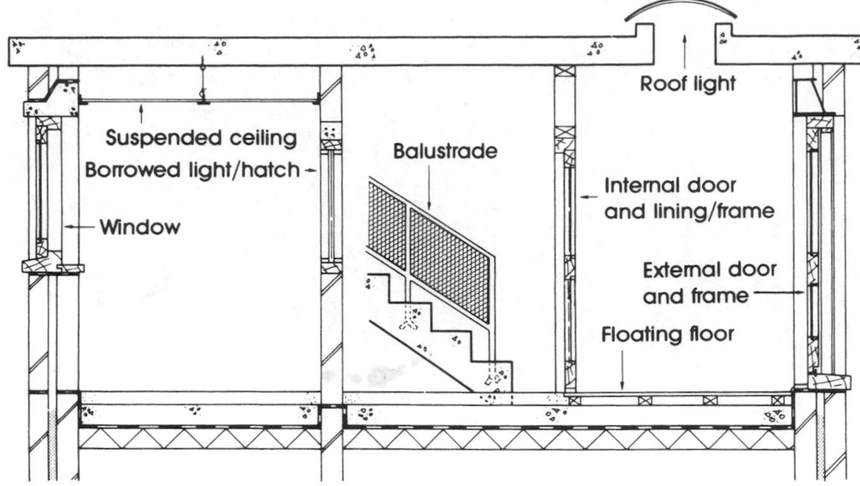

Secondary elements

Doors are moveable barriers used to cover an opening in a structure. Their main function is to allow access in a building and passage between its interior spaces. Other functional requirements include weather protection, fire resistance, sound and thermal insulation, security, privacy, ease of operation and durability. They may be classified by their method of construction and method of operation.

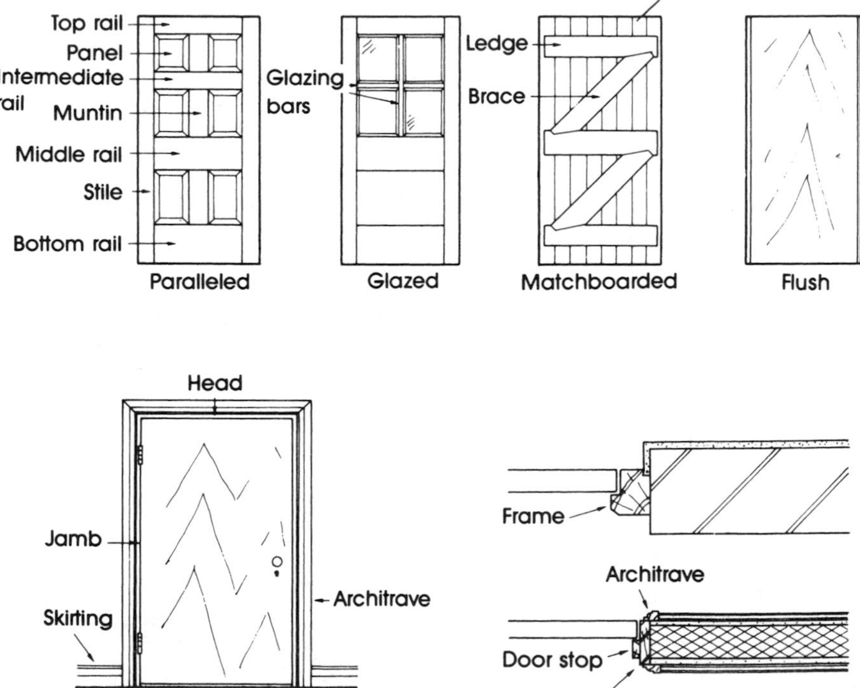

The surround to the wall opening on which doors are hung may be either a *frame* or *lining*.

Windows are glazed openings in a wall used to allow daylight and air in and give occupants an outside view.

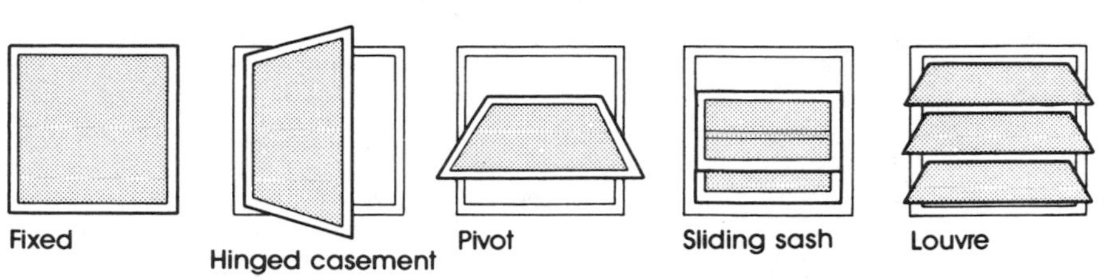

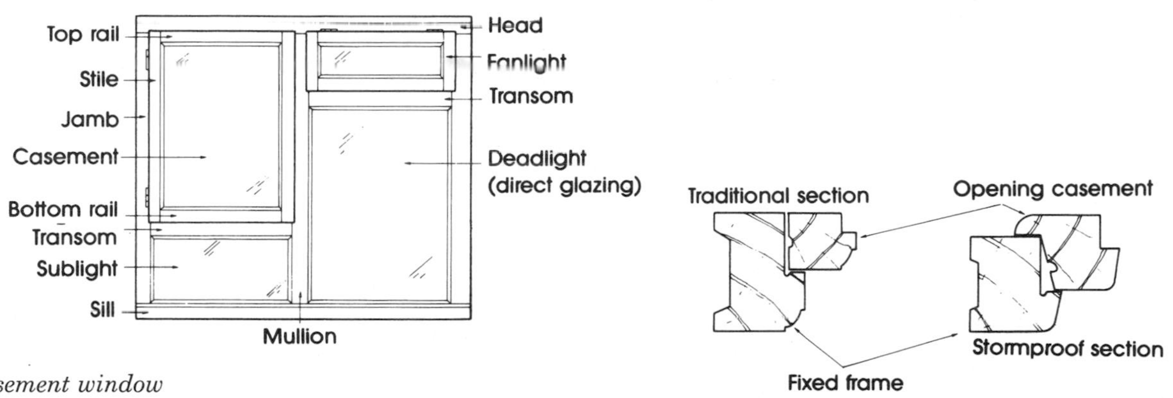

Casement window

Finishing elements – the final surface of an element that may be a self-finish such as concrete and face brickwork, or an applied finish such as plaster, paint and timber trim.

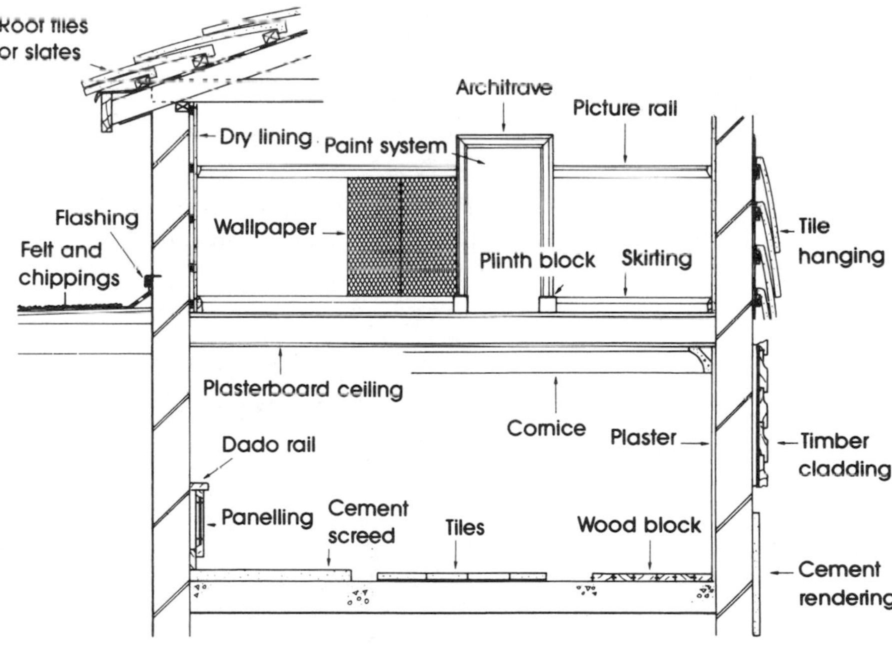

Components – the various parts or materials that are combined to form the elements of a building.

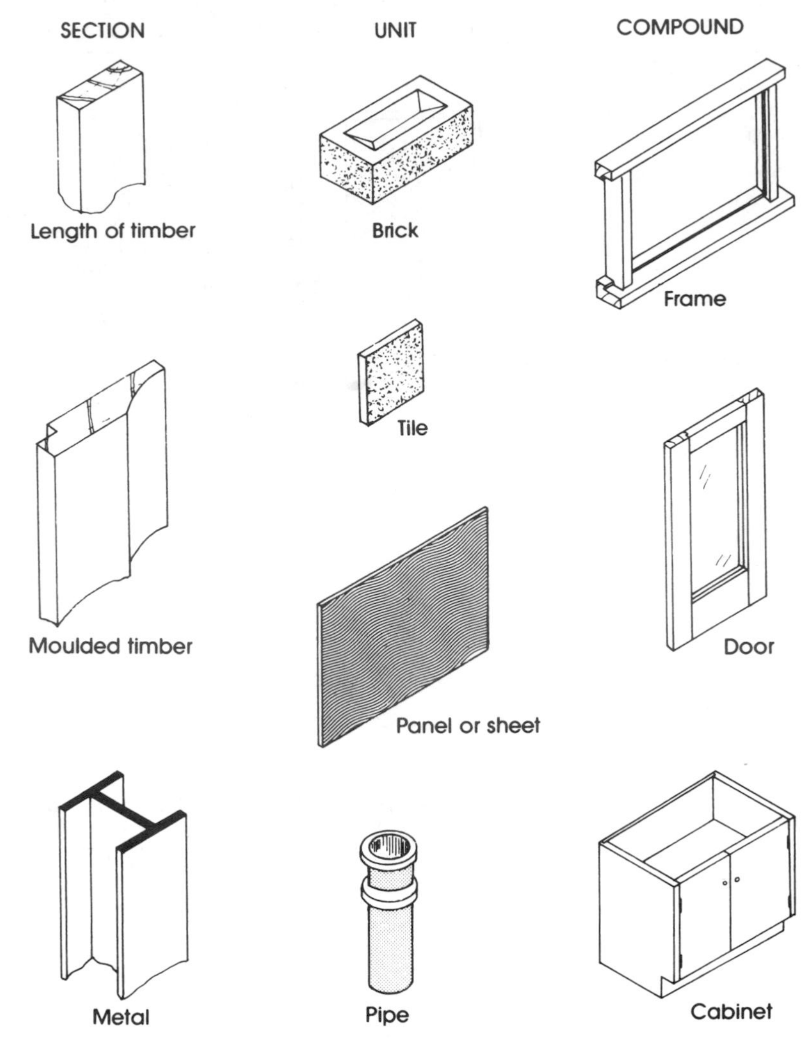

Services – the electrical, plumbing, mechanical and specialist installations in a building, normally piped, wired or ducted into or within a building.

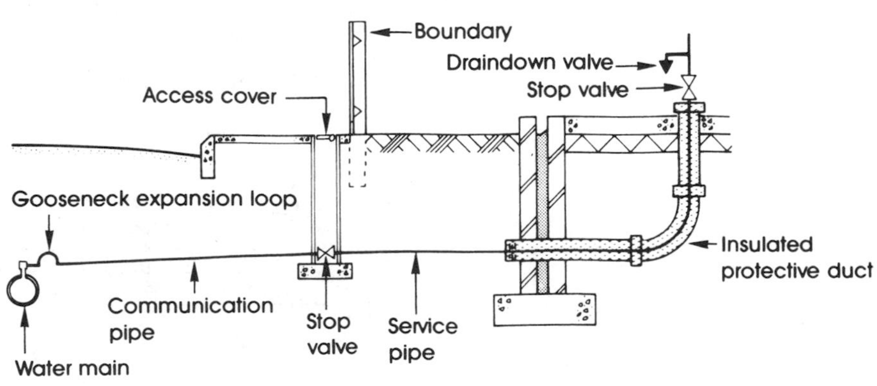

Water supply

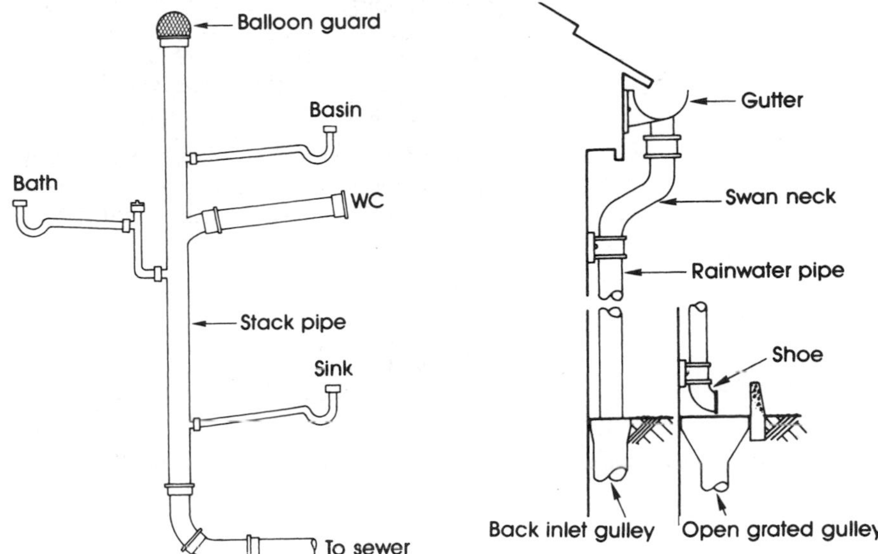

Above ground drainage

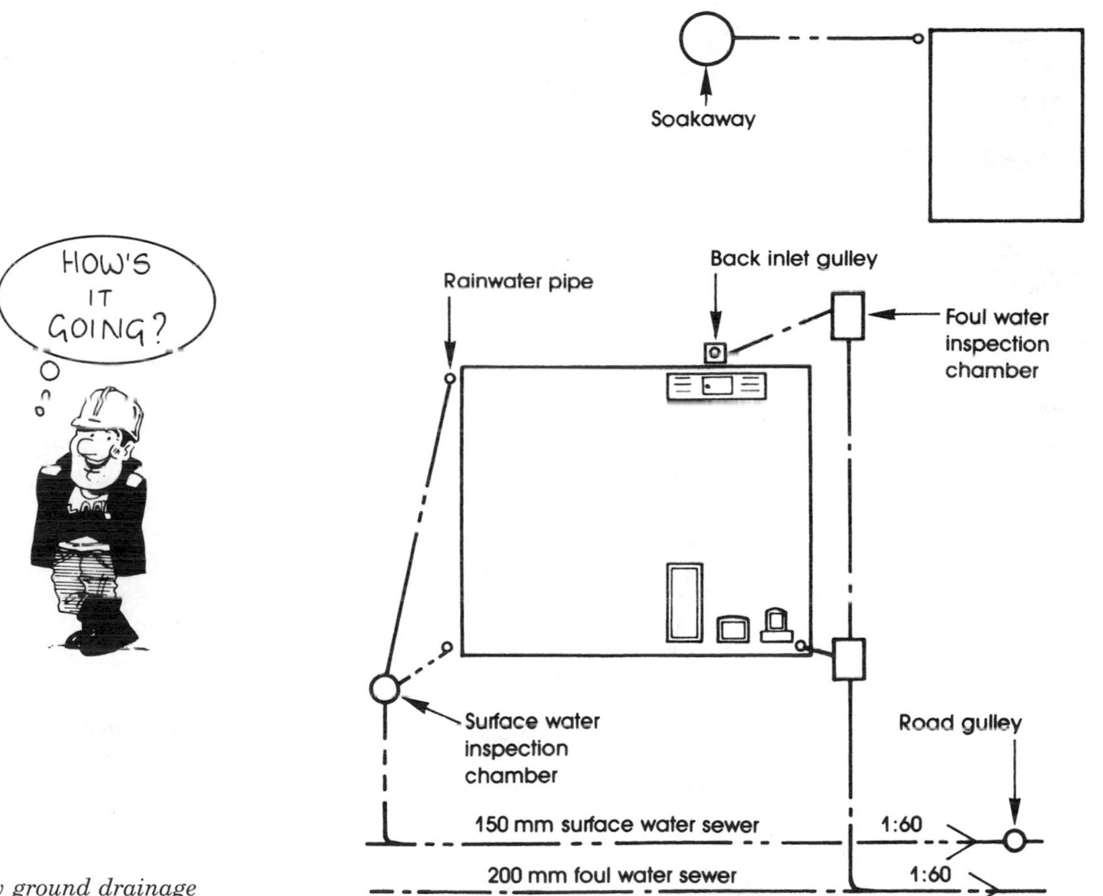

Below ground drainage

Site works

Excavation – the process of removing earth to form a hole in the ground can be dug manually using a shovel or mechanically using a digger excavator.

Oversite excavation is the removal of topsoil and vegetable matter from a site prior to the commencement of building work. Excavation depth is typically between 150 and 300 mm.

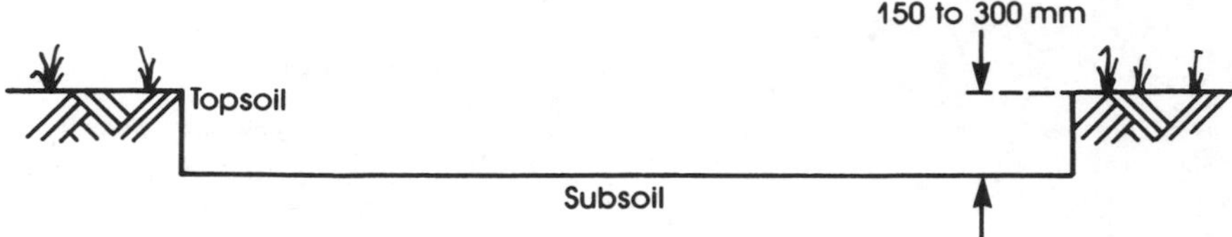

Reduced level excavation is a process carried out after oversite excavation on undulating ground. It consists of cutting and filling operations to produce a level surface, called the formation level.

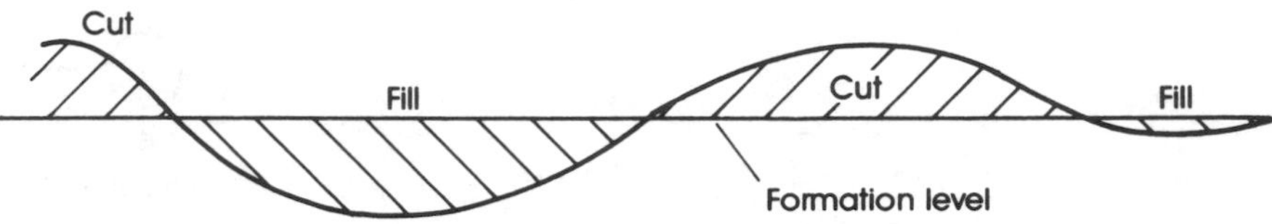

Trench excavations are long narrow holes in the ground, to accommodate strip foundations or underground services. Deep trenches may be battered or timbered to prevent the sides from caving in.

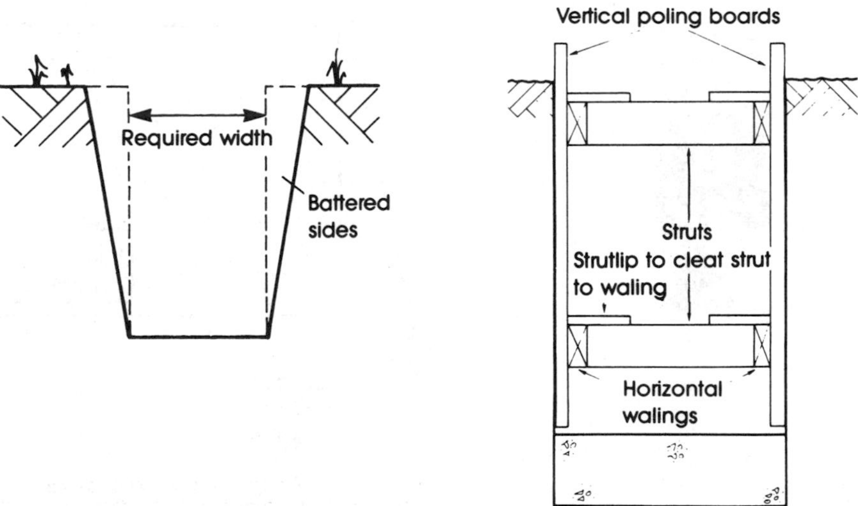

Timbering to sides of trench

Pit excavations are deep rectangular holes in the ground, normally for column base pad foundations. Larger holes may be required for basements etc. Sides may be battered or timbered depending on depth.

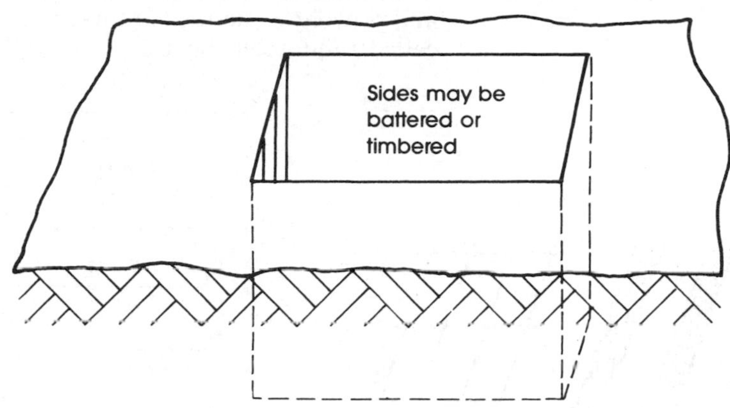

Road construction – in this context the scope is limited to small estate roads, access roads and driveways. Once excavated and scraped down to the formation level construction of the surface can commence on the sub-grade.

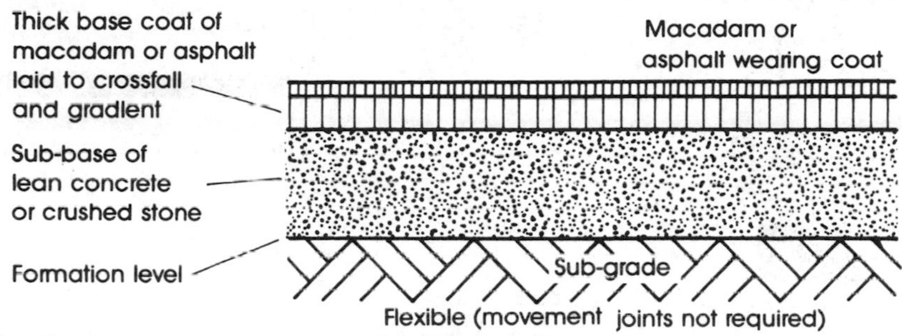

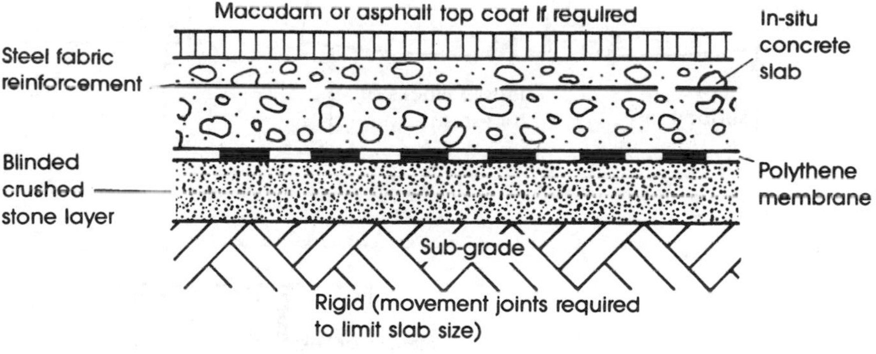

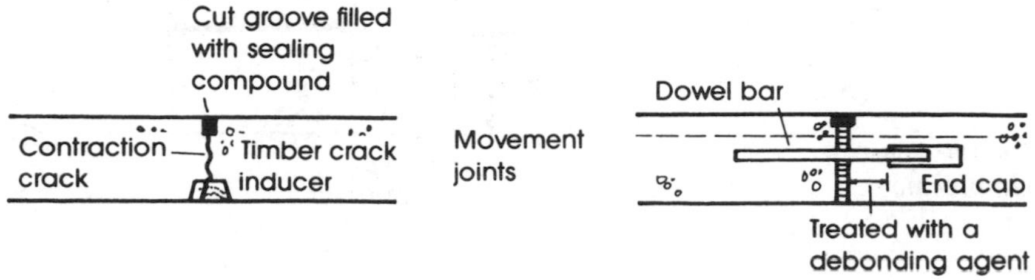

Steel fixing – Concrete is very strong when being compressed (squashed) but comparatively weak in tension (stretched or bent). Few structures are subjected to loadings which are totally compressive. Thus steel reinforcement is normally introduced to increase strength and prevent structural failure, thereby producing a composite material called reinforced concrete.

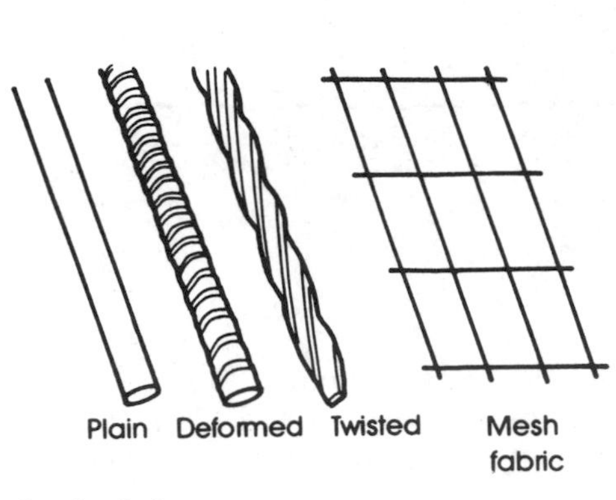

Steel reinforcement

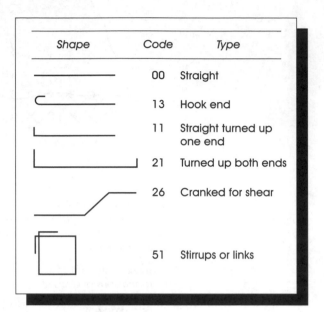

Shape	*Code*	*Type*
	00	Straight
	13	Hook end
	11	Straight turned up one end
	21	Turned up both ends
	26	Cranked for shear
	51	Stirrups or links

Typical standard shape codes

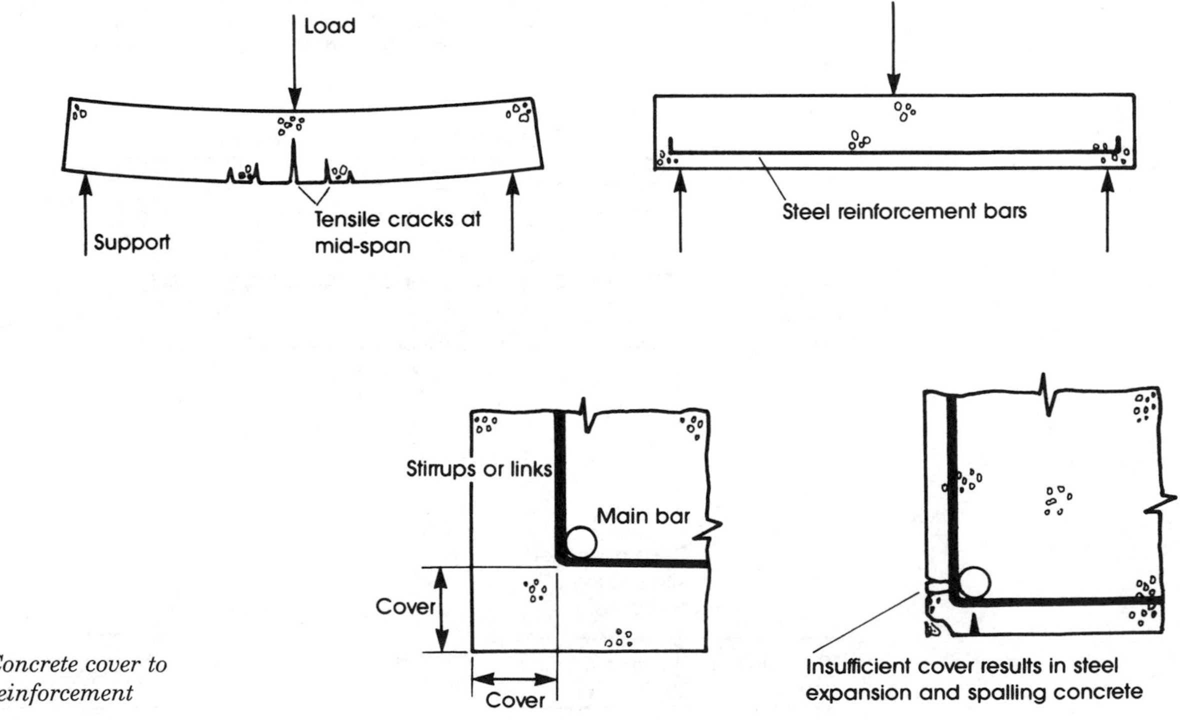

Concrete cover to reinforcement

Formwork – a structure which is usually temporary but can be partly or wholly permanent, designed to contain fresh, fluid concrete, form it into the required shape and dimension and support it until it cures (hardens) sufficiently to become self-supporting. The surface in contact with the concrete is known as the form face whilst the supporting structure can be referred to as forcework.

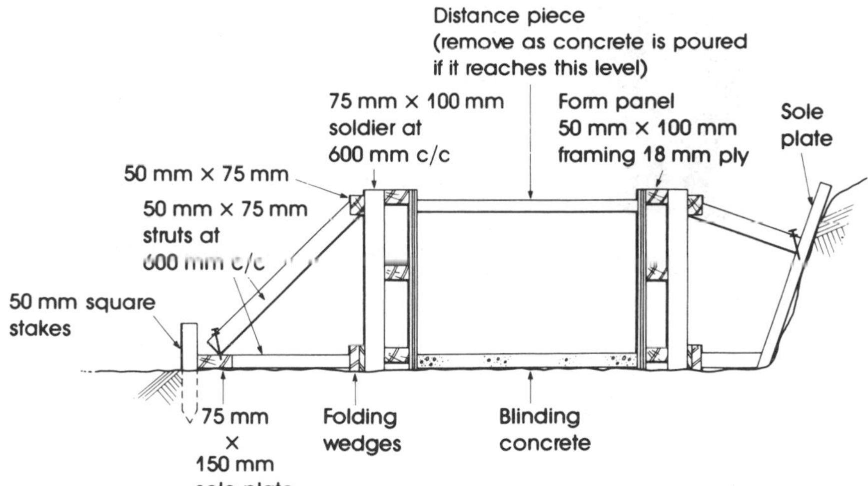

Column pad base formwork

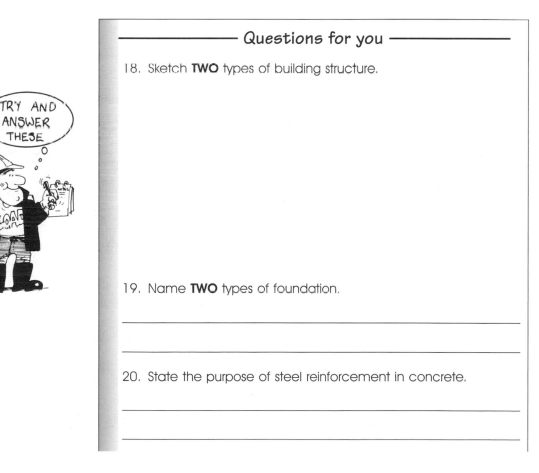

Questions for you

18. Sketch **TWO** types of building structure.

19. Name **TWO** types of foundation.

____

____

20. State the purpose of steel reinforcement in concrete.

____

____

21. Sketch and label a typical cavity wall.

22. Sketch and label a typical timber upper floor construction.

23. State the purpose of formwork.

24. Name **TWO** methods of excavation.

25. List **FOUR** finishing elements.

26. State two types of road surface.

27. Name any *six* of the lettered features shown.

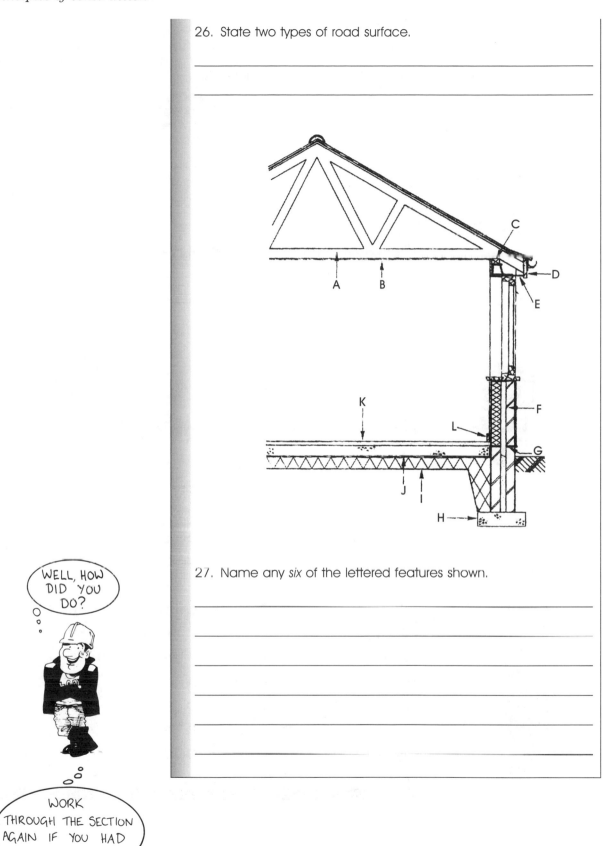

WELL, HOW DID YOU DO?
WORK THROUGH THE SECTION AGAIN IF YOU HAD ANY PROBLEMS

WORD-SQUARE SEARCH

Hidden in the word square are the following 20 words associated with '*The Construction Industry*'. You may find the words written forwards, backwards, up, down or diagonally.

Public sector	Carpenter
Services	Component
Wall	Formwork
Architect	Floors
Foundation	Tiler
Engineer	Window
Ganger	Restoration
New	Excavation
Surveyor	Roof
Substructure	Element

Draw a ring around the words, or line in using a highlight pen thus:

EXAMPLE

EXAMPLE

COMPLETE THE WORD SQUARE

P	U	B	L	I	C	S	E	C	T	O	R	P	E	T	E	R	E
A	H	R	I	S	E	R	V	I	C	E	S	R	E	T	T	E	F
C	R	L	S	L	L	A	F	D	N	D	E	R	E	G	N	A	G
F	I	C	A	V	W	A	L	L	O	P	P	E	I	V	E	O	H
O	A	O	H	A	Y	A	T	A	I	O	C	E	E	B	W	R	A
R	U	M	R	I	T	W	T	R	T	T	C	N	I	L	E	T	T
M	R	P	E	N	T	A	L	E	A	T	D	I	A	W	L	E	S
W	T	O	T	J	N	E	D	A	D	L	F	G	H	A	M	D	U
O	A	N	T	F	E	S	C	D	N	E	O	N	T	R	E	B	B
R	C	E	I	N	M	T	R	T	U	T	I	E	N	V	T	C	S
K	C	N	F	C	E	C	O	N	O	T	R	N	I	H	S	A	T
S	B	T	P	F	L	R	O	O	F	L	A	F	A	V	T	I	R
R	Q	S	O	V	E	W	A	C	A	R	P	E	N	T	E	R	U
O	E	T	H	N	O	P	D	E	A	N	E	D	I	C	C	A	C
O	A	R	S	D	R	E	S	T	O	R	A	T	I	O	N	N	T
L	A	U	N	I	O	L	A	E	D	E	R	S	O	D	C	D	U
F	A	I	T	I	L	E	R	R	O	Y	E	V	R	U	S	I	R
C	W	N	O	I	T	A	N	O	I	T	A	V	A	C	X	E	E

WELL, HOW MANY DID YOU GET?

2 Health and safety

Accidents

Definition

An accident is often described as a chance event or an unintentional act. This description is not acceptable, as accidents do not 'just happen' they do not 'come out of the blue', they are caused! A better definition of an accident is therefore:

An accident is an event causing injury, damage or loss that might have been avoided by following correct methods and procedures.

Accident statistics

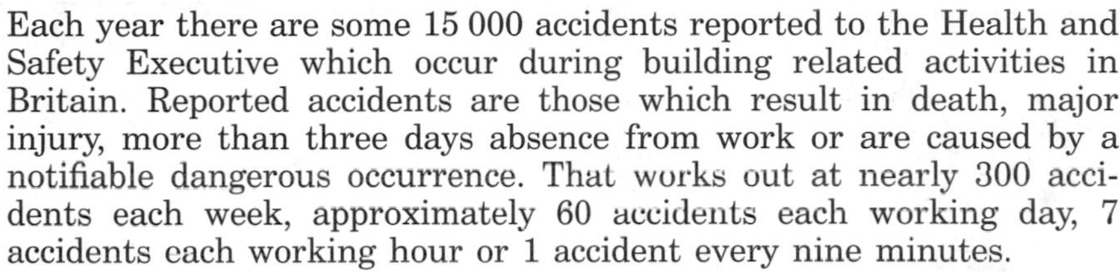

Each year there are some 15 000 accidents reported to the Health and Safety Executive which occur during building related activities in Britain. Reported accidents are those which result in death, major injury, more than three days absence from work or are caused by a notifiable dangerous occurrence. That works out at nearly 300 accidents each week, approximately 60 accidents each working day, 7 accidents each working hour or 1 accident every nine minutes.

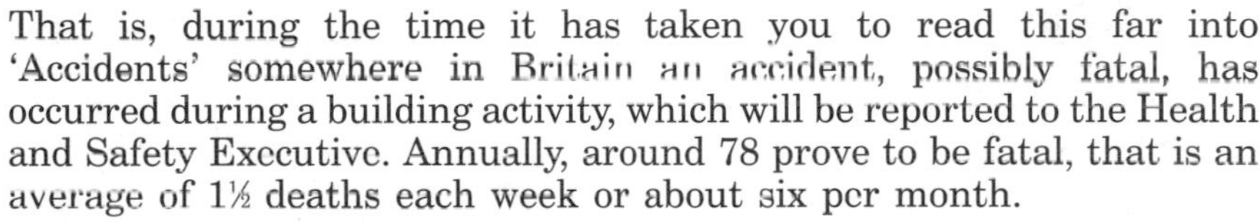

That is, during the time it has taken you to read this far into 'Accidents' somewhere in Britain an accident, possibly fatal, has occurred during a building activity, which will be reported to the Health and Safety Executive. Annually, around 78 prove to be fatal, that is an average of 1½ deaths each week or about six per month.

These figures are not intended to frighten you or put you off a future career in the building industry, but simply to make you aware of the hazards involved so that all concerned will make a conscious effort to improve them.

Causes of accidents

The Health and Safety Executive break down these reported accident figures into the type of accident and the occupations within the industry of those involved. The illustrations on the following three pages show further information about the causes of fatal accidents and the breakdown of reported accidents by occupation. More than 50 per cent of fatal accidents involve falls of persons. Of the reported accidents by occupation, carpenters and joiners are most at risk with 10 per cent of the total followed by bricklayers at 5.6 per cent, then electricians at 5 per cent.

WHERE'S THIS ACCIDENT, THEN?

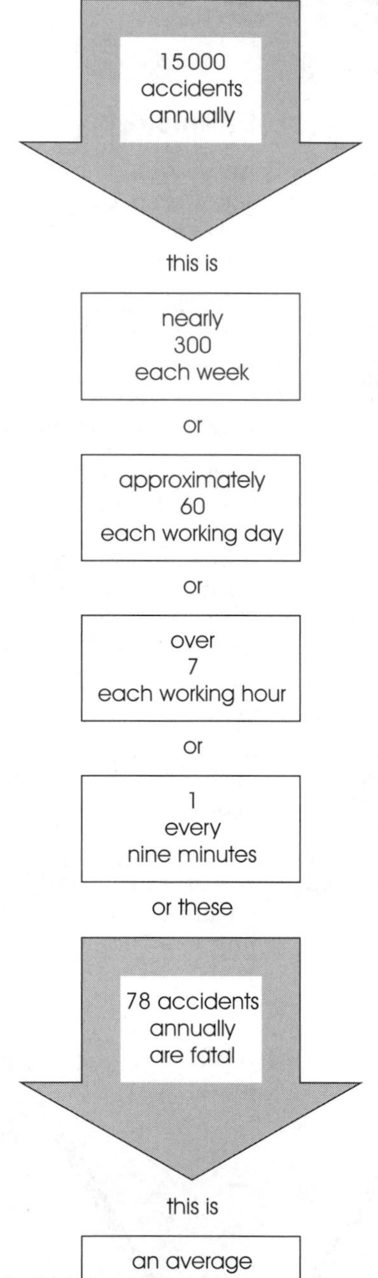
15000
accidents
annually
this is
nearly
300
each week
or
approximately
60
each working day
or
over
7
each working hour
or
1
every
nine minutes
or these
78 accidents
annually
are fatal
this is
an average
of 1½ deaths
each week

Reported accidents

STUDY THIS DIAGRAM

THINK
SAFETY
DON'T BECOME
A STATISTIC!

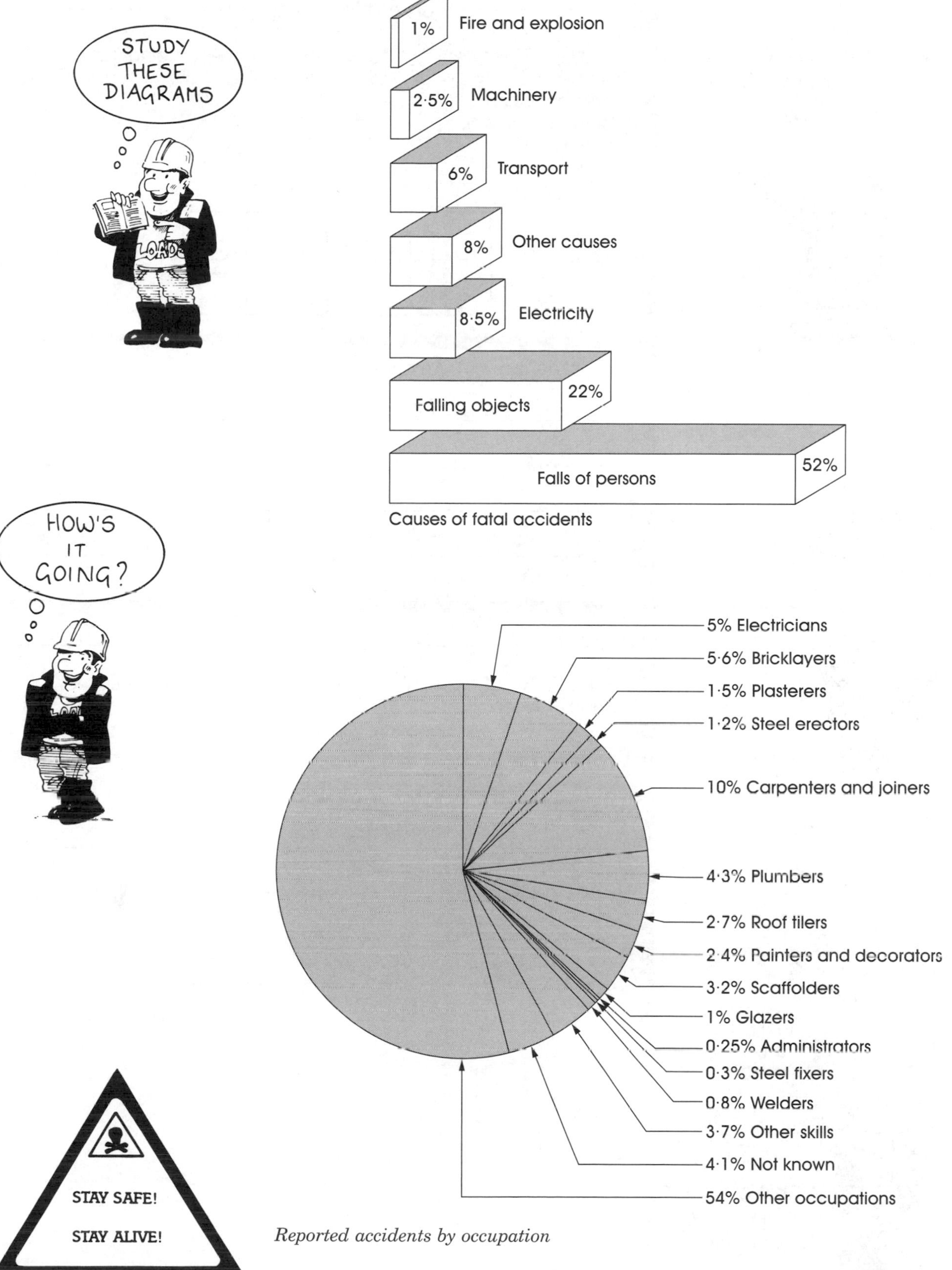

Reported accidents by occupation

TRY AND ANSWER THESE

Questions for you

1. Over 50 per cent of fatal accidents in the building industry involve:
(a) machinery
(b) electric shock
(c) falls
(d) transport

a	b	c	d
[ ]	[ ]	[ ]	[ ]

2. Define the term 'accident'

____

____

____

____

____

____

3. Define what constitutes a reported accident and state to whom it is reported.

____

____

____

____

____

____

WELL, HOW DID YOU DO?
WORK THROUGH THE SECTION AGAIN IF YOU HAD ANY PROBLEMS

Health and safety controls

In the mid 1970s the Health and Safety at Work Act (HASAWA) was introduced. HASAWA was seen as an enabling umbrella. It introduced the main statutory legislation, completely covering the health and safety of all persons at their place of work and protecting other people from risks occurring through work activities.

It has overseen the gradual replacement of previous piecemeal health and safety requirements by revised and up-to-date measures prepared in consultation with industry and its workers. As a member of the construction industry you are required to know about your responsibilities with regards to the following legislation:

- The Health and Safety at Work Act
- The Management of Health and Safety at Work Regulations
- The Manual Handling Operations Regulations
- The Personal Protective Equipment at Work Regulations
- The Health and Safety (Safety Signs and Signals) Regulations
- The Construction (Design & Management) Regulations
- The Construction (Health, Safety & Welfare) Regulations
- The Construction (Head Protection) Regulations
- The Provision and Use of Work Equipment Regulations
- The Control of Substances Hazardous to Health Regulations

The Health and Safety at Work Act (HASAWA)

The four main objectives of the HASAWA are as follows:

1) To secure the health, safety and welfare of all persons at work.
2) To protect the general public from risks to health and safety arising out of work activities.
3) To control the use, handling, storage and transportation of explosives and highly flammable substances.
4) To control the release of noxious or offensive substances into the atmosphere.

These objectives can be achieved only by involving everyone in health and safety matters. This includes:

- Employers and management
- Employees (and those undergoing training)
- Self-employed
- Designers, manufacturers and suppliers of equipment and materials.

Employers' and management duties

Employers have a general duty to ensure the health and safety of their employees, visitors and the general public. This means that the employer must:

1) Provide and maintain a safe working environment.
2) Ensure safe access to and from the workplace.
3) Provide and maintain safe machinery, equipment and methods of work.
4) Ensure the safe handling, transport and storage of all machinery, equipment and materials.
5) Provide their employees with the necessary information, instruction, training and supervision to ensure safe working.
6) Prepare, issue to employees and update as required a written statement of the firm's safety policy.
7) Involve trade union safety representatives (where appointed) with all matters concerning the development, promotion and maintenance of health and safety requirements.

Note: An employer is not allowed to charge an employee for anything done, or equipment provided, to comply with any health and safety requirement.

Employees' duties

An employee is an individual who offers his or her skill and experience etc. to his or her employer in return for a monetary payment. It is the duty of all employees while at work to comply with the following:

1) Take care at all times and ensure that their actions do not put at 'risk' themselves, their workmates or any other person.
2) Co-operate with their employers to enable them to fulfil the employers' health and safety duties.
3) Use the equipment and safeguards provided by the employers.
4) Never misuse or interfere with anything provided for health and safety.

Self-employed duties

The self-employed person can be thought of as both the employer and employee; therefore their duties under the Act are a combination of those of the employer and employee.

Designers', manufacturers' and suppliers' duties

Under the Act, designers, manufacturers and suppliers as well as importers and hirers of equipment, machinery and materials for use at work have duty to:

1) Ensure that the equipment machinery or material is designed, manufactured and tested so that when it is used correctly no hazard to health and safety is created.

2) Provide information or operating instructions as to the correct use, without risk, of their equipment, machinery or material. **Note:** Employers should ensure this information is passed on to their employees.
3) Carry out research so that any risk to health and safety is eliminated or minimised as far as possible.

a
WARNING
CONTAINS
ASBESTOS
Breathing asbestos dust is dangerous to health
Follow safety instructions

Enforcement of Safety Legislation

Under the HASAWA a system of control was established, aimed at reducing death, injury and ill health. This system of control consists of the Health and Safety Executive (HSE). The Executive is divided into a number of specialist inspectorates or sections which operate from local offices situated throughout the country. From the local office, inspectors visit individual workplaces.

Note: The section with the main responsibility for the building industry is the Factory Inspectorate.

The Health and Safety Executive inspectors have been given wide powers of entry, examination and investigation in order to assist them in the enforcement of the HASAWA and other safety legislation. In addition to giving employers advice and information on health and safety matters, an inspector can do the following:

1) *Enter premises* in order to carry out investigations, including the taking of measurements, photographs, recordings and samples. The inspector may require the premises to be left undisturbed while the investigations are taking place.
2) *Take statements.* An inspector can ask anyone questions relevant to the investigation and also require them to sign a declaration as to the truth of the answers.
3) *Check records.* All books, records and documents required by legislation must be made available for inspection and copying.
4) *Give information.* An inspector has a duty to give employees or their safety representative information about the safety of their workplace and details of any action he/she proposes to take. This information must also be given to the employer.
5) *Demand.* The inspector can demand the seizure, dismantling, neutralising or destruction of any machinery, equipment, material or substance that is likely to cause immediate serious personal injury.
6) *Issue an improvement notice.* This requires the responsible person (employer or manufacturer, etc.) to put right within a specified period of time any minor hazard or infringement of legislation.
7) *Issue a prohibition notice.* This requires the responsible person to stop immediately any activities which are likely to result in serious personal injury. This ban on activities continues until the situation is corrected. An appeal against an improvement or prohibition notice may be made to an industrial tribunal.
8) *Prosecute.* All persons, including employers, employees, self-employed, designers, manufacturers and suppliers who fail to comply with their safety duty may be prosecuted in a magistrates' court or in certain circumstances in the higher court system. Conviction can lead to unlimited fines, or a prison sentence, or both.

BBS Construction Services

STATEMENT OF COMPANY POLICY ON HEALTH AND SAFETY

The Directors accept that they have a legal and moral obligation to promote health and safety in the workplace and to ensure the co-operation of employees in this. This duty of care extends to all persons who may be affected by any operation under the control of BBS Construction Services.

Employees also have a statutory duty to safeguard themselves and others and to co-operate with management to secure a safe work environment.

The directors shall ensure, so far as reasonably practicable, that:

- Adequate resources and competent advice are made available in order that proper provision can be made for health and safety.
- Safe systems of working are devised and maintained.
- All employees are provided with all information, instructions, training ad supervision. required to secure the safety of all persons.
- All plant, machinery and equipment is safe and without risk to health.
- All places of work are maintained in a safe condition with safe means of access and egress.
- Arrangements are made for safe use, handling, storage and transport of all articles and substances.
- The working environment is maintained in a condition free of risks to health and safety and that adequate welfare facilities are provided.
- Assessment of all risks are made and control measures put in place to reduce or eliminate them.
- All arrangements are monitored and reviewed periodically.

These statements have been adopted by directors of the company and form the basis of our approach to health and safety matters.

Ivor Carpenter, Finance Director

Christine Whiteman, Human Resource Director

James Brett, Managing Director

The Director with responsibility for Health and Safety

Peter Brett, Chief Executive

BBS Construction Services

SITE SAFETY INDUCTION

As part of the company's commitment to safety the following checklist is provided for management, when employing new staff or on the transfer of existing staff to your site.

MANAGEMENT MUST:

- Issue and explain the company's Safety Policy.
- Introduce your site safety advisor controller.
- Discuss and record any previous safety training and experience.
- Issue and discuss appropriate safety method statement.
- Emphasis the following points:
  - Emergency and first aid procedures applicable to the site
  - Personal safety responsibilities, house keeping, hygiene and PPE
  - Need to report accidents 'near misses' and unsafe conditions
  - Need for authorisation and training, before use of all plant machinery and powered hand tools
  - Location of all welfare facilities
- Explain the procedure to be followed in the event of a health and safety dispute (consult safety advisor/controller in first instances).
- Show the site notice board (in the rest room) where safety notices and information are displayed.
- Finally inform of company's key phrase for all matters 'IF IN DOUBT ASK' then invite questions.

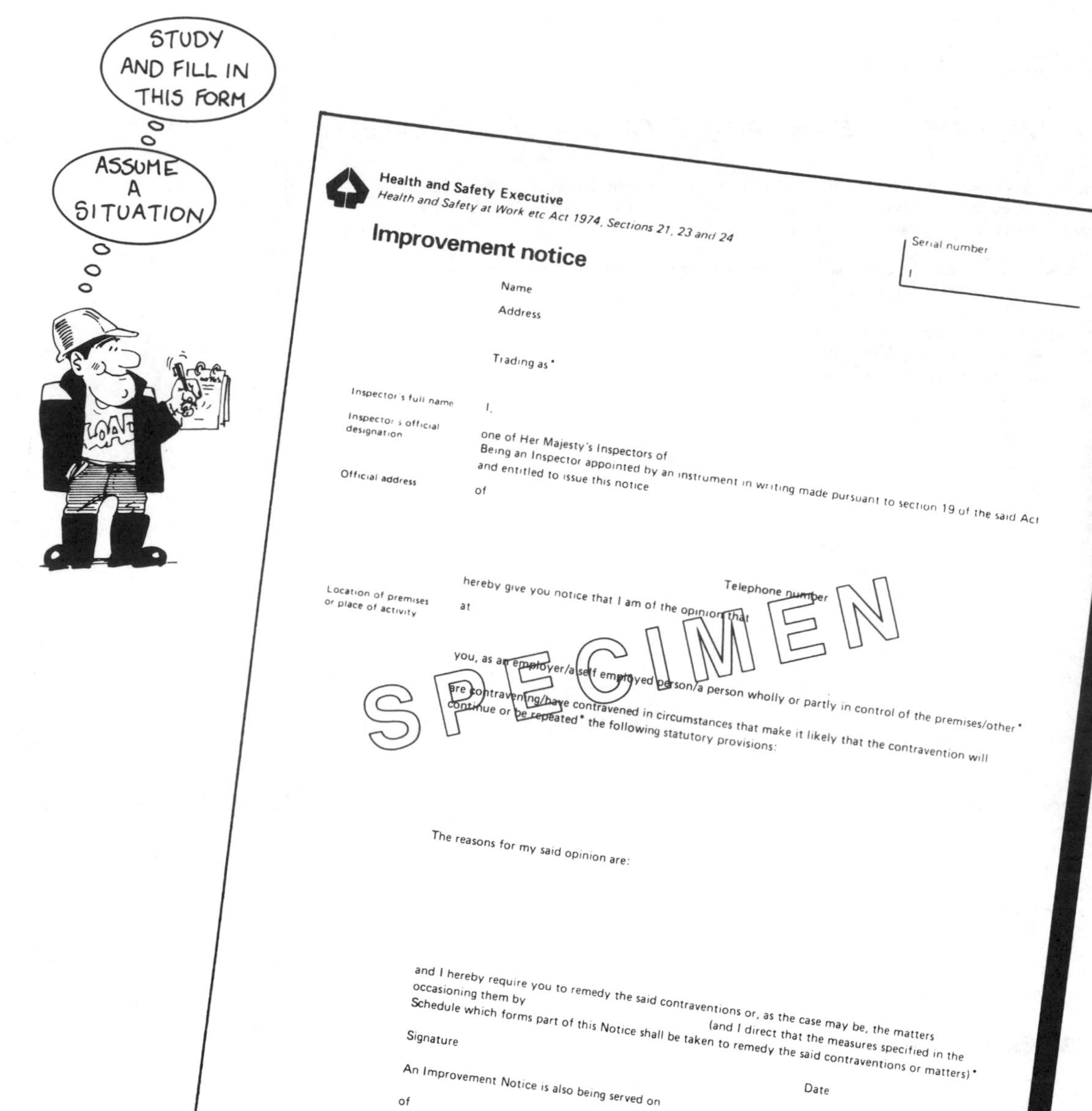

STUDY AND FILL IN THIS FORM
ASSUME A SITUATION

Health and Safety Executive
Health and Safety at Work etc Act 1974, Sections 21, 23 and 24

Improvement notice

Serial number I

Name

Address

Trading as*

Inspector's full name — I,

Inspector's official designation — one of Her Majesty's Inspectors of
Being an Inspector appointed by an instrument in writing made pursuant to section 19 of the said Act and entitled to issue this notice

Official address — of

Telephone number

hereby give you notice that I am of the opinion that

Location of premises or place of activity — at

you, as an employer/a self employed person/a person wholly or partly in control of the premises/other*
are contravening/have contravened in circumstances that make it likely that the contravention will continue or be repeated* the following statutory provisions:

SPECIMEN

The reasons for my said opinion are:

and I hereby require you to remedy the said contraventions or, as the case may be, the matters occasioning them by
(and I direct that the measures specified in the Schedule which forms part of this Notice shall be taken to remedy the said contraventions or matters)*

Signature — Date

An Improvement Notice is also being served on
of
related to the matters contained in this notice

Environment and Safety Information Act 1988 — This is a relevant notice for the purposes of the Environment and Safety Information Act 1988 YES/NO*
This page only will form the register entry*.

Signature — Date

LP1 (rev 12/88)

See notes overleaf

* delete as appropriate

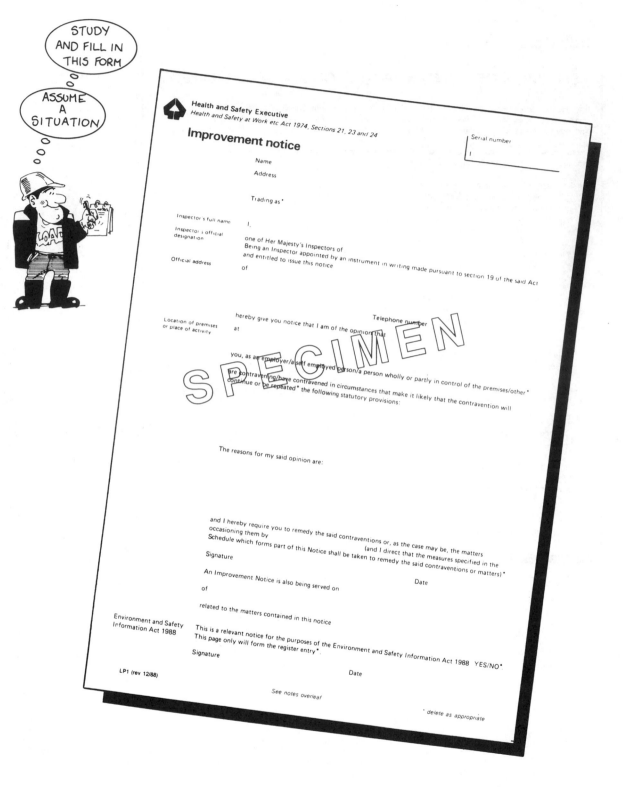

Improvement notice

STUDY AND FILL IN THIS FORM
ASSUME A SITUATION
LOAD

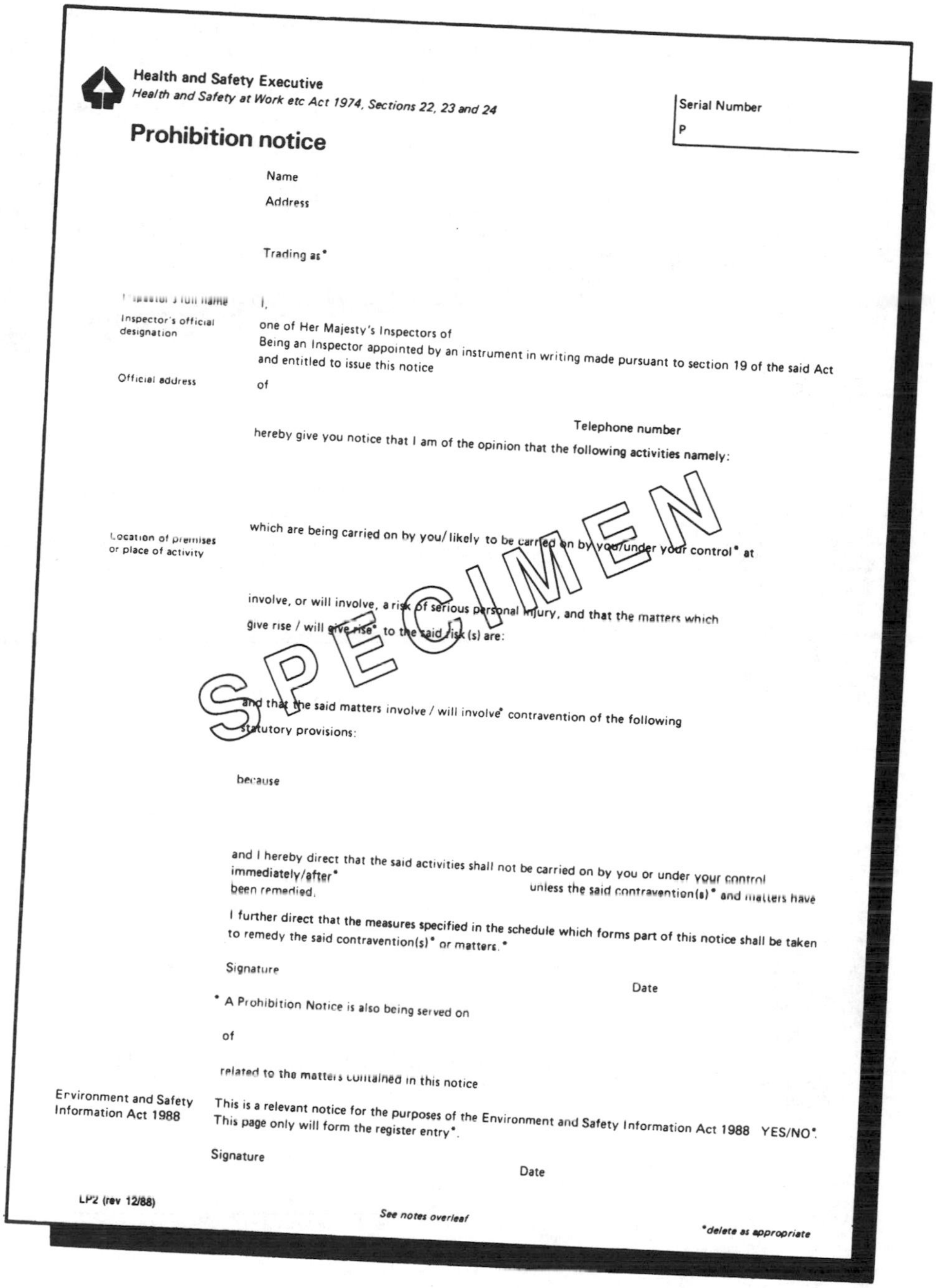

Health and Safety Executive
Health and Safety at Work etc Act 1974, Sections 22, 23 and 24

Serial Number
P

Prohibition notice

Name

Address

Trading as*

Inspector's full name — I,

Inspector's official designation — one of Her Majesty's Inspectors of

Being an Inspector appointed by an instrument in writing made pursuant to section 19 of the said Act and entitled to issue this notice

Official address — of

Telephone number

hereby give you notice that I am of the opinion that the following activities namely:

Location of premises or place of activity — which are being carried on by you/ likely to be carried on by you/under your control* at

involve, or will involve, a risk of serious personal injury, and that the matters which give rise / will give rise* to the said risk (s) are:

and that the said matters involve / will involve* contravention of the following statutory provisions:

because

and I hereby direct that the said activities shall not be carried on by you or under your control immediately/after* unless the said contravention(s)* and matters have been remedied.

I further direct that the measures specified in the schedule which forms part of this notice shall be taken to remedy the said contravention(s)* or matters.*

Signature

Date

* A Prohibition Notice is also being served on

of

related to the matters contained in this notice

Environment and Safety Information Act 1988 — This is a relevant notice for the purposes of the Environment and Safety Information Act 1988 YES/NO*. This page only will form the register entry*.

Signature

Date

LP2 (rev 12/88)

See notes overleaf

**delete as appropriate*

SPECIMEN

Prohibition notice

BUILDING AND CONSTRUCTION

Harry Whiteman is Safety Consultant to BBS Contracts

SAFETY NEWS. June. PSB

Maidstone Crown Court fined a building contractor £25 000 recently over an incident where an 18-year-old trainee lost both hands and feet when the scaffold tube he was unloading touched an overhead 33 000 volt electric cable. The Health and Safety Executive had asked in the magistrates court, where the maximum penalty is £2000, for this case to be referred to the Crown Court for sentence.

A major contractor in Birmingham city centre was recently fined £500 for supplying only two safety helmets for the 20 people who were employed on-site.

A 22-year-old site operative who lost the sight in one eye as a result of a grinding wheel accident, was fined £250 in Northampton Magistrates' Court this week. The Health and Safety Executive who brought the prosecution claimed that the operative had failed to take notice of the safety sign or wear the safety goggles which had been supplied by the employer.

TRY AND ANSWER THESE

WELL, HOW DID YOU DO?

WORK THROUGH THE SECTION AGAIN IF YOU HAD ANY PROBLEMS

Questions for you

4. State **TWO** of the main objectives of the Health and Safety at Work Act.

____

____

____

____

5. State **TWO** duties of each of the following under the Health and Safety at Work Act.

(a) Employers ______________________________

(b) Employees ______________________________

6. State **THREE** main powers of a Health and Safety Executive Inspector.

____

____

____

Management of H&S legislation

Health and Safety at Work Regulations (HSWR)

These regulations apply to everyone at work. They require your employer and the self-employed to plan, control, organise, monitor and review their work. In doing this, they must:

- assess the risks associated with the work being undertaken
- have access to competent health and safety advice
- provide employees with health and safety information and training
- appoint competent persons in their workforce to assist them in complying with obligations under health and safety legislation
- make arrangements to deal with serious and imminently dangerous situations
- co-operate in all health and safety matters with others who share the workplace.

Your duties as an employee under the regulations are:

- To use all machinery, equipment, dangerous substances, means of production, transport equipment and safety in accordance with the training and instructions given.
- Inform your employer or named competent person of dangerous situations and/or shortcomings in the health and safety arrangements.

Risk assessment

Risk management is the key part of the regulations, in order to put in place control measures. It involves employers identifying the hazards involved in their work, assessing the likelihood of any harm arising and deciding on adequate precautionary control measures.

Risk assessment is a five-step process:

1) *Looking for the hazards.* Consider the job to be undertaken:
   - How will it be done?
   - Where is it done?
   - What equipment and materials will be used?
2) *Decide who might be at risk and how.* Consider:
   - employees
   - the self-employed
   - other companies working on the job
   - visitors to the job
   - the general public who may be on or near the job.
3) *Evaluate the risks and decide on the action to be taken.* Typical questions are:
   - Can the hazard be completely removed?
   - Can the job be done in another safer way?
   - Can a different, less hazardous material be used?

BBS Construction Services

RISK ASSESSMENT

Activity covered by assessment: ____________________

Location of activity: ____________________

Persons involved: ____________________

Date of assessment: ____________________

Tick appropriate box ☑

- Does the activity involve a potential risk? YES ☐ NO ☐
- If YES can the activity be avoided? YES ☐ NO ☐
- If NO what is the level of risk? LOW ☐ MEDIUM ☐ HIGH ☐
- What remedial action can be taken to control or protect against the risk?

1 ____________________

2 ____________________

3 ____________________

4 ____________________

5 ____________________

MANAGEMENT SUMMARY:

Priority for action: LOW ☐ MEDIUM ☐ HIGH ☐

Action to be taken: ____________________

Date action to be taken by: ____________________

Date for reassessment: ____________________

Assessor's name and signature: ____________________

ASSESS THE RISK – PUT IN CONTROLS – CHECK THEY WORK

If any of the answers are yes, change the job to eliminate the risk. If risks cannot be eliminated:

- Can they be controlled?
- Can protective measures be taken?

4) *Record the findings.* Employers should make a record of the risk assessment and pass it on to their employees. This should include details of significant risks involved and the measures taken to remove or control them.

5) *Review the findings.* Periodic reviews of the findings are required to ensure that they are still effective. New assessments will be required:
   - when the risks or conditions change
   - when new risks or conditions are encountered for the first time.

The regulations

READ THIS PAGE
LOADS

The Manual Handling Operations Regulations

These regulations require employers and the self-employed to avoid the need to undertake manual handling operations that might create a risk of injury. Where avoidance is not reasonably practical, they have to make an assessment with the aim of (re)moving hazards and minimising potential risk of injury by:

- Avoiding all unnecessary manual handling.
- Mechanising or automating handling tasks e.g. by the use of cranes, hoists, fork-lift trucks and conveyor belts, etc.
- Arranging for heavy or awkward loads to be shared when finally lifting or moving into position by hand.
- Ordering materials in easily handled sizes e.g. bagged sand, cement and plaster etc., are all available in 25 kg bags.
- Positioning all loads mechanically as near as possible to where they will be used, so as to reduce the height they have to be manually lifted and the distance they have to be carried.
- Providing employees with advice and training in safe lifting techniques and sensible handling of loads.

As an employee, you are required to make full and proper use of anything put into place by your employer, to reduce the risk of injury during manual handling operations

REFER TO THESE REGULATIONS
REGS.
LOADS

The Personal Protective Equipment at Work Regulations

Personal protective equipment (PPE) means all pieces of equipment, additions or accessories designed to be worn or held by a person at work to protect against one or more risks. Typical items of PPE are safety footwear, waterproof clothing, safety helmets, gloves, high visibility clothing, eye protection, dust masks, respirators and safety harnesses.

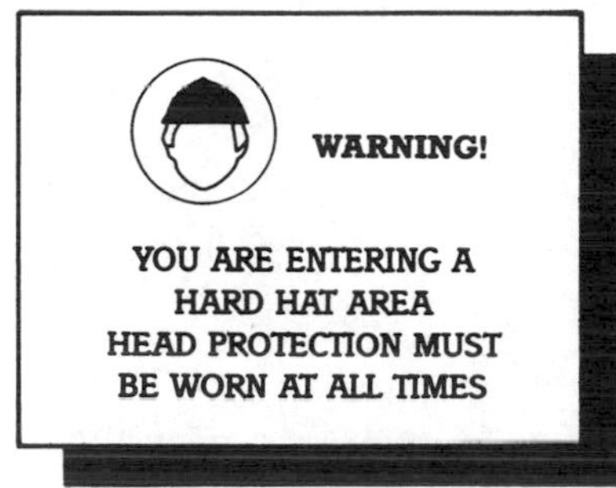
WARNING!
YOU ARE ENTERING A
HARD HAT AREA
HEAD PROTECTION MUST
BE WORN AT ALL TIMES

The use of PPE is seen as the ***last*** not the ***first*** resort. The first consideration is to undertake a risk assessment, with a view to preventing or controlling any risk at its source, by making machinery or work processes safer.

PPE requirements –

- All items of PPE must be suitable for the purpose it is being used for and provision must be made for PPE maintenance, replacement and cleaning.
- Where more than one item is being worn, they must be compatible.
- Training must be provided in the correct use of PPE and its limitations.
- Employers must ensure that appropriate items are provided and are being properly used.
- Employees and the self-employed must make full use of PPE provided and in accordance with the training given. In addition any defect or loss must be reported to their employers.

The Health and Safety (Safety Signs and Signals) Regulations

Safety signals and signals legislation requires employers to provide safety signs in a variety of situations that do or could affect health and safety. There are four types of safety signs in general use. Each of these signs have a designated shape and colour, to ensure that health and safety information is presented to employees in a consistent, standard way, with the minimum use of words.

Details of these signs and typical examples of use are given in the table.

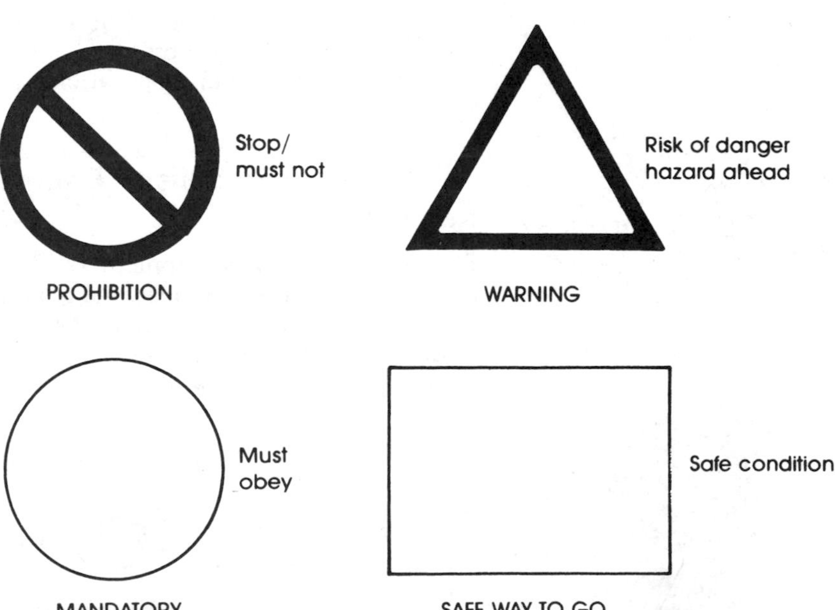

Safety signs

Purpose	*Sign*	*Definition*	*Examples for use*
Prohibition	white red	A sign prohibiting certain behaviour	No smoking; Smoking and naked flames prohibited; Do not extinguish with water; Not drinking water; Pedestrians prohibited
Caution	yellow black	A sign giving warning of certain hazards	Caution, risk of fire; Caution, toxic hazard; Caution, corrosive substance; General warning caution, risk of danger; Caution, risk of electric shock; Perimeter of hazard
Safe condition	green	A sign providing information about safe confitions	First aid; Indication of direction; Indication of direction
Mandatory	blue	A sign indicating that a special course of action is required	Head protection must be worn; Eye protection must be worn; Hearing protection must be worn; Foot protection must be worn; Hand protection must be worn; Respiratory protection must be worn
Supplementary	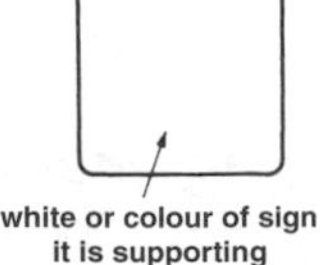 white or colour of sign it is supporting	A sign with text. Can be used in conjunction with a safety sign to provide additional information	IMPORTANT REPORT ALL ACCIDENTS IMMEDIATELY; SCAFFOLDING INCOMPLETE; SAFETY HELMETS ARE PROVIDED FOR YOUR SAFETY AND MUST BE WORN; PETROLEUM MIXURE HIGHLY FLAMMABLE NO SMOKING OR NAKED LIGHTS; WARNING HIGH VOLTAGE CABLES OVERHEAD; 

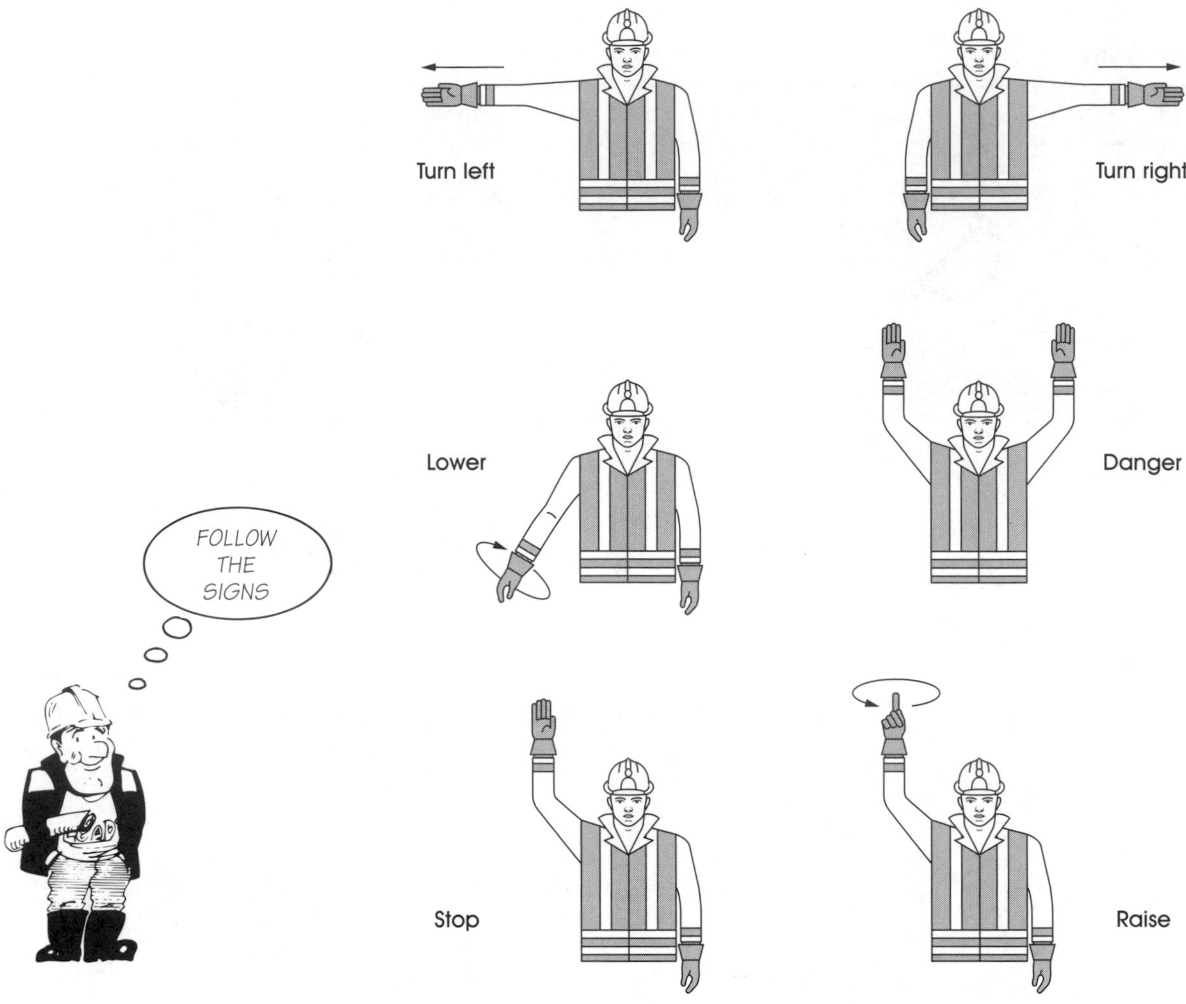

In addition the following points, which are of particular concern to construction work, are highlighted in the regulations:

- In order to avoid confusion, too many signs should not be placed together.
- Signs should be removed when the situation they refer to ceases to exist.
- Fire fighting equipment and its place of storage, must be identified by being red in colour.
- Traffic routes should be marked out using yellow.
- Acoustic fire evacuation signals must be continuous and sufficiently loud to be heard above other noises on site.
- Anyone giving hand signals must wear distinctive brightly coloured clothing and use the standard arm and hand movements.

The Construction (Design & Management) Regulations (CDM)

Design and management legislation requires that health and safety is taken into account and managed during all stages of a construction project. From its conception, design and planning, throughout the actual construction process and afterwards during maintenance and repair.

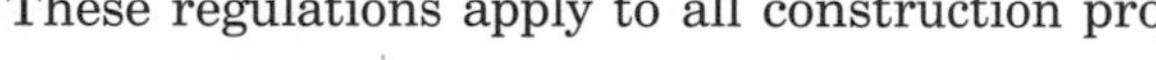

These regulations apply to all construction projects that:

- include any demolition work, or
- will last for more than 30 days, or
- will involve more than 500 person days of work, or
- will involve more than 5 workers on site at anyone time.

Health and safety plan and file – The regulations require the client, the designers and the building contractors to play their part in improving on-site health and safety. In doing this they will have to draw up a two-stage health and safety plan:

- *Stage 1.* A design plan, which highlights any particular risks of the project, and the equipment and the level of health and safety competence that a prospective contractor will require.
- *Stage 2.* A construction plan, which sets out how health and safety will be managed during the project.

They will also have to draw up a health and safety file. This is to be produced at the end of a project and passed on to the client or building user. It should contain details of health and safety risks that will have to be managed during future maintenance and cleaning work.

Client responsibilities

- Appoint a planning supervisor to draw up the first stage safety plan. Co-ordinate with the principle contractor with regards the second stage. Compile the safety file.
- Appoint the principle contractor.
- Be satisfied that the planning supervisor, principle contractor, designers and any nominated sub-contractors are competent to deal with the health and safety aspects of the project.
- Ensure that construction work does not start until a suitable safety plan is in place.
- Keep the health and safety file available for inspection.

Designer responsibilities – The term 'designer' is used to describe everyone who prepares drawings or specifications for a product and thus includes architects, structural engineers, surveyors, the planning supervisor and other designers on health and safety matters.

- Consider at the design stage, the foreseeable health and safety risks associated with the project. Not only during construction, but also during later maintenance and cleaning.
- Provide information with their design, on any aspect which might affect the health and safety of contractors, cleaners or anyone else who might be affected by their work.

Principal contractor responsibilities – This is the main contractor who has been awarded the contract to undertake the construction work. They may appoint other contractors and sub-contractors, to undertake specific parts of the construction work.

- Prepare and maintain the second stage of the safety plan and supply all relevant information to the planning supervisor for inclusion in the safety file.
- Ensure co-operation between all contractors on health and safety matters.
- Ensure all contractors and employees comply with the requirements of the health and safety plan.

- Ensure that only authorised persons are permitted to enter areas where construction work is being carried out.

Contractor responsibilities – The term 'contractor' is used to include sub-contractors and self-employed persons working on site.

- Co-operate with the principal contractor and other contractors in order to achieve safe and healthy site conditions.
- Provide health and safety information for inclusion as required, into both the health and safety plan and file.

The Construction (Health, Safety and Welfare) Regulations (CHSWR)

The main objective of CHSWR is to promote the health and safety of employees, the self-employed and others who may be affected by construction activities. The issues covered in the regulations include:

- provision of welfare facilities
- provision of working platforms
- prevention of falls
- support of excavations
- provision of guard rails and barriers
- procedures in the event of fire or other emergency
- use of vehicles and transport routes on site
- inspections and reports.

The main specific requirements of CHSWR are as follows.

Toilets – No specific number, but must be clean, with adequate ventilation and lighting. Men and women can use the same toilet provided each is in a separate lockable room.

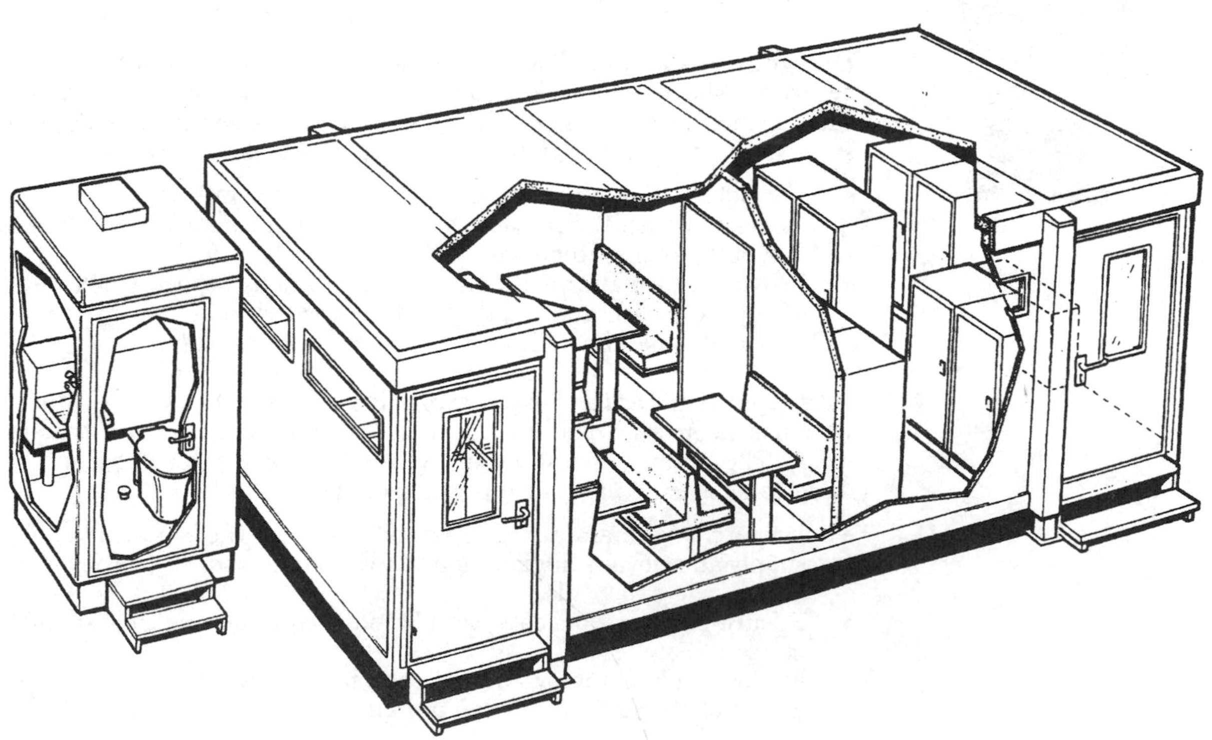

Washing facilities – Wash hand basins with hot and cold or warm water, to be provided in the immediate vicinity of toilets and changing rooms. These must include soap and towels or other drying facility. Where the work is particularly dirty or involves exposure to toxic or corrosive substances, showers may be required. All rooms containing washing facilities must have adequate ventilation and lighting. Unisex facilities are suitable for washing of hands, faces and arms, otherwise in a separate room which is used by one person at a time and can be locked from the inside.

Drinking water – (wholesome) to be readily accessible in suitable places and clearly marked. Cups or other drinking containers must be provided, unless the water is supplied via a drinking fountain.

Storage and changing of clothing – Secure accommodation must be provided for normal clothing not worn at work and for protective clothing not taken home. Separate lockers may be required where there is a risk of protective clothing contaminating normal clothing. This accommodation should include changing facilities and a means of drying wet clothing.

Rest facilities – Accommodation must be provided for taking breaks and meals. These facilities must include tables and chairs, a means of boiling water and a means of preparing food.

Working platforms – Where it is possible for a person to fall two metres or more, the working platform, must be inspected by a competent person: before its first use; after alteration; after strong winds or other events likely to affect its stability and at least once every seven days.

Prevention of falls – Edge protection is required to all working platforms and other exposed edges where it is possible to fall two metres or more. They should be:

- Sufficiently rigid for the purpose
- include a guard rail at least 910 mm above the edge.
- Include a toe board at least 150 mm high.
- be subdivided with intermediate guard rails, additional toe boards or brick guards etc., so that the maximum unprotected gap is 470 mm.

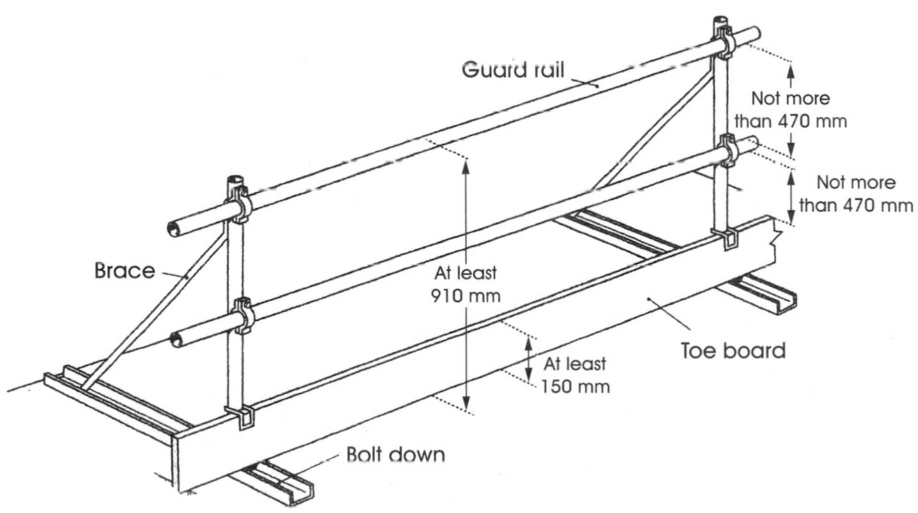

YOU WILL BE INFORMED OF YOUR EMPLOYER'S CHSWR PROVISIONS AT A SAFETY INDUCTION

Other types of barrier may be used to protect edges, provided that they give the equivalent standard of protection against falls of persons and rolling or kicking of materials over the edge.

Support of excavations – Measures must be taken to prevent injury by collapsing excavations, falling materials or contact with buried underground services. Support for excavations is to be provided at an early stage. Sides of excavations must either be battered back to a safe angle or be supported with timbering or a proprietary system. All support work is to be carried out or altered by or under the supervision of a competent person. Measures must be taken to prevent people, materials or vehicles falling into excavations. For example: by the use of edge protection guard rails; not storing materials, waste or plant items near excavations; keeping traffic routes clear of excavations.

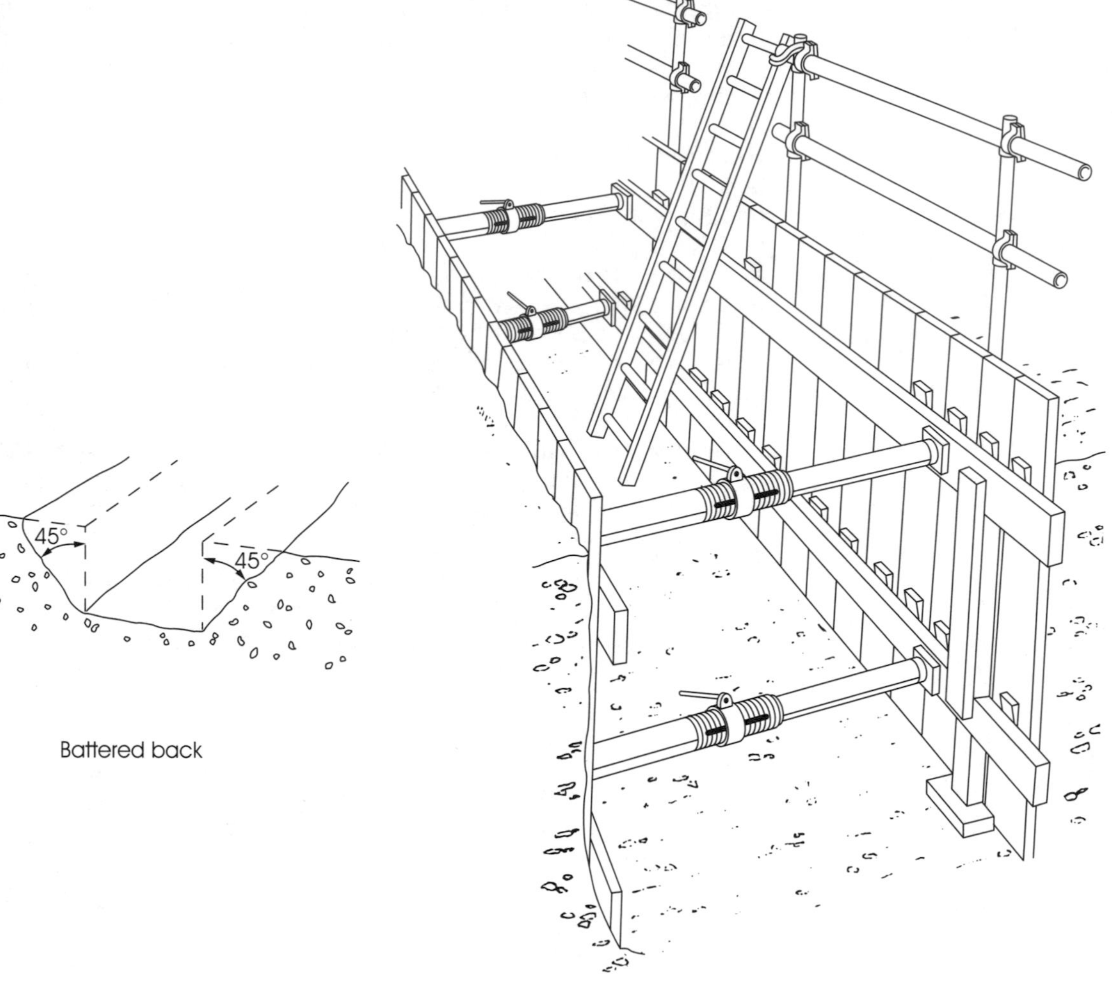

Battered back

Shoring

Emergency procedures – These are the arrangements made to deal with any unforeseen emergency, including fire, flooding, explosion and asphyxiation. These procedures must include the following:

- Provision of emergency signals, routes and exits for evacuation. These should be kept clear, be marked and illuminated.
- Provision to notify the emergency services.
- Provision of first aid and other facilities for treating and recovering injured persons.

These procedures should be co-ordinated by a trained person, who will take responsibility and control.

Use of vehicles –

- All drivers must be trained and visiting drivers informed of site transport rules.
- Suitable traffic routes must be provided, clearly marked avoiding sharp bends and blind corners including safe entry and exit points.
- Pedestrians and vehicles should be separated as far as possible.
- Reversing should be avoided wherever possible. Audible alarms are advisable where reversing is necessary.
- Provide trained signallers, wearing high visibility clothing to assist drivers.

Construction (Health, Safety and Welfare) Regulations 1996

INSPECTION REPORT

Report of results of every inspection made in pursuance of regulation 29(1)

1. Name and address of person for whom inspection was carried out.

2. Site address.

3. Date and time of inspection.

4. Location and description of workplace (including any plant, equipment or materials inspected)

5. Matters which give rise to any health and safety risks.

6. Can work be carried out safely? Y / N

7. If not, name of person informed.

8. Details of any other action taken as a result of matters identified in 5 above.

9. Details of any further action considered necessary.

10. Name and position of person mak[ing] the report.

11. Date report handed over.

A COMPETENT PERSON MUST CARRY OUT INSPECTIONS AND MAKE REPORTS

Construction (Health, Safety and Welfare) Regulations 1996

INSPECTION REPORTS: NOTES

Place of work requiring inspection	Timing of frequency of inspection					
	Before being used for the first time.	After substantial addition, dismantling or alteration.	After any event likely to have affected its strength or stability.	At regular intervals not exceeding 7 days.	Before work at the start of every shift.	After accidental fall of rock, earth or any material.
Any working platform or part thereof or any personal suspension equipment.	✓	✓	✓	✓		
Excavations which are supported in pursuit of paragraphs (1), (2) or (3) of regulation 12.			✓		✓	✓
Cofferdams and caissons.			✓		✓	

NOTES

General

1. The inspection report should be completed before the end of the relevant period.
2. The person who prepares the report should, within 24 hours, provide either the report or a copy to the person on whose behalf the inspection was carried out.
3. The report should be kept on site until work is complete. It should then be retained for three months at the office of the person for whom the inspection was carried out.

Working platforms only

1. An inspection is only required where a person is liable to fall more than 2 metres from a place of work.
2. Any employer or any other person who controls the activities of persons using a scaffold shall ensure that it is stable and of sound construction and that the relevant safeguards are in place before his employees or persons under his control first use the scaffold.
3. No report is required following the inspection of any mobile tower scaffold which remains in the same place for less than 7 days.
4. Where an inspection of a working platform or part thereof or any personal suspension equipment is carried out.
   i. before it is taken into use for the first time; or
   ii. after any substantial addition, dismantling or other alterations;

   not more than one report is required for any 24 hour period.

Excavations only

1. The duties to inspect and prepare a report apply only to any excavation which needs to be supported to prevent any person being trapped or buried by an accidental collapse, or dislodgement of material from its sides, roof or area adjacent to it. Although an excavation must be inspected at the start of every shift, only one report of such inspections is required every 7 days. Reports must be completed for all inspections carried out during this period for other purposes, e.g. after accidental fall material.

Checklist of typical scaffolding faults

Footings	Standards	Ledgers	Bracing	Putlogs and transoms	Couplings	Bridles	Ties	Boarding	Guard-rails and toe-boards	Ladders
Soft and uneven	Not plumb	Not level	Some missing	Wrongly spaced	Wrong fitting	Wrong spacing	Some missing	Bad boards	Wrong height	Damaged
No base plates	Jointed at same height	Joints in same bay	Loose	Loose	Loose	Wrong couplings	Loose	Trap boards	Loose	Insufficient length
No safe plates	Wrong spacing	Loose	Wrong fittings	Wrongly supported	Damaged	No check couplers	Not enough	Incomplete	Some missing	Not bad
Undermined	Damaged	Damaged	–	–	No check couplers	–	–	Insufficient supports	–	–

Inspections and reports – Competent persons must carry out the following inspections and make written reports:

- *Working platforms* – inspect before use, after alteration, after any event which may have affected its stability and at least once every seven days.
- *Excavations* – inspect at the start of each shift before work commences and after any fall of material, rock or earth.

No work is to commence unless the competent person is satisfied that work can be carried out safely. The workplace must not be used until defects have been put right.

Notifications reports and records – These are also required for the following actions and incidents and are usually submitted on standard forms obtainable from the relevant authority. A record can be kept on-site by making a photocopy of the completed form before submitting it.

- Notification of a construction project, which will last more than 30 days or 500 person days or have more than 5 workers on site at a time.
- Notification and record of accidents resulting in death or major injuries or notifiable dangerous occurrences, or more than three days absence from work, or for a specified disease associated with the work. Major injuries can be defined as most fractures, amputations, loss of sight or any other injury involving a stay in hospital. Many incidents can be defined as notifiable dangerous occurrences but in general they include the collapse of a crane, hoist, scaffolding or building, an explosion or fire, or the escape of any substance that is liable to cause a health hazard or major injury to any person.
- A record of all accidents and first aid treatments.

See pp. 60 and 61.

The Construction (Head Protection) Regulations

This legislation places a duty on employers and the self-employed to provide and ensure that suitable head protection is worn on site. Employees and the self-employed are obliged to wear them. Employers must ensure that suitable head protection is properly worn at all times on site, unless there is no foreseeable risk or injury to the head. Site rules should be set down giving guidance to employees and the self-employed and for site visitors.

The Provision and Use of Work Equipment Regulations (PUWER)

The main objectives of PUWER are to ensure that all equipment used in the workplace is:

- suitable for its use
- properly maintained
- provided with all appropriate safety devices and warning notices
- all users and supervisors of equipment are given health and safety information, training and written instructions.

HSE Health & Safety Executive

Notification of project

NOTES
1 This form can be used to notify any project covered by the Construction (Design and Management) Regulations 1994 which will last longer than 30 days or 500 person days. It can also be used to provide additional details that were not available at the time of initial notification of such projects (any day on which construction work is carried out (including holidays and weekends) should be counted, even if the work on that day is of short duration. A person day is one individual, including supervisors and specialists, carrying out construction work for one normal working shift.)

2 The form should be completed and sent to the HSE area office covering the site where construction work is to take place. You should send it as soon as possible after the planning supervisor is appointed to the project.

3 The form can be used by contractors working for domestic clients. In this case only parts 4–8 and 11 need to be filled in.

HSE - For official use only

Client V PV NV — Planning supervisor V PV NV

Focus serial number — Principal contractor V PV NV

1 Is this the initial notification of this project or are you providing additional information that was not previously available?

Initial notification ☐ Additional notification ☐

2 Client: name, full address, postcode and telephone number *(if more than one client, please attached details on separate sheet)*

Name: Telephone number:
Address:
Postcode:

3 Planning Supervisor: name, full address, postcode and telephone number

Name: Telephone number:
Address:
Postcode:

4 Principal Contractor: *(or contractor when project for domestic*

Name:
Address:
Postcode:

5 Address of site: where construction is to be carried out

Address:
Postcode:

F10 (rev0.3 95)

6 Local Authority: name of the local government district council or island council within whose district the operations are to be carried out

7 Please give your estimates on the following: Please indicate if these estimates are original ☐ revised ☐ *(tick relevant box)*

a. The planned date for the commencement of the construction work

b. How long the construction work is expected to take *(in weeks)*

c. The maximum number of people carrying out construction work on site at any one time

d. The number of contractors expected to work on site

8 Construction work: give brief details of the type of construction work that will be carried out

9 Contractors: name full address and postcode of those who have been chosen to work on the project *(if required continue on a separate sheet). (Note this information is only required when it is known at the time notification is first made to HSE. An update is not required)*

Declaration of planning supervisor

10 I hereby declare that..*(name of organisation)* has been appointed as planning supervisor for the project

Signed by or on behalf of the organisation..*(print name)*..

Date..

Declaration of principal contractor

11 I hereby declare that..*(name of principal contractor)* has been appointed as principal contractor *(or contractor undertaking project for domestic client)*

Signed by or on behalf of the organisation..*(print name)*..

Date..

TO BE COMPLETED BEFORE WORK STARTS ON SITE

Notification of a construction project

HSE Health & Safety Executive

Health and Safety at Work etc Act 1974
The Reporting of Injuries, Diseases and Dangerous Occurrences Regulations 1995

Report of an injury or dangerous occurrence

Filling in this form
This form must be filled in by an employer or other responsible person.

Part A

About you

1 What is your full name?

2 What is your job title?

3 What is your telephone number?

About your organisation

4 What is the name of your organisation?

5 What is the address and postcode?

6 What type of work does the organisation do?

Part B

About the incident

1 On what date did the incident happen?

/ /

2 At what time did the incident happen?
(Please use the 24-hour clock e.g. 0600)

3 Did the incident happen at the above address?

Yes ☐ Go to question 4

No ☐ Where did the incident happen?

☐ elsewhere in your organisation – give the name, address and postcode

☐ at someone else's premises – give the name, address and postcode

☐ in a public place – give details of where it happened

If you do not know the postcode, what is the name of the local authority?

4 In which department, or where on the premises, did the incident happen?

F2508 (01/96)

Part C

About the injured person

If you are reporting a dangerous occurrence, go to Part F.
If more than one person was injured in the same incident, please attach the details asked for in Part C and Part D for each injured person.

1 What is their full name?

2 What is their home address and postcode?

3 What is their home phone number?

4 How old are they?

5 Are they

☐ male?

☐ female?

6 What is their job title?

7 Was the injured person (tick only one box)

☐ one of your employees?

☐ on a training scheme? Give details:

☐ on work experience?

☐ employed by someone else? Give details of this employer:

3 Was the injury (tick the one box that applies)

☐ a fatality?

☐ a major injury or condition? (see accompanying notes)

☐ an injury to an employee or self-employed person which prevented them doing their normal work for more than 3 days?

☐ an injury to a member of the public which meant they had to be taken from the scene of the accident to a hospital for treatment?

4 Did the injured person (tick the boxes that apply)

☐ become unconscious?

☐ need resuscitation?

☐ remain in hospital for more than 24 hours?

☐ none of the above?

Part E

About the kind of accident

Please tick the one box that best describes what happened, then go to Part G.

☐ Contact with moving machinery or material being machined

☐ Hit by a moving flying or falling object

☐ Hit by a moving vehicle

☐ Hit something fixed or stationary

☐ Injured while handling, lifting or carrying

☐ Slipped, tripped or fell on the same level

☐ Fell from a height
How high was the fall

metres

☐ Trapped by something collapsing

☐ Drowned or asphyxiated

☐ Exposed to, or in contact with , a harmful substance

☐ Exposed to fire

☐ Exposed to an explosion

☐ Contact with electricity or an electrical discharge

☐ Injured by an animal

☐ Physically assaulted by a person

☐ Another kind of accident (describe it in Part G)

Part F

Dangerous occurrences

Enter the number of the dangerous occurrence you are reporting. (The numbers are given in the Regulations and in the notes which accompany this form)

Part G

Describing what happened

Give as much detail as you can. For instance;

- the name of any substance involved
- the name and type of any machine involved
- the events that led to the incident
- the part played by any people

If it was a personal injury, give details of what the person was doing. Describe any action that has since been taken to prevent a similar incident. Use a separate piece of paper if you need to.

Part H

Your signature

Signature

Date

/ /

Where to send the form
Please send it to the Enforcing Authority for the place where it happened. If you do not know the Enforcing Authority, send it to the nearest HSE office.

For official use
Client number Location number Event number

☐ INV REP ☐ Y ☐

REPORT ALL ACCIDENTS, INJURIES AND DANGEROUS OCCURRENCES

Regulation of hazardous substances

The Control of Substances Hazardous to Health Regulations (COSHH)

Where people use or are exposed to hazardous substances, the COSHH regulations require:

- the assessment of risks involved
- the prevention of exposure to risk
- measures taken to adequately control it
- monitoring of the effectiveness of the measures taken.

Identification – People may be exposed to risks either because they handle a hazardous substance, or because during the work a hazardous substance is created. Manufacturers and suppliers of such substances are required to provide safety data sheets for reference purposes.

ALWAYS READ THE SAFETY DATA SHEET AND TAKE APPROPRIATE ACTION
REGS.

BBS: Panel Products
33 Stafford Thorne Street
Nottingham NG22 3RD
Tel. 011594000

SAFETY DATA SHEET

Chemical Name:	Interior Medium Density Fibreboard (MDF)	
Trade Name:	MeDFit	
Chemical Family:	Wood Based Panel Product	
Formula:	Mixture	
Ingredients:	Mixed Softwoods	82%
	Urea Formaldehyde Resins	8–10%
	Paraffin Wax	0.5%
	Water	6–8%
	Silica	<0.05%
	Free Formaldehyde	<0.04%

Physical and Chemical Characteristics:
Specific Gravity: 0.65–0.99
Appearance/colour: Cream to light brown, solid wood texture.

Fire and Explosion Hazard:
Extinguisher Media: Water
Explosion Hazard: None for the sheet material. However airbourne dust produced during re-manufacturing operations may cause an explosion hazard. Dust should be continuously removed from processing machinery. Smoking should not be permitted in the working area.

Health Hazards: During re-manufacture wood dust may:
- increase mucosal output
- cause reddening and itching of the skin
- irritation of the throat and eyes.

Most of the effects are readily reversible after the end of exposure.

Personal Protection: During re-manufacturing operations:
- wear dust mask and eye protection
- apply a barrier cream (replenish after washing)
- wash before eating, drinking, smoking and going to the toilet.

Special controls: During re-manufacturing operations the use of high efficiency dust collection equipment is strongly recommended to ensure compliance with the COSHH regulations.

First Aid:
Inhalation of dust: Clean nasal passages, take in plenty of fresh air

Contamination of eyes: Flush with an approved eye wash solution for a prolonged period.

EXPLOSIVE

HIGHLY INFLAMMABLE

IRRITANT

CORROSIVE

TOXIC

HARMFUL

OXIDIZING

Assessment – Must be undertaken by employers (see risk assessment). They must look at the ways people may be exposed to hazardous substances in the particular type of work undertaken. For example:

- breathing in dust, fumes or vapours
- swallowing or eating contaminated materials
- contact with skin or eyes.

Prevention – Where harm from a substance is likely, the first course of action should be prevention of exposure by:

- Doing the job in a different way so that the substance is removed or not created e.g. rodding out blocked drains rather than using hazardous chemicals.
- Using a less hazardous substitute substance, e.g. using water-based paints rather than more hazardous spirit-based ones.

Control – Where the substance has to be worked with because either there is no choice, or because alternatives also present equal risks, exposure must be controlled by:

- Using the substance in a less hazardous form, e.g. use a sealed surface glass fibre insulation quilt rather than an open fibre one, to reduce the risk of skin contact or the inhalation of fine strands.
- Using a less hazardous method of working with the substance. Wet rubbing down of old lead based painted surfaces rather than dry rubbing down which causes hazardous dust; or applying spirit-based products by brush or roller rather than by spraying.
- Limiting the amount of substance used.
- Limiting the amount of time people are exposed.
- Keeping all containers closed when not in use.
- Providing good ventilation to the work area. Mechanical ventilation may be required in confined spaces.
- When cutting or grinding use tools fitted with exhaust ventilation or water suppression to control dust.

Study the table of recognised hazards in various construction jobs and suggested methods of control. See p. 64.

Protection – If exposure cannot be prevented, or adequately controlled using any of the above, also use personal protective equipment (PPE).

- Always wear protective clothing. Overalls, gloves (for protection and anti-vibration), boots, helmets, ear protection, eye protection goggles or visors and dust masks or respirators as appropriate.
- The use of barrier and after-work creams is recommended to protect skin from contact dermatitis.
- Ensure items of PPE are kept clean, so that they do not themselves become a source of contamination.
- All items of PPE should be regularly maintained, checked for damage and stored in clean dry conditions.
- Replacement items of PPE and spare parts, must be available for use when required.

Personal hygiene – Protection does not stop with PPE. Hazardous substance can be easily transferred from contaminated clothing and unwashed hands and face.

Hazardous substances in construction

Substances	*Health risk*	*Jobs*	*Controls*
DUSTS:			
Cement (Also when wet)	SK I ENT	Masonry, rendering	Prevent spread. Protective clothing, respirator when handling dry, washing facilities, barrier cream.
Gypsum	SK I ENT	Plastering	
Man-made mineral fibre	I SK ENT	Insulation	Minimise handling/cutting, respirator, one piece overall, gloves, eye protection.
Silica	I	Sand blasting, grit blasting: scrabbling granite, polishing	Substitution – e.g. with grit, silica-free sand; wet methods; process enclosure/extraction; respirator.
Wood dust (Dust from treated timber e.g. with pesticide may present extra hazards)	I SK ENT	Power tool use in carpentry, especially sanding	Off-site preparation; on-site – enclosures with exhaust ventilation; portable tools – dust extraction; washing facilities; respirator.
Mixed dusts (Mineral and biological)	I SK ENT	Demolition and refurbishment	Minimise dust generation; use wet methods where possible; segregate or reduce number of workers exposed; protective clothing, respirator; good washing facilities/showers. Tetanus immunisation.
FUMES/GASES:			
Various welding fumes from metals or rods	I	Welding/cutting activities	Mechanical ventilation in enclosed spaces; air supplied helmet; elsewhere good general ventilation.
Hydrogen sulphide	I ENT	Sewers, drains, excavations, manholes	All work in confined spaces – exhaust and blower ventilation; self contained breathing equipment confined space procedures.
Carbon monoxide/nitrous oxide	I	Plant exhausts	Position away from confined spaces. Where possible maintain exhaust filters; forced ventilation and extraction of fumes.
SOLVENTS: In many construction products – paints, adhesives, strippers, thinners, etc.	I SK SW	Many trades, particularly painting, tile fixing. Spray application is high risk. Most brush/roller work less risk. Regulation exposure increases risks	Breathing apparatus for spraying, particularly in enclosed spaces; use of mistless/airless methods. Otherwise ensure good general ventilation. Washing facilities, barrier cream.
RESIN SYSTEMS:			
Isocynates (MDI:TDI)	I ENT SK SW	Thermal insulation	Mechanical ventilation where necessary; respirators; protective clothing, washing facilities. Skin checks, respirators checks.
Polyurethane paints	I ENT SK SW	Decorative surface coatings	Spraying – airline/self contained breathing apparatus; elsewhere good general ventilation. One-piece overall, gloves, washing facilities.
Epoxy	I SK SW	Strong adhesive applications	Good ventilation, personal protective equipment (respirator; clothing) washing facilities, barrier cream. As above.
Polyester	I SK ENT SW	Glass fibre claddings and coatings	
PESTICIDES: (e.g. timber preservatives, fungicides, weed killers)	I SK ENT SW	Particularly in-situ timber treatment. Handling treated timber	Use least toxic material. Mechanical ventilation, respirator, impervious gloves, one-piece overall and head cover. In confined spaces – breathing apparatus. Washing facilities, skin checks. If necessary biological checks. Handle only dry material.
ACIDS/ALKALIS:	SK ENT	Masonry cleaning	Use weakest solutions. Protective clothing, eye protections. Washing facilities (first aid including eye bath and copious water for splash removal).
MINERAL OIL:	SK I	Work near machines, compressors, etc. Mould release agents	Filters to reduce mist. Good ventilation. Protective clothing. Washing facilities; barrier creams. Skin checks.
SITE CONTAMINANTS: e.g. Arsenic. Phanols; heavy metals; Micro organisms etc. e.g. Wells disease, tetanus, hepatitis B	I SK SW	Site re-development of industrial premises or hospitals – particularly demolition ground work and drain/sewers	Thorough site examination and clearance procedures. Respirators, protective clothing. Washing facilities/showers. Immunisation for tetanus.

Health risk
SK = skin; I = inhalation; ENT = irritant eyes, nose, throat; SW = ingestion
Table extracted from Control of Substances Hazardous to Health (COSHH) regulations

Questions for you

7. Name **THREE** notices or certificates that must be displayed at a building site.

____

____

____

____

____

____

____

8. Name **TWO** parts of the Construction Regulations.

____

____

9. A safety sign that is contained in a yellow triangle with a black border is:
(a) prohibiting certain behaviour
(b) warning of certain hazards
(c) providing information about safety
(c) indicating that safety equipment must be worn.

a [ ]	b [ ]	c [ ]	d [ ]

- Always wash hands and face before eating, drinking or smoking and also at the end of the working shift.
- Never eat, drink or smoke near to the site of exposure.
- Change out of contaminated work wear into normal clothes before travelling home.
- Have contaminated work wear regularly laundered.

Monitoring and health surveillance – Must be carried out to ensure exposure to hazardous substances is being adequately controlled.

- Monitoring of the workplace is required to ensure exposure limits are not being exceeded e.g. regular checks on noise levels and dust or vapour concentrations.
- Health surveillance is a legal duty for a limited range of work exposure situations (for example asbestos). However many employers operate a health surveillance programme for all their employees. This gives medical staff the opportunity to check the general health of workers, as well as giving early indications of illness, disease

and loss of sensory perception. Simple checks can be made on a regular basis including blood pressure, hearing, eyesight and lung peak flow. Any deterioration over a time indicates the need for further action.

Information – Employers must provide their employees, who are or may be exposed to hazardous substances with:

- Safety information and training for them to know the risks involved.
- A safe working method statement, including any precautions to be taken or PPE to be worn.
- Results of any monitoring and health surveillance checks.

BBS: Shopfitting Services
33 Stafford Thorne Street
Nottingham NG22 3RD
Tel. 0115 94000

SAFETY METHOD STATEMENT

Process: **The re-manufacture of MDF panel products.** During this process a fine airborne dust is produced. This may cause skin, eye, nose and throat irritation. There is also a risk of explosion. The company has controls in place to minimise any risk. However, for your own safety and the protection of others, you must play your part by observing the following requirements.

General Requirements: At all times observe the following safety method statements and the training you have received from the company.

- Manual Handling
- Use of Woodworking Machines
- Use of Powered Hand Tools
- General House Keeping

Specific Requirements:

- When handling MDF, always wear gloves or barrier cream as appropriate. Barrier cream should be replenished after washing.
- When sawing, drilling, routing or sanding MDF, always use the dust extraction equipment and wear dust masks and eye protection.
- Always brush down and wash thoroughly to remove all dust, before eating, drinking, smoking, going to the toilet and finally at the end of the shift.
- Do not smoke outside the designated areas.
- If you suffer from skin irritation or other personal discomfort seek first aid treatment or consult the nurse.

IF IN DOUBT ASK

ALWAYS READ SAFETY METHOD STATEMENTS AND TAKE APPROPRIATE ACTION

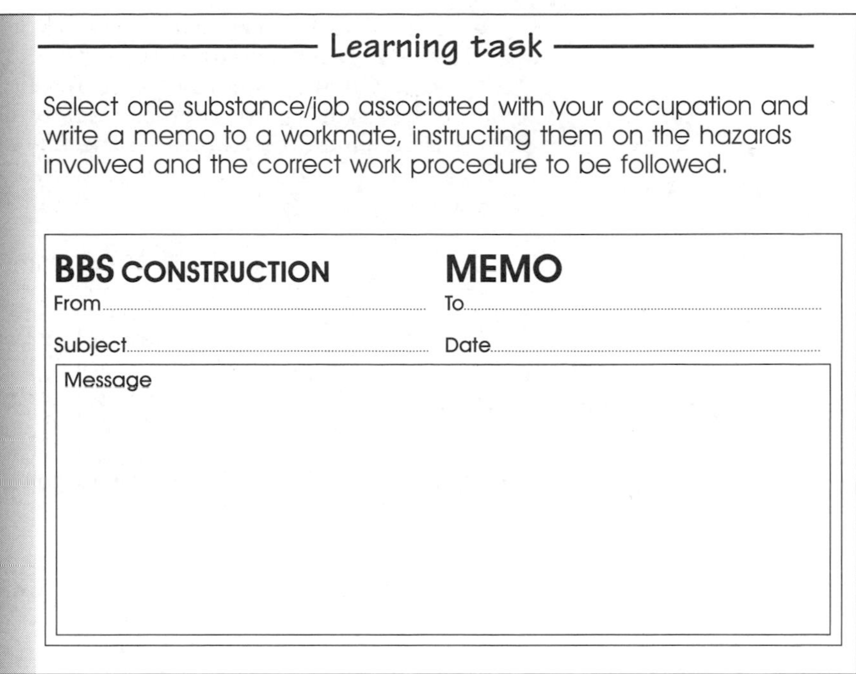

Learning task

Select one substance/job associated with your occupation and write a memo to a workmate, instructing them on the hazards involved and the correct work procedure to be followed.

BBS CONSTRUCTION **MEMO**

From.......................... To..........................

Subject.......................... Date..........................

Message

General safety

It should be the aim of everyone to prevent accidents. Remember, you are required by law to be aware and fulfil your duties under the Health and Safety at Work Act and other regulations.

The main contribution you as an operative can make towards the prevention of accidents is to work in the safest possible manner at all times, thus ensuring that your actions do not put at risk yourself, your workmates or the general public.

Safety: on site and in the workshop

A safe working area is a tidy working area. All unnecessary obstructions, which may create a hazard should be removed, e.g. off-cuts of material, unwanted materials, disused items of plant, and the extraction or flattening of nails from discarded pieces of timber. Therefore:

WARNING!
FIRE EXTINGUISHERS CAN GIVE OFF DANGEROUS FUMES

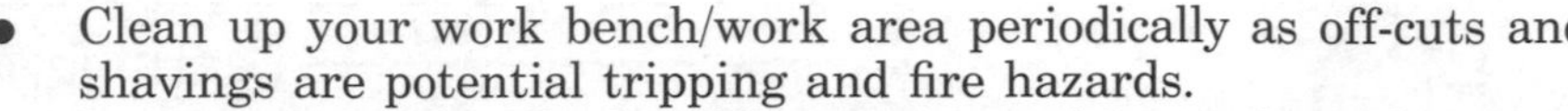

- Clean up your work bench/work area periodically as off-cuts and shavings are potential tripping and fire hazards.
- Learn how to identify the different types of fire extinguishers and what type of fire they can safely be used on. Staff in each work area should be trained in the use of fire extinguishers, see table.

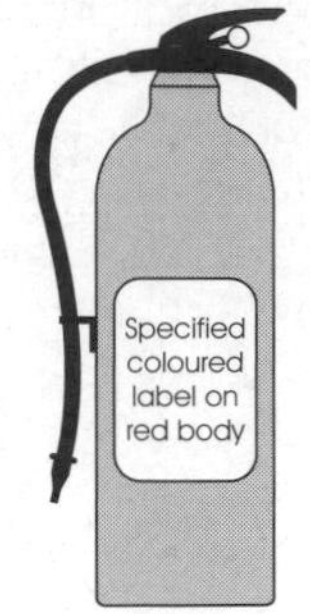

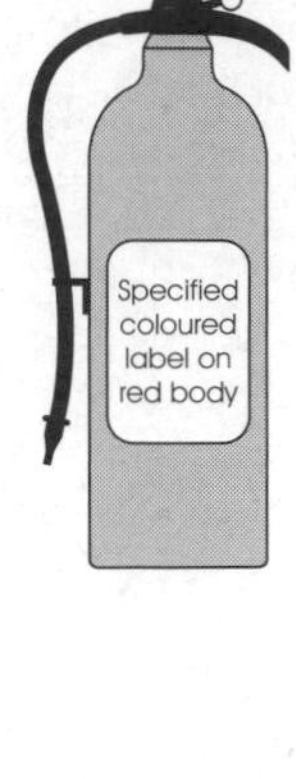

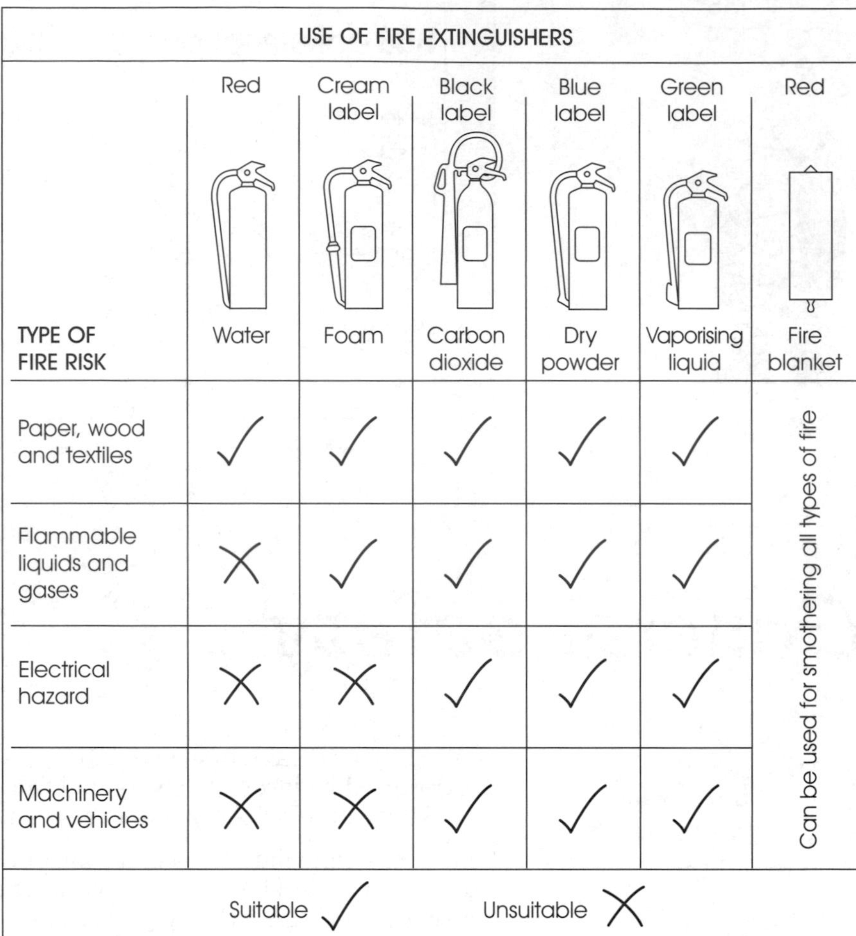

USE OF FIRE EXTINGUISHERS						
	Red	Cream label	Black label	Blue label	Green label	Red
TYPE OF FIRE RISK	Water	Foam	Carbon dioxide	Dry powder	Vaporising liquid	Fire blanket
Paper, wood and textiles	✓	✓	✓	✓	✓	Can be used for smothering all types of fire
Flammable liquids and gases	✗	✓	✓	✓	✓	
Electrical hazard	✗	✗	✓	✓	✓	
Machinery and vehicles	✗	✗	✓	✓	✓	
Suitable ✓			Unsuitable ✗			

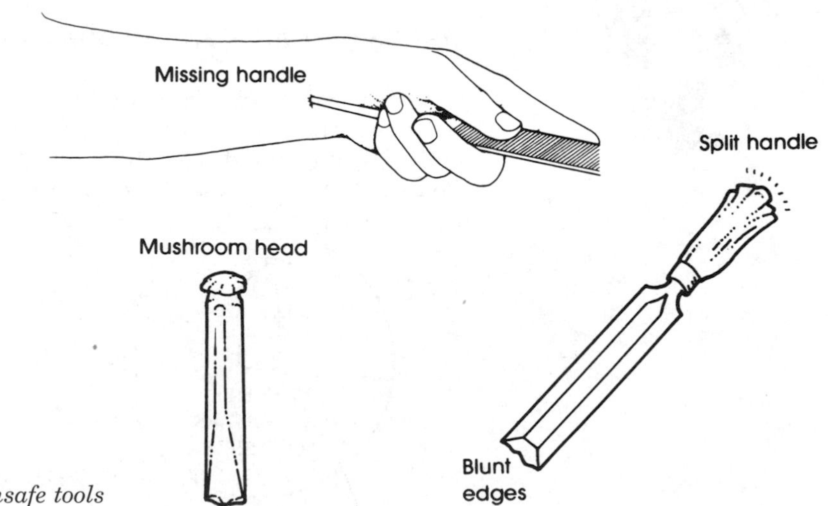

Unsafe tools

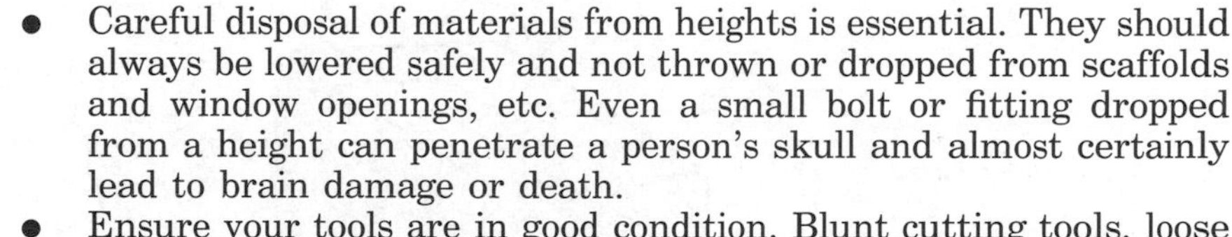

- Careful disposal of materials from heights is essential. They should always be lowered safely and not thrown or dropped from scaffolds and window openings, etc. Even a small bolt or fitting dropped from a height can penetrate a person's skull and almost certainly lead to brain damage or death.
- Ensure your tools are in good condition. Blunt cutting tools, loose hammer heads, broken or missing handles and mushroom heads must be repaired immediately or the use of the tool discontinued.
- When moving materials and equipment always look at the job first; if it is too big for you then get help. Look out for splinters, nails and sharp or jagged edges on the items to be moved. Always lift with your back straight, elbows tucked in, knees bent and feet slightly apart. When putting an item down ensure that your hands and fingers will not be trapped.

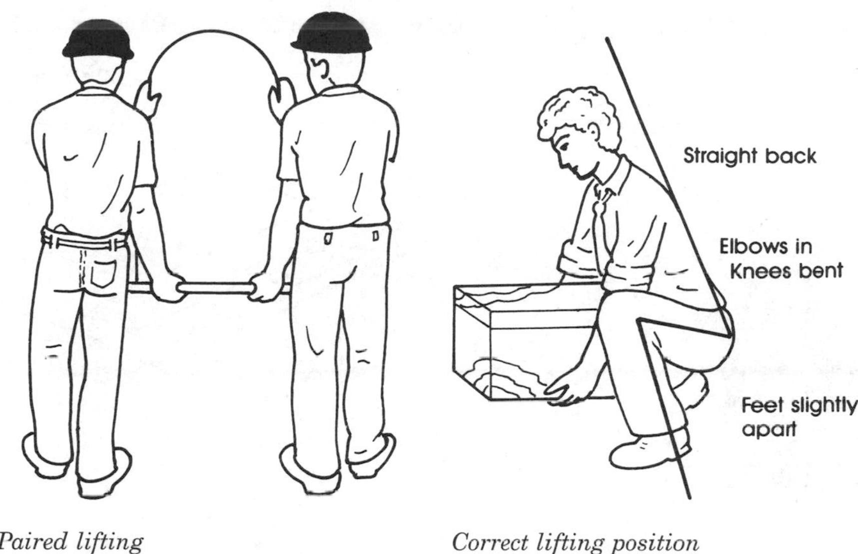

Paired lifting

Correct lifting position

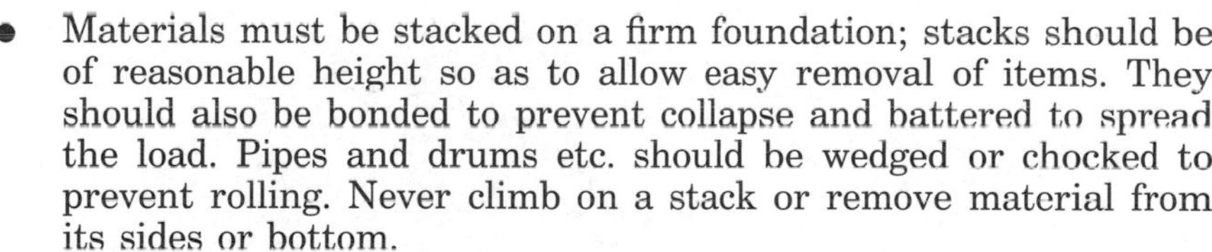

- Materials must be stacked on a firm foundation; stacks should be of reasonable height so as to allow easy removal of items. They should also be bonded to prevent collapse and battered to spread the load. Pipes and drums etc. should be wedged or chocked to prevent rolling. Never climb on a stack or remove material from its sides or bottom.

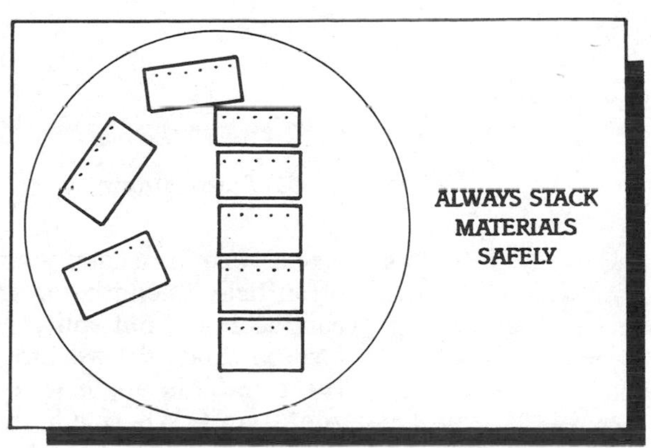

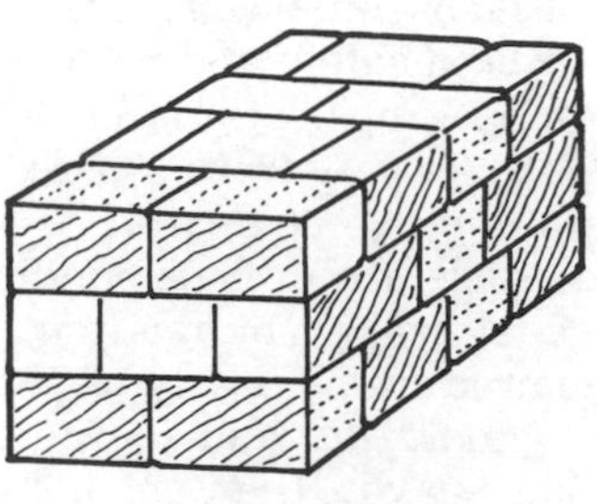

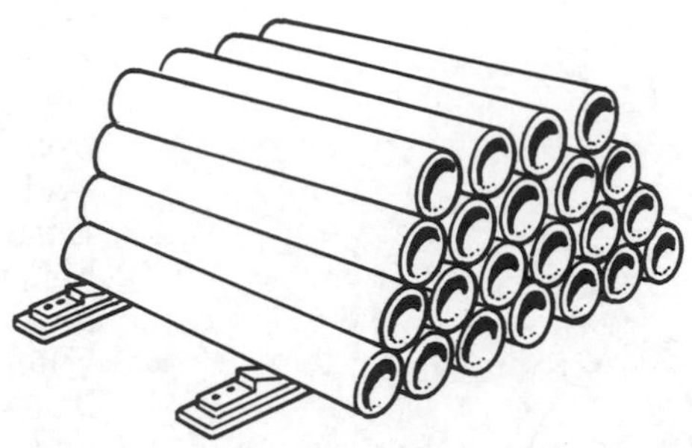

Bonded material storage *Chocked material storage*

- Excavations and inspection chambers should be either protected by a barrier or covered over completely to prevent people carelessly falling into them.

Protection of excavations

- Extra care is needed when working at heights. Ladders should be of sufficient length for the work in hand and should be in good condition and not split, twisted or with rungs missing. They should also be used at a working angle of 75 degrees and securely tied at the top. This angle is a slope of four vertical units to one horizontal unit. Where a fixing at the top is not possible, an alternative is the stake-and-guy rope method illustrated.

- Scaffolds should be inspected before working on them. Check to see that all components are there and in good condition, not bent, twisted, rusty, split, loose or out of plumb and level. Also ensure that the base has not been undermined or is too close to excavations. If in doubt do not use, and have it looked at by an experienced scaffolder. For further information see 'Scaffolding' (page 203).

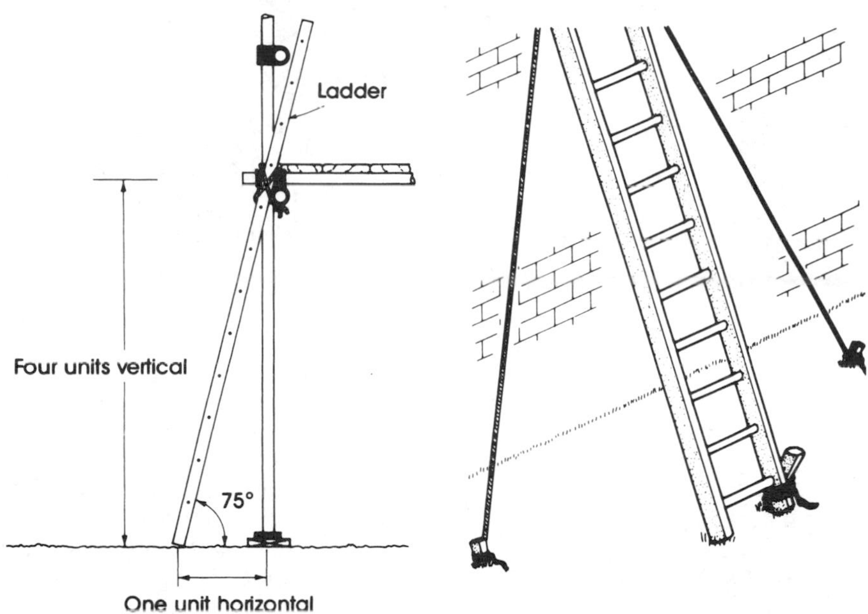

Ladder access *Stakes and guys*

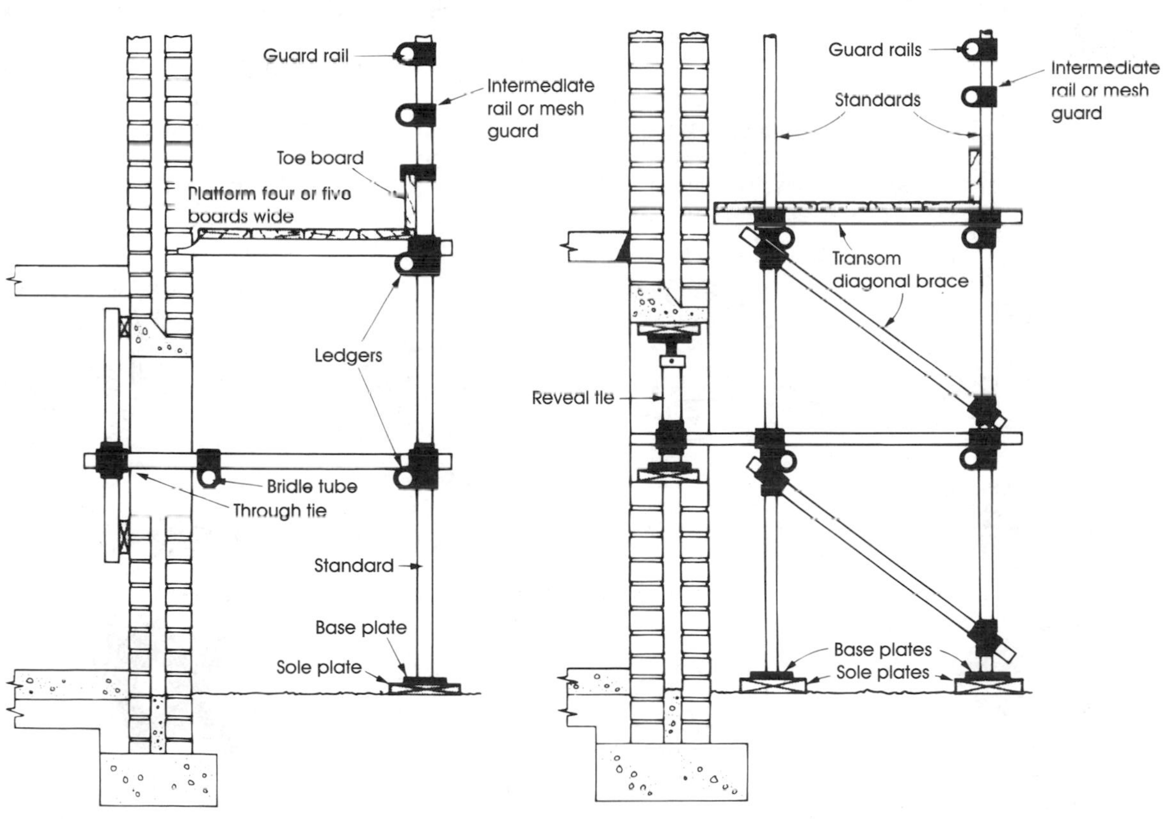

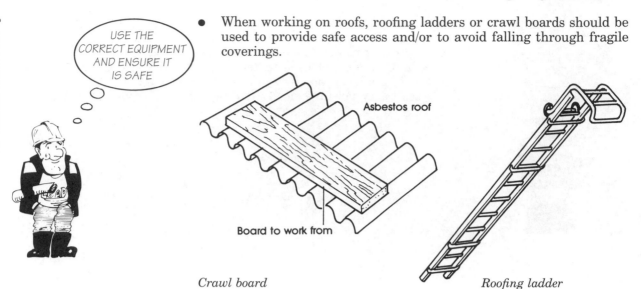

- When working on roofs, roofing ladders or crawl boards should be used to provide safe access and/or to avoid falling through fragile coverings.

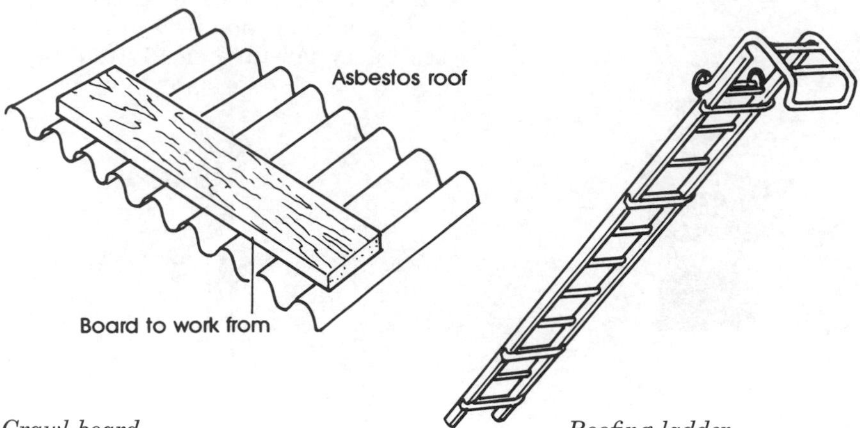

Crawl board *Roofing ladder*

- Working with electrical and compressed-air equipment brings additional hazards as they are both potential killers. Installations and equipment should be checked regularly by qualified personnel; if anything is incomplete, damaged, frayed, worn or loose, do not use it, but return it to stores for attention. Ensure cables and hoses are kept as short as possible and routed safely out of the way to prevent risk of tripping and damage, or in the case of electric cables, from lying in damp conditions.

Protective clothing and equipment – *Always wear the correct protective equipment* for the work in hand. **Safety helmets** and **safety footwear** should be worn at all times. Wear **ear protectors** when carrying out noisy activities, and **safety goggles** when carrying out any operation that is likely to produce dust, chips or sparks etc. **Dust masks** or **respirators** should be worn where dust is being produced or fumes are present, and **gloves** when handling materials. **Wet weather clothing** is necessary for inclement conditions. *Many of these items must be supplied free of charge by your employer.*

Personal hygiene – Care should be taken with personal hygiene which is just as important as physical protection. Some building materials have an irritant effect on contact with the skin. Some are poisonous if swallowed, while others can result in a state of unconsciousness (narcosis) if their vapour or powder is inhaled. These harmful effects can be avoided by taking proper precautions: follow the manufacturer's instructions; avoid inhaling fumes or powders; wear a barrier cream; thoroughly wash your hands before eating, drinking, smoking and after work.

First aid – First aid is the treatment of persons with the purpose of preserving life until medical help is obtained and also the treatment of minor injuries for which no medical help is required.

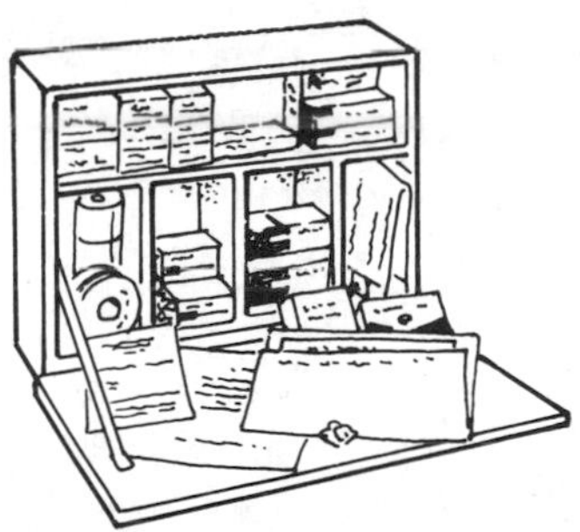

In all cases first aid should only be administered by a trained first-aider. Take care not to become a casualty yourself! Send for the nearest first-aider and/or medical assistance (phone 999) immediately. Even minor injuries where you may have applied a simple plaster or sterilised dressing to yourself may require further attention.

Note: you are strongly recommended to seek medical attention if a minor injury becomes inflamed, painful or festered.

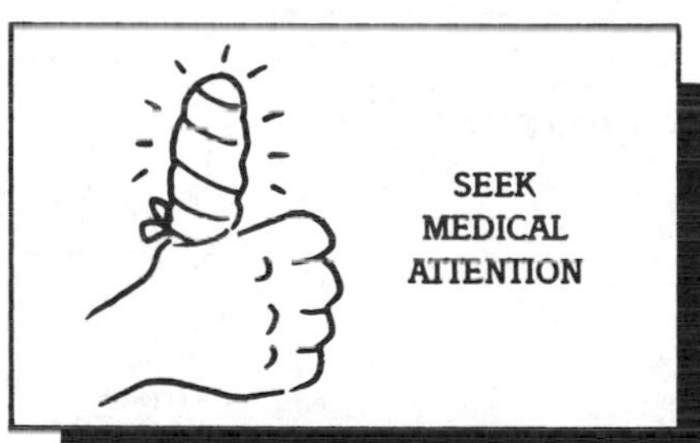

For further information see the General First Aid Guidance Leaflet shown on page 74, which should be found in every first aid box.

(b) *Chemical burns* Remove any contaminated clothing which shows no sign of sticking to the skin and flush all affected parts of the body with plenty of clean, cool water ensuring that all the chemical is so diluted as to be rendered harmless. Apply a sterilised dressing to exposed, damaged skin and clean towels to damaged areas where the clothing cannot be removed. (N.B. Take care when treating the casualty to avoid contamination).

(c) *Foreign bodies in the eye* If the object cannot be removed readily with a clean piece of moist material, irrigate with clean, cool water. People with eye injuries which are more than minimal must be sent to hospital with the eye covered with an eye pad from the container.

(d) *Chemical in the eye* Flush the open eye at once with clean, cool water; continue for at least 5 to 10 minutes and, in any case of doubt, even longer. If the contamination is more than minimal, send the casualty to hospital.

(e) *Electric shock* Ensure that the current is switched off. If this is impossible, free the person, using heavy duty insulating gloves (to BS 697/1977) where these are provided for this purpose near the first aid container, or using something made of rubber, dry cloth or wood or a folded newspaper; use the casualty's own clothing if dry. *Be careful* not to touch the casualty's skin before the current is switched off. If breathing is failing or has stopped, start resuscitation and continue until breathing is restored or medical, nursing or ambulance personnel take over.

(f) *Gassing* Move the casualty to fresh air *but make sure that whoever does this is wearing suitable respiratory protection.* If breathing has stopped, start resuscitation and continue until breathing is restored or until medical, nursing or ambulance personnel take over. If the casualty needs to go to hospital make sure a note of the gas involved is sent with him.

General

(a) *Hygiene* When possible, wash your hands before treating wounds, burns or eye injuries. Take care in any event not to contaminate the surfaces of dressings

(b) *Treatment position* Casualties should be seated or lying down while being treated

(c) *Record-keeping* An entry must be made in the accident book (for example B1 510 Social Security Act Book) of each case

(d) *Minor injuries* Casualties with minor injuries, of a sort they would attend to themselves if at home, may wash their hands and apply a small sterilised dressing from the container

(e) *First aid materials* Each article used from the container should be replaced as soon as possible

Health and Safety (First Aid) Regulations 1981

General first aid guidance for first aid boxes

Note: Take care not to become a casualty yourself while administering first aid. Be sure to use protective clothing and equipment where necessary. If you are not a trained first-aider, send immediately for the nearest first-aider where one is available.

Advice on treatment

If the assistance of medical or nursing personnel will be required, send for a doctor or nurse (where they are employed at the workplace) or ambulance immediately. When an ambulance is called, arrangements should be made for it to be directed to the scene without delay.

Priorities

(1) *Breathing* If the casualty has stopped breathing, resuscitation must be started at once *before any other treatment is given* and should be continued until breathing is restored until medical, nursing or ambulance personnel take over.

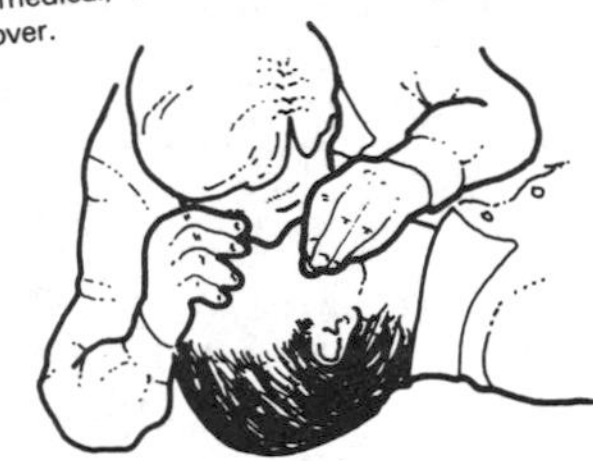

Mouth-to-mouth resuscitation

(2) *Bleeding* If bleeding is more than minimal, control it by direct pressure – apply a pad of sterilised dressing or, if necessary, direct pressure with fingers or thumb on the bleeding point. Raising a limb if the bleeding is sited there will help reduce the flow of blood (unless the limb is fractured).

(3) *Unconsciousness* Where the patient is unconscious, care must be taken to keep the airway open. This may be done by clearing the mouth and ensuring that the tongue does not block the back of the throat. Where possible, the casualty should be placed in the recovery position.

Recovery position

(4) *Broken bones* Unless the casualty is in a position which exposes him to further danger, do not attempt to move a casualty with suspected broken bones or injured joints until the injured parts have been supported. Secure so that the injured parts cannot move.

(5) *Other injuries*

(a) *Burns and scalds* Small burns and scalds should be treated by flushing the affected area with plenty of clean cool water before applying a sterilised dressing or a clean towel. Where the burn is large or deep, simply apply a dry sterile dressing. (N.B. Do not burst blisters or remove clothing sticking to the burns or scalds).

TRY AND ANSWER THESE

Questions for you

10. The correct angle for an access ladder to a scaffold is:
(a) 1 unit horizontal, 2 units vertical
(b) 1 unit horizontal, 4 units vertical
(c) 2 units horizontal, 1 unit vertical
(d) 4 units horizontal, 1 unit vertical

a [ ]	b [ ]	c [ ]	d [ ]

11. State **FOUR** building site operations where you would insist on the use of protective equipment. Name the item of protective equipment in each case.

12. Describe **FOUR** general procedures to be followed, which would aid either site or workshop safety.

WELL, HOW DID YOU DO?
WORK THROUGH THE SECTION AGAIN IF YOU HAD ANY PROBLEMS

READ THE INSTRUCTIONS AND COMPLETE THE TASK

Learning task

Hazard spotting

The following is intended to reinforce the work undertaken in this book. It gives you an opportunity to use your newly acquired safety awareness.

From the illustration (pages 78–79) of an unsafe building site, you are required to identify safety hazards, breaches of regulations and general bad practices, etc.

There are at least 20 to be found, which relate to areas covered as part of this package.

How many hazards can you spot?

Circle each hazard etc. and number it as in the sketch below:

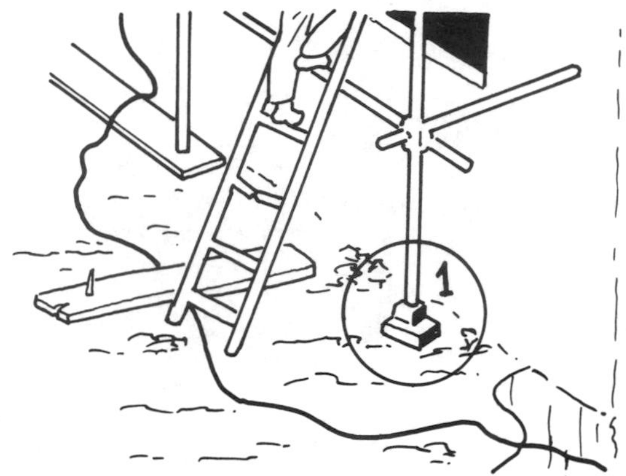
1

Then describe each hazard, etc. like this:

(1) Sole plate missing from under scaffold standard.

Hazards spotted:

LIST THE HAZARDS

(1) ______________________

(2) ______________________

(3) ______________________

(4) ______________________

(5) ______________________

(6) ______________________

(7) ______________________

(8) ____________________

(9) ____________________

(10) ____________________

(11) ____________________

(12) ____________________

(13) ____________________

(14) ____________________

(15) ____________________

(16) ____________________

(17) ____________________

WELL, HOW MANY DID YOU GET?

(18) ____________________

(19) ____________________

(20) ____________________

(21) ____________________

(22) ____________________

(23) ____________________

(24) ____________________

(25) ____________________

BBS
ON HIRE

SPOT THE HAZARDS
NO WAY WOULD I WORK HERE
THINK SAFETY DON'T BECOME A STATISTIC
CONTRACT

COMPLETE THE WORD SQUARE

WORD-SQUARE SEARCH

Hidden in the word square are 12 words associated with safety. You may find the words written forwards, backwards, up, down and diagonally. Solve the clues and then see if you can find the words.

Clues

This package has raised your _ _ _ _ _ _ awareness.

Used at a working angle of 75 degrees _ _ _ _ _ _.

Safety _ _ _ _ _ _ _ should be worn at all times when working on site.

An event causing injury or damage _ _ _ _ _ _ _ _.

A notice to stop work immediately _ _ _ _ _ _ _ _ _ _ _ _.

Carried out before eating _ _ _ _.

Main statutory (umbrella) legislation _ _ _ _ _ _ (abbreviation).

Should be covered _ _ _ _ _ _ _ _ _ _ _.

About 78 each year are _ _ _ _ _.

A HSE inspector may _ _ _ _ _ premises to carry out investigations.

Notifiable dangerous occurrences must be _ _ _ _ _ _ _ _ to the HSE.

Working platforms are covered under the _ _ _ _ _ _ _ _ _ _ _ _ _ regulations.

Draw a ring around the words, or line in using a highlight pen as shown on the left.

EXAMPLE

EXAMPLE

WELL, HOW MANY DID YOU GET?

A	C	R	I	B	Z	F	A	L	B	A	D	P	E	T	E	R	E
C	H	R	I	S	I	H	S	A	L	E	B	R	E	T	T	E	F
C	R	L	S	L	L	A	F	D	A	D	E	O	R	I	P	P	G
S	I	C	A	V	T	S	A	D	L	P	P	H	I	V	E	O	H
D	A	D	D	A	Y	A	T	A	P	O	C	I	E	B	H	R	A
B	U	I	L	N	T	W	T	R	R	T	C	B	I	L	E	T	T
C	R	E	T	N	E	A	L	E	O	T	D	I	A	W	L	E	E
Q	T	W	V	J	F	G	D	A	B	L	F	T	H	A	M	D	S
T	A	T	E	F	A	S	Y	D	I	E	O	I	T	R	E	B	E
W	C	C	O	N	S	T	R	U	C	T	I	O	N	V	T	C	W
X	C	O	B	C	A	C	O	N	S	T	G	N	I	H	S	A	W
U	B	N	E	F	S	R	T	R	S	L	A	F	A	V	T	I	N
R	Q	S	P	V	R	D	A	C	C	I	D	E	N	T	P	E	R
O	E	T	V	N	N	P	D	C	A	N	E	D	I	C	C	A	C
L	A	R	D	E	R	B	N	H	I	B	I	T	T	I	L	N	A
C	A	U	T	I	O	L	A	D	D	E	R	S	O	D	O	D	V
F	A	C	T	A	L	B	M	L	A	D	E	D	E	R	S	I	E
C	S	N	O	I	T	A	V	A	C	X	E	C	I	D	E	N	T

3 Communications

Scale

A **scale rule** has a series of marks used for measuring purposes.

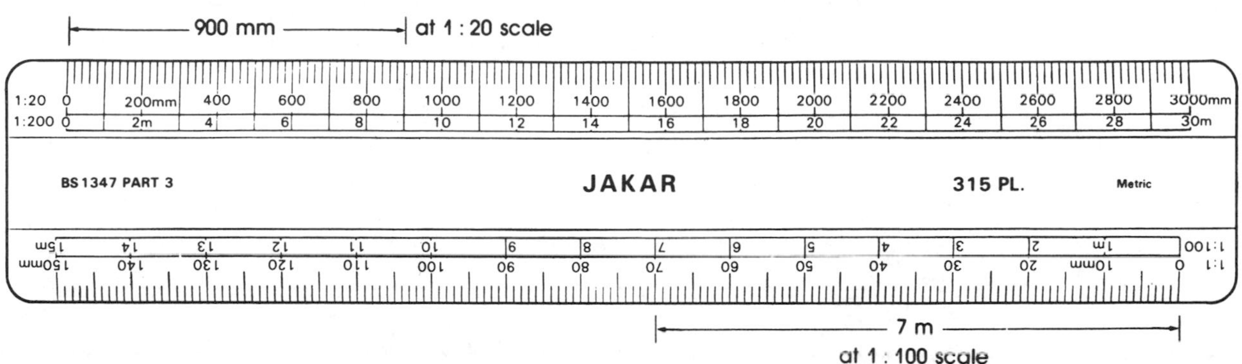

Scales use ratios to relate measurements on a drawing or model to the real dimensions of the actual job. It is impractical to draw buildings, plots of land and most parts of a building to their full size, as they simply will not fit on a piece of paper.

Instead they are normally drawn to a smaller size which has a known ratio to the real thing. These are then called **scale drawings**.

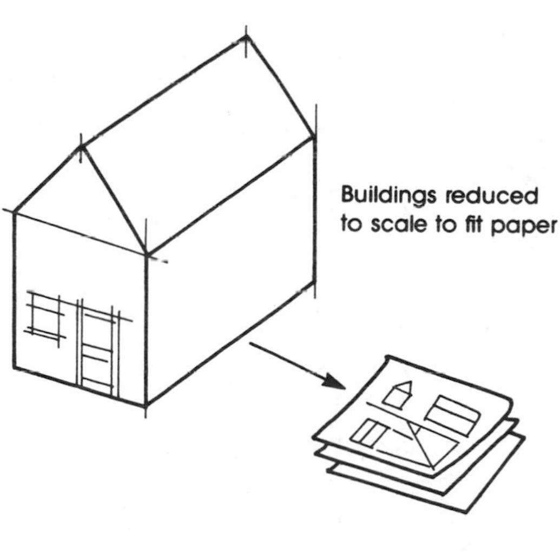

The main scales (ratios) used in the construction industry are:

1:1 1:5 1:10 1:20 1:50 1:100 1:200 1:500 1:1250 1:2500

The ratio shows how much smaller the plan or model is to the original. In a house drawing to a scale of 1:20, 1 mm would stand for 100 mm and so on.

An object drawn 100 mm long would represent a much bigger size.

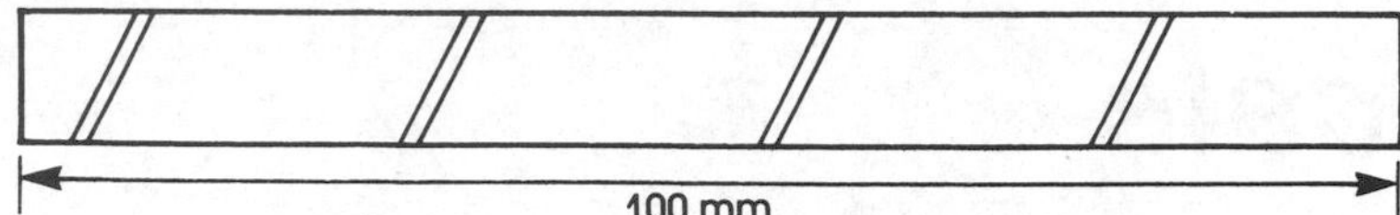

To a scale of 1:10 it would represent 1000 mm or 1 m. Whereas to a scale of 1:50 it would represent 5000 mm or 5 m.

It is simply a matter of multiplying the scale measurement by the scale ratio.

Example

10 mm at a 1:50 scale equals 500 mm

Learning task

Complete the table below.

Scale ratio of job	*Size drawn*	*Actual size*
1:1	100 mm	100 mm
1:5	250 mm	1250 mm
1:10	100 mm	1000 mm
1:20	75 mm	___________
1:50	125 mm	___________
1:100	150 mm	___________
1:200	125 mm	___________
1:1500	45 mm	___________
1:1250	25 mm	___________
1:2500	50 mm	___________

Scale rules can be used for both preparing and reading scale drawings.

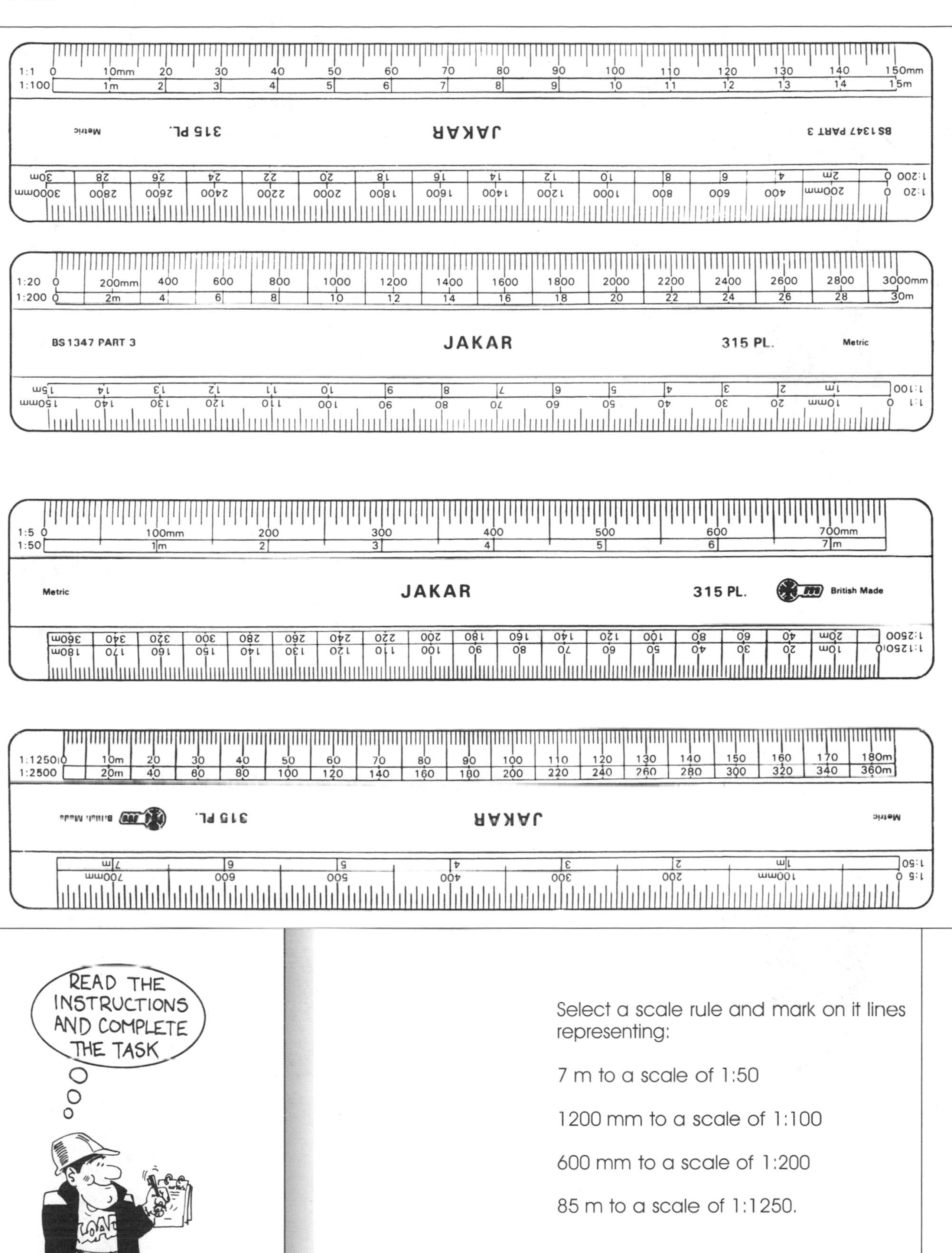

READ THE INSTRUCTIONS AND COMPLETE THE TASK

Select a scale rule and mark on it lines representing:

7 m to a scale of 1:50

1200 mm to a scale of 1:100

600 mm to a scale of 1:200

85 m to a scale of 1:1250.

READ THE INSTRUCTIONS AND COMPLETE THE TASK

Use a scale rule to measure the following lines and then complete the table.

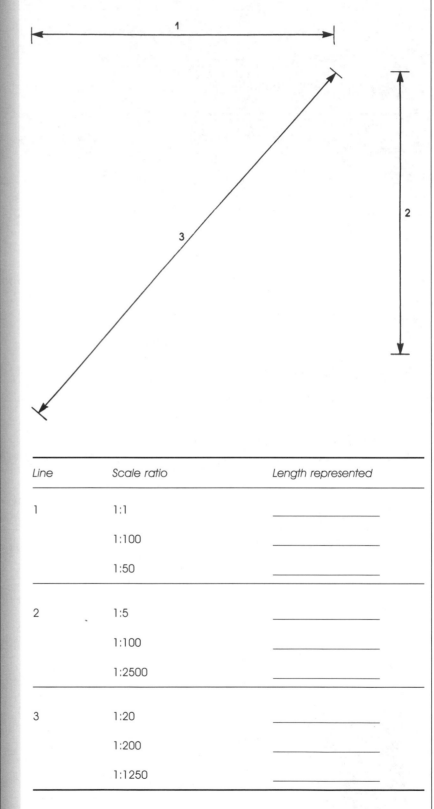

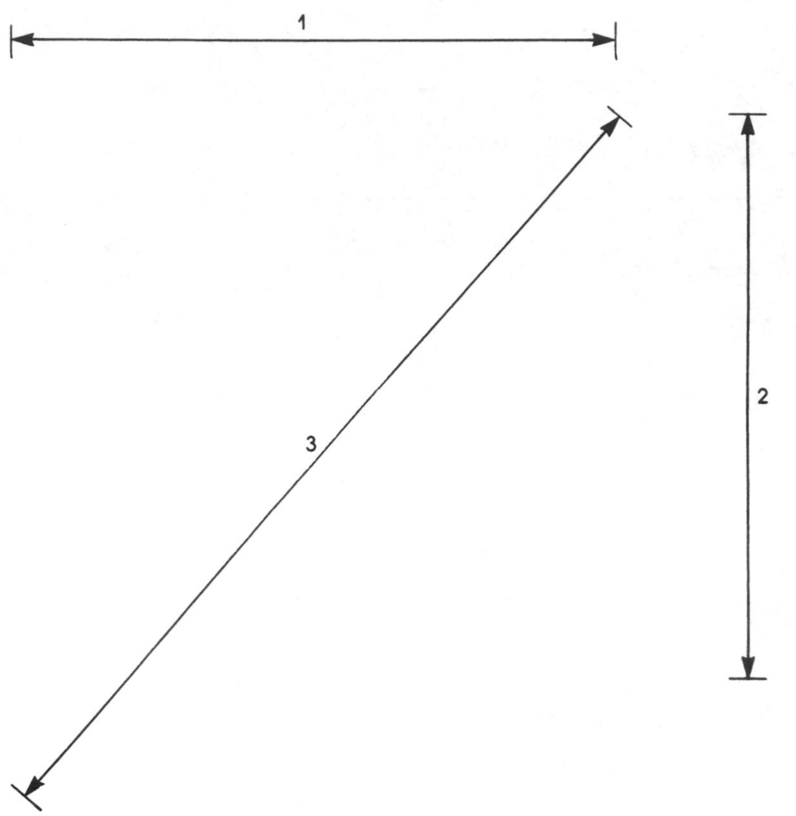
1
2
3

COMPLETE THIS TABLE

Line	*Scale ratio*	*Length represented*
1	1:1	__________
	1:100	__________
	1:50	__________
2	1:5	__________
	1:100	__________
	1:2500	__________
3	1:20	__________
	1:200	__________
	1:1250	__________

The outline ground floor plan of a house is drawn to a scale of 1:50.

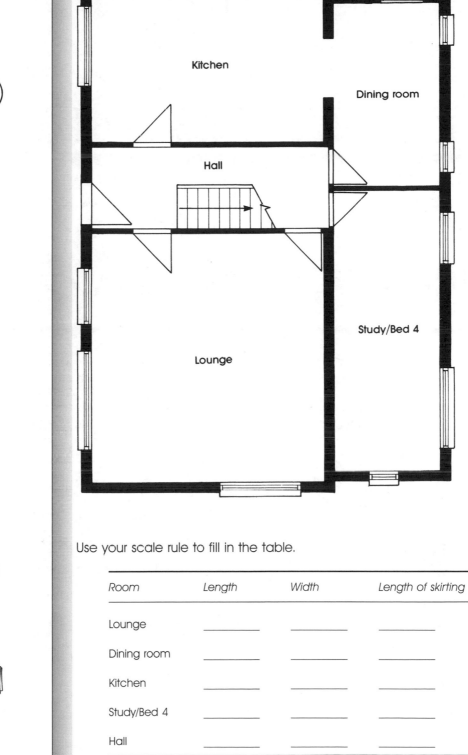

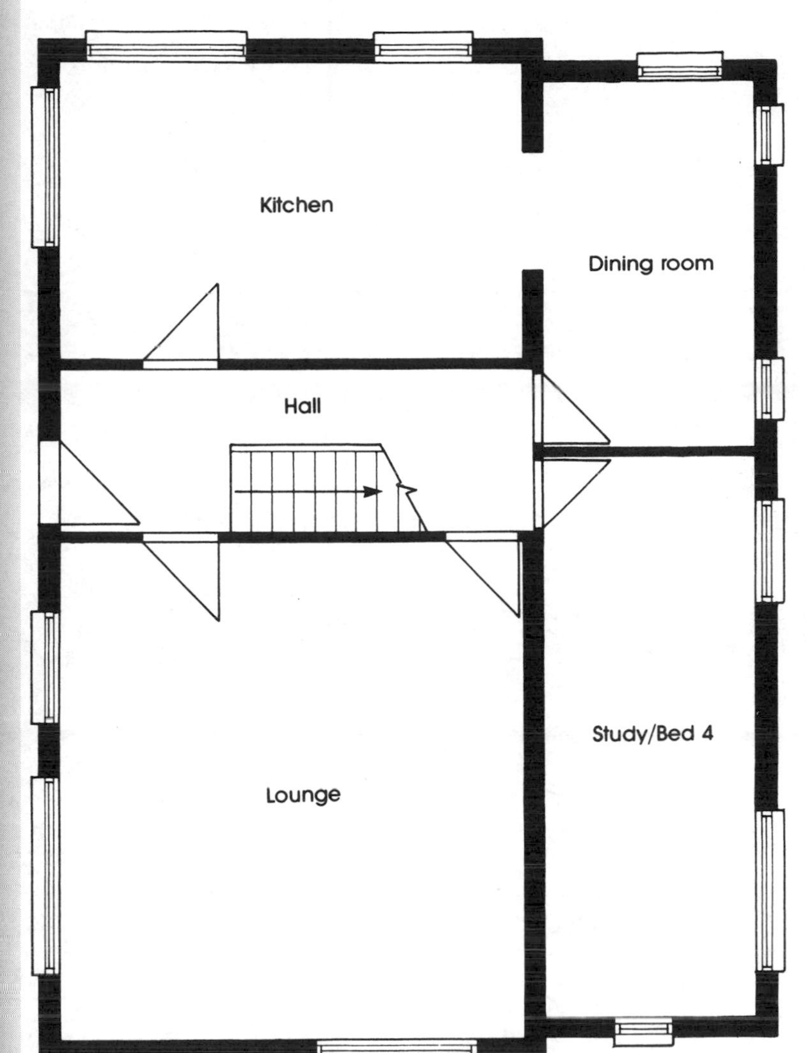

Use your scale rule to fill in the table.

Room	*Length*	*Width*	*Length of skirting*
Lounge	______	______	______
Dining room	______	______	______
Kitchen	______	______	______
Study/Bed 4	______	______	______
Hall	______	______	______

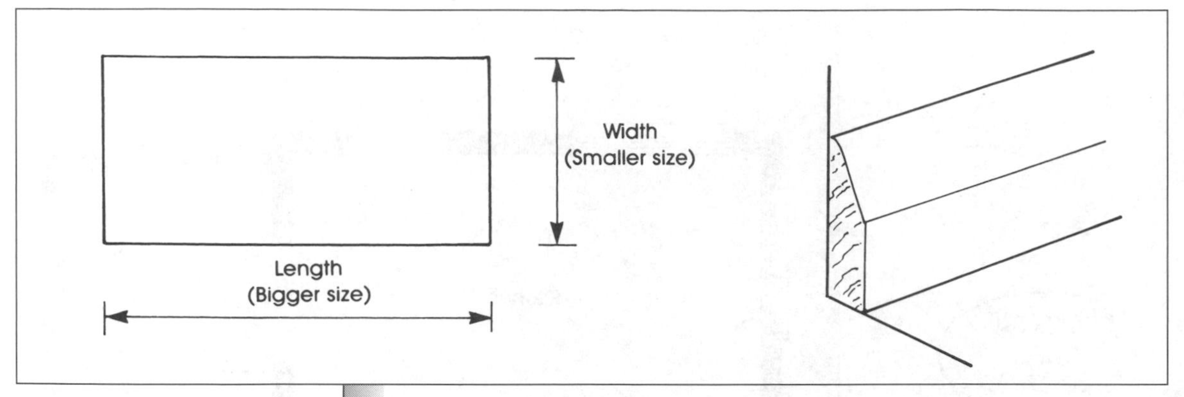

Skirting is the piece of wood fixed at the junction of the floor and wall joint. (Refers to learning task on previous page.)

Drawings, symbols and abbreviations

Drawings

These are the major means used to communicate technical information between all parties involved in the building process. They must be clear, accurate and easily understood by everyone who uses them. In order to achieve this architects and designers will use standardised methods for layout, symbols and abbreviations.

Line	Use
Thick	Main outlines
Medium	General details and outlines
Thin	Construction and dimension lines
Breakline	Breaks in the continuity of a drawing
Thick chain	Pipe lines, drains and services
Thin chain	Centre lines
Section line	Showing the position of a cut (the pointers indicate the direction of view)
Broken line	Showing details which are not visible
Dimension line	Showing the distance between two points

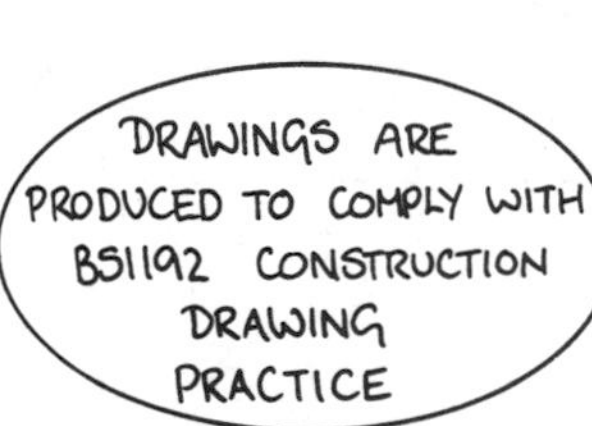

DRAWINGS ARE PRODUCED TO COMPLY WITH BS1192 CONSTRUCTION DRAWING PRACTICE

Symbols

These are graphical illustrations used to represent different materials and components in a building drawing.

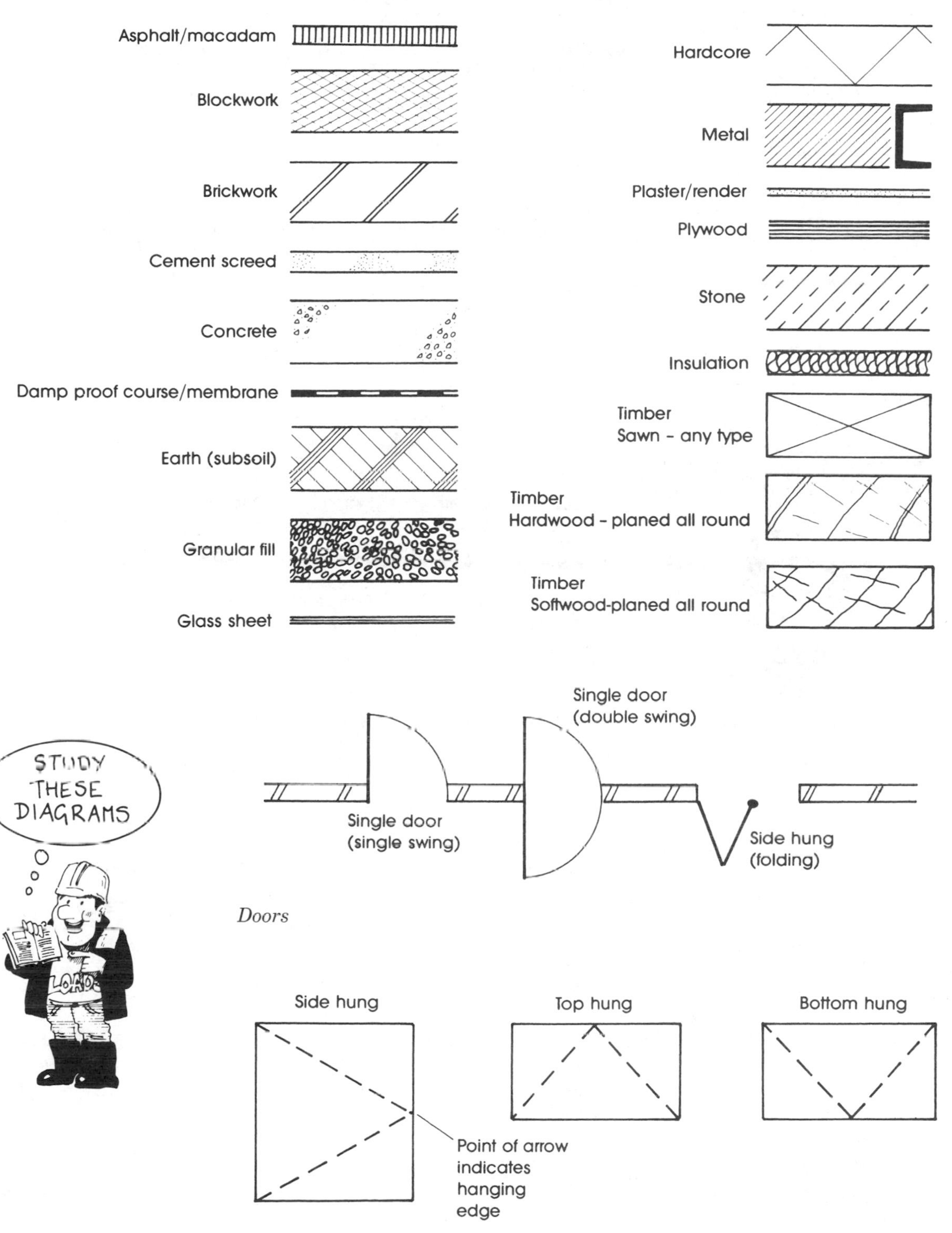

Doors

Casement windows

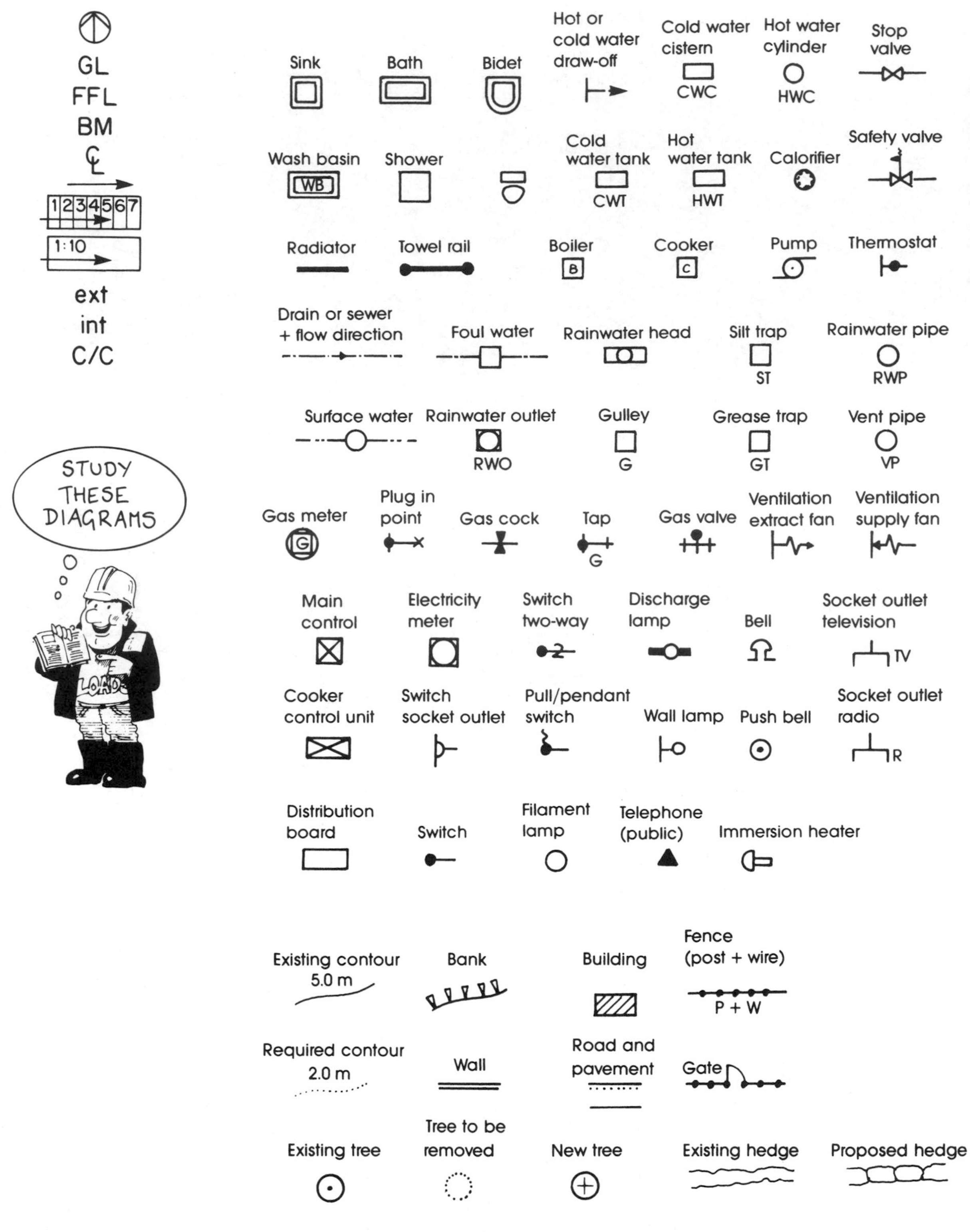
GL
FFL
BM
1 2 3 4 5 6 7
1:10
ext
int
C/C
STUDY THESE DIAGRAMS
Sink
Bath
Bidet
Hot or cold water draw-off
Cold water cistern
CWC
Hot water cylinder
HWC
Stop valve
Wash basin
WB
Shower
Cold water tank
CWT
Hot water tank
HWT
Calorifier
Safety valve
Radiator
Towel rail
Boiler
B
Cooker
C
Pump
Thermostat
Drain or sewer + flow direction
Foul water
Rainwater head
Silt trap
ST
Rainwater pipe
RWP
Surface water
Rainwater outlet
RWO
Gulley
G
Grease trap
GT
Vent pipe
VP
Gas meter
G
Plug in point
Gas cock
Tap
G
Gas valve
Ventilation extract fan
Ventilation supply fan
Main control
Electricity meter
Switch two-way
2
Discharge lamp
Bell
Socket outlet television
TV
Cooker control unit
Switch socket outlet
Pull/pendant switch
Wall lamp
Push bell
Socket outlet radio
R
Distribution board
Switch
Filament lamp
Telephone (public)
Immersion heater
Existing contour
5.0 m
Bank
Building
Fence (post + wire)
P + W
Required contour
2.0 m
Wall
Road and pavement
Gate
Existing tree
Tree to be removed
New tree
Existing hedge
Proposed hedge

Abbreviations

These are a short way of writing a word or group of words. They allow maximum information to be included in a concise way. Here are some abbreviations commonly used in the building industry.

THESE
ARE USED ON
DRAWINGS

Aggregate	agg	Foundation	fdn
Air brick	AB	Fresh air inlet	FAI
Aluminium	al	Glazed pipe	GP
Asbestos	abs	Granolithic	grano
Asbestos cement	absct	Hardcore	hc
Asphalt	asph	Hardboard	hdbd
Bitumen	bit	Hardwood	hwd
Boarding	bdg	Inspection chamber	IC
Brickwork	bwk	Insulation	insul
BS* Beam	BSB	Invert	inv
BS Universal beam	BSUB	Joist	jst
BS Channel	BSC	Mild steel	MS
BS equal angle	BSEA	Pitch fibre	PF
BS unequal angle	BSUA	Plasterboard	pbd
BS tee	BST	Polyvinyl acetate	PVA
Building	bldg	Polyvinylchloride	PVC
Cast iron	CI	Rainwater head	RWH
Cement	ct	Rainwater pipe	RWP
Cleaning eye	CE	Reinforced concrete	RC
Column	col	Rodding eye	RE
Concrete	conc	Foul sewers	FS
Copper	Copp cu	Sewers surface water	SWS
Cupboard	cpd	Softwood	swd
Damp proof course	DPC	Tongue and groove	T & G
Damp proof membrane	DPM	Unglazed pipe	UGP
Discharge pipe	DP	Vent pipe	VP
Drawing	dwg	Wrought iron	WI
Expanding metal lathing	EML		

*BS = British Standard

Methods of projection

Drawings can be produced as either a series of flat views called orthographic projection or in a form which closely resembles their actual appearance called pictorial projection.

Orthographic projection – a method used for working drawing plans, elevations and sections. A separate drawing of all the views of an object is produced in a systematic manner on the same drawing sheet.

First angle – a form of orthographic projection used for building drawings where in relation to the front view the other views are arranged as follows: the view from above is drawn below; the view from below is drawn above; the view from the left is drawn to the right; the view from the right is drawn to the left; the view from the rear is drawn to the extreme right. A sectional view may be drawn to the left or the right.

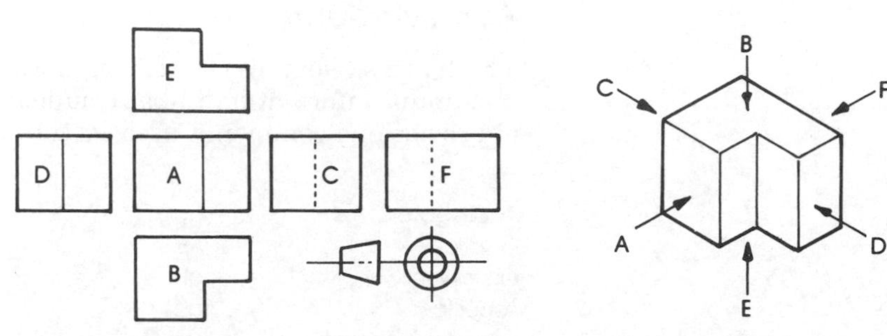

First angle *Isometric view*

Third angle – a form of orthographic projection used for engineering drawings. It is also termed American projection. In relation to the front elevation the other views are arranged as follows: the view from above is drawn above; the view from below is drawn below; the view from the left is drawn left; the view from the right is drawn right; the view from the rear is drawn to the extreme right. A sectional view may be drawn to the left or the right.

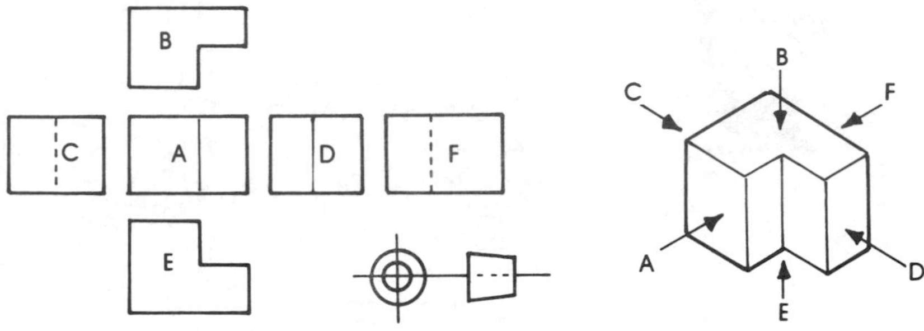

Third angle *Isometric view*

Plan – the shape or surface of an object when looked down on vertically. Floor plans in building drawings are normally taken just above window sill height.

Elevation – the view of an object from either side, front or rear.

Section – the cut surface produced when an object is imagined as if cut through with a saw.

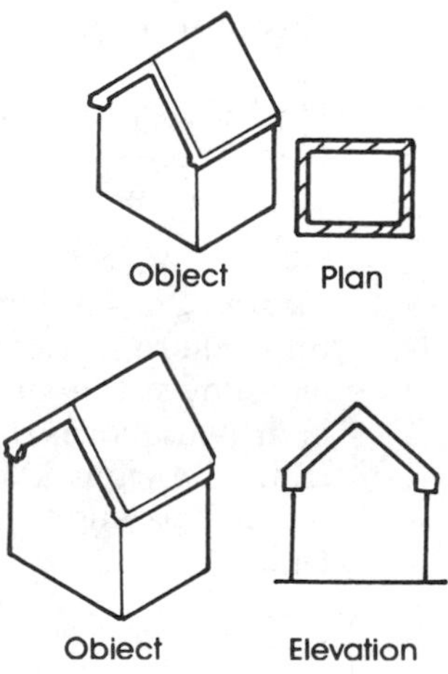

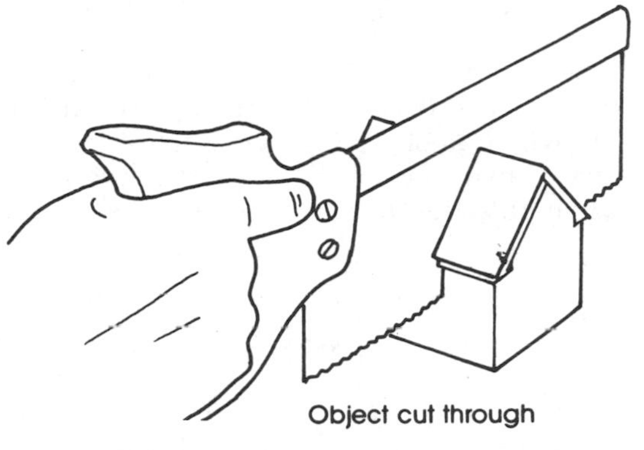
Object cut through

Section

Pictorial projection – a method of drawing objects in a three-dimensional form. Often used for design and marketing purposes, as the finished appearance of the object can be more readily appreciated.

Axonometric – a form of pictorial drawing where all vertical lines are drawn vertical, while horizontal lines are drawn at 45 degrees to the horizontal, giving a true plan shape (below). Often used in kitchen design.

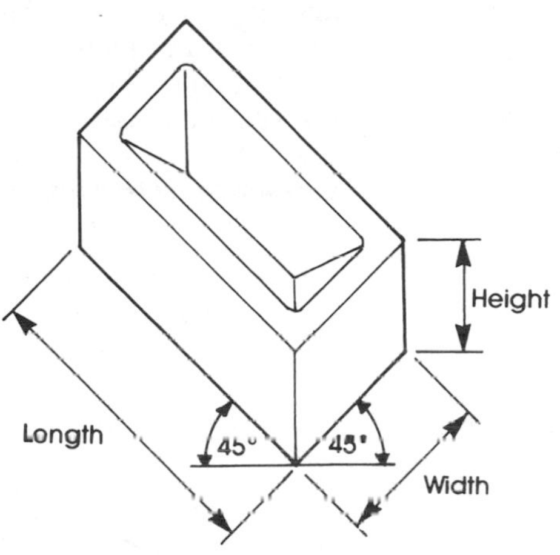

Isometric – a form of pictorial drawing where all vertical lines are drawn vertical, while all horizontal lines are drawn at an angle of 30 degrees to the horizontal (below).

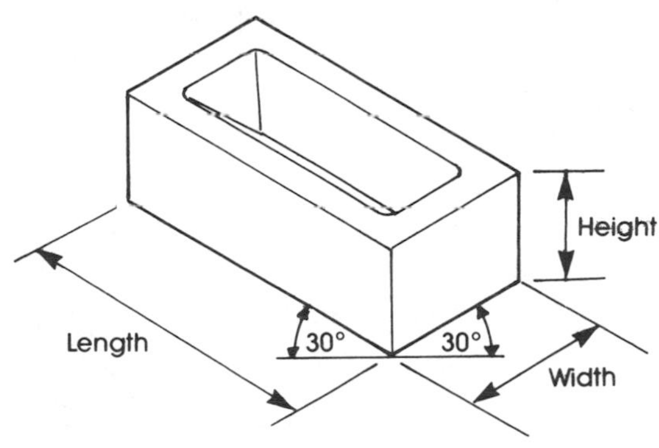

Oblique – a form of pictorial projection which can take the form of either cabinet or cavalier (below). All vertical lines are drawn vertical and all horizontal lines in the front elevation are drawn horizontal to give a true front view. All horizontal lines in the side elevations are drawn at 45 degrees to the horizontal. In cavalier these 45 degree lines are drawn to their full length, while in cabinet they are drawn to half their full length.

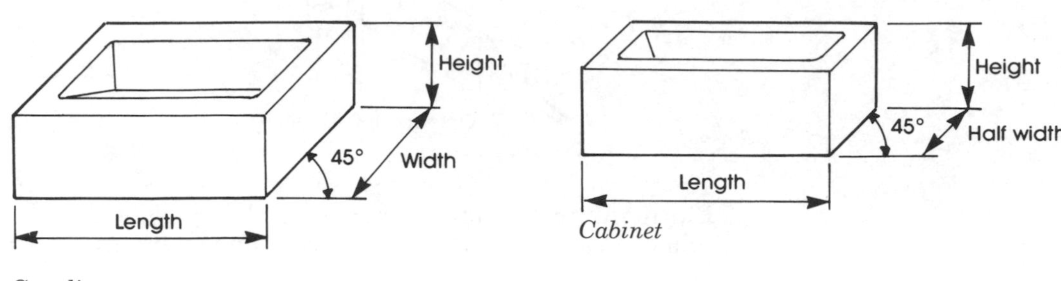

Cabinet

Cavalier

Sketches

These are rough outlines or initial drafts of an idea before full working drawings are made. Alternatively sketches may be prepared to convey thoughts and ideas. It is often much easier to produce a sketch of your intentions rather than to describe in words or produce a long list of instructions. Sketches can be produced either freehand, that is without the use of any equipment, or be more accurately produced using a ruler and set square to give basic guidelines. Methods of projects follow those used for drawing, e.g. orthographic or pictorial.

Pictorial sketching is made easy if you imagine the object you wish to sketch with a three-dimensional box around it. Draw the box first, lightly, with a 2H pencil, then draw in the object using an HB pencil.

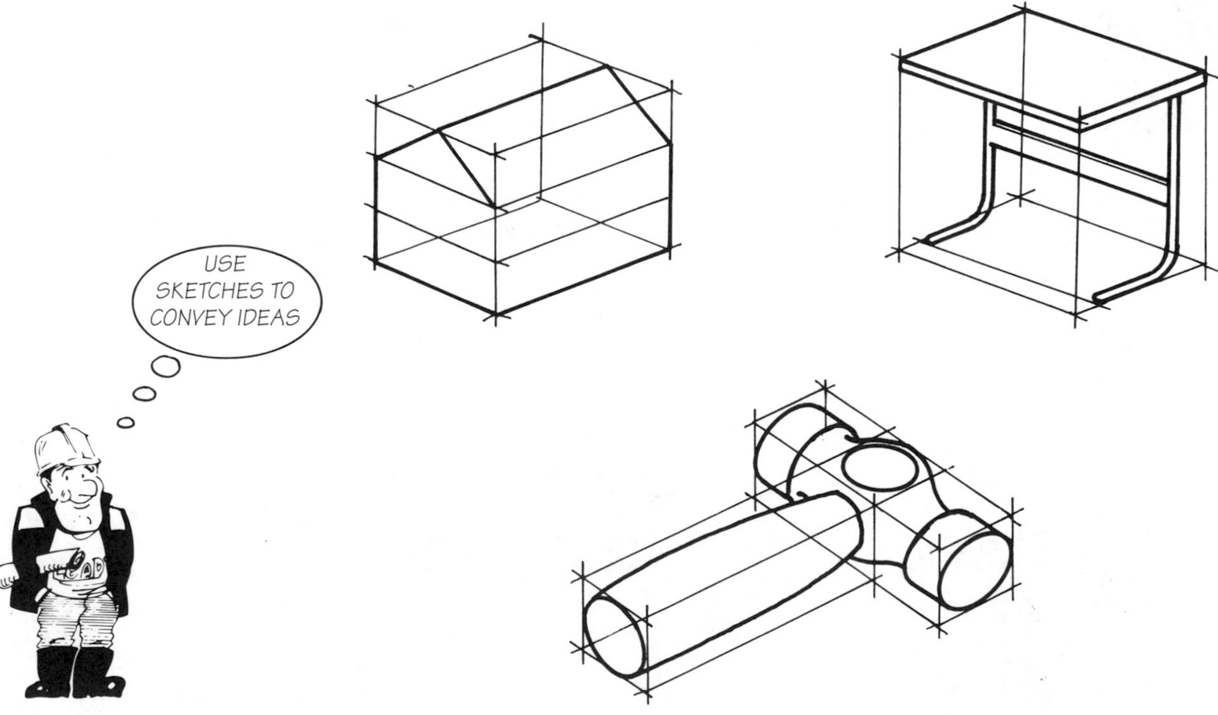

Learning task

Use the box method to sketch a hand tool and a component or element associated with your occupation.

Construction activity documents

Architects' drawings

A range of scale working drawings showing plans, elevations, general arrangements, layouts and details of a proposed construction, the main types being:

Location drawings

Block plans identify the proposed site in relation to the surrounding area.

The scales most commonly used are 1:2500 and 1:1250.

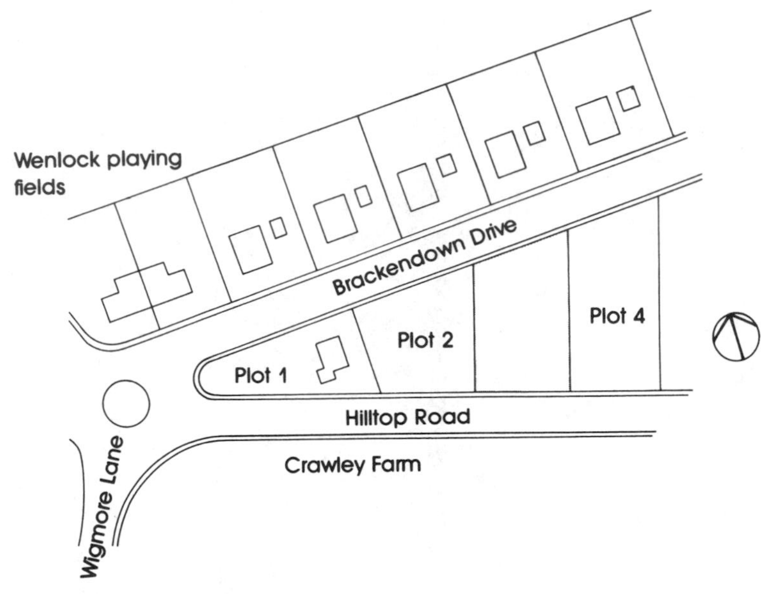

Site plans show the position of a proposed building and the general layout of the road services and drainage etc. on the site. The scales most commonly used are 1:500 and 1:200.

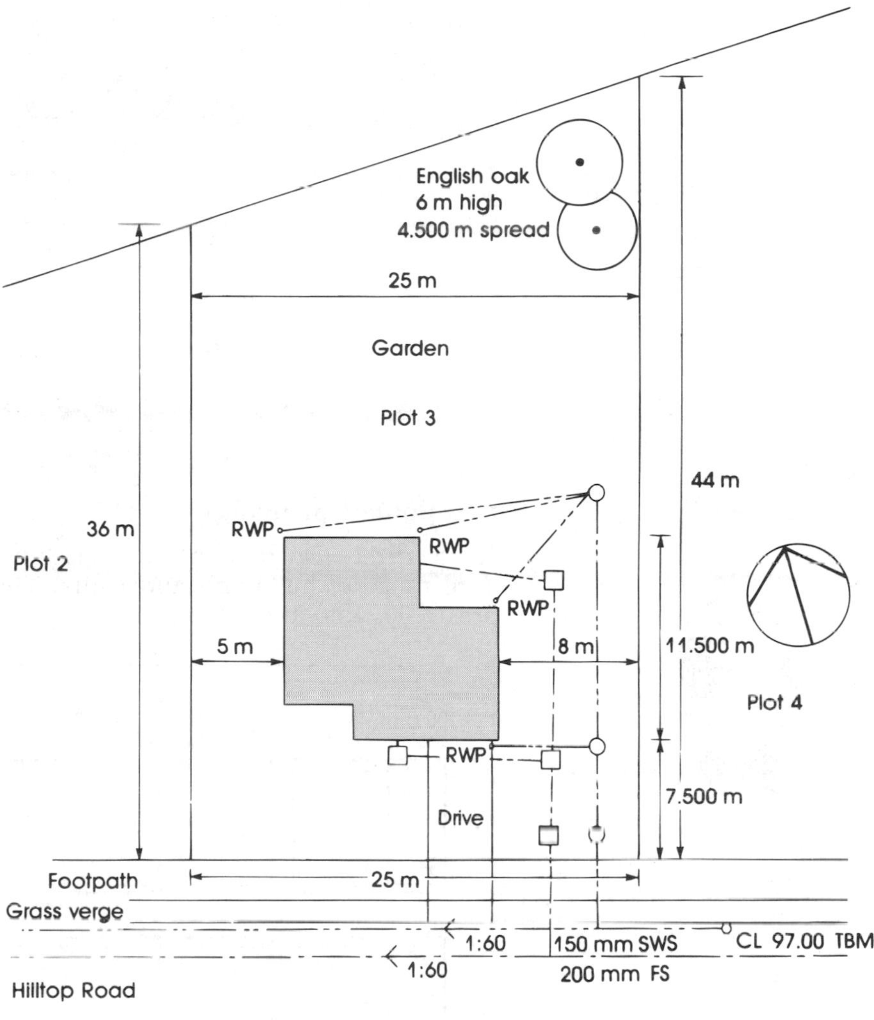

Site plan

General location plans show the positions occupied by the various areas within a building and identify the locations of principal elements and components. The scales most commonly used are 1:200, 1:100 and 1:50.

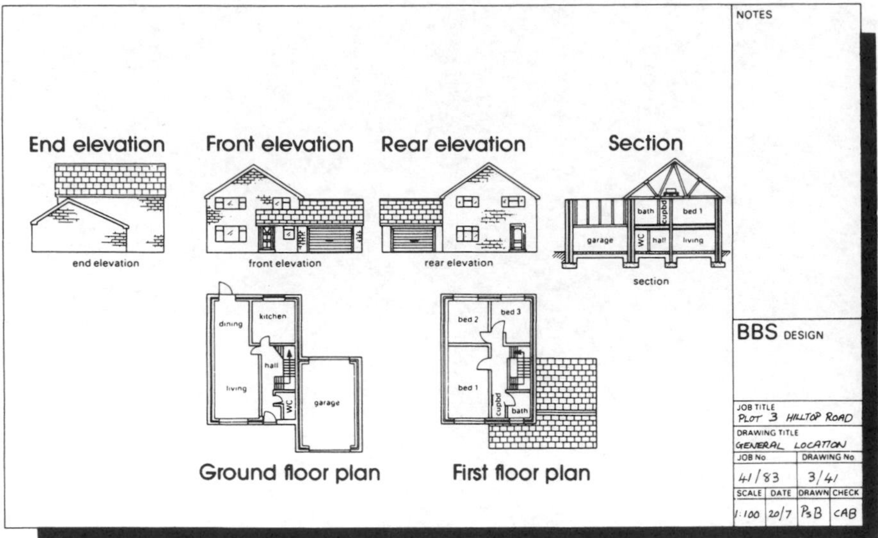

Component drawings

Range drawings show the basic sizes in a reference system of a standard range of building components. The scales most commonly used are 1:100, 1:50 and 1:20.

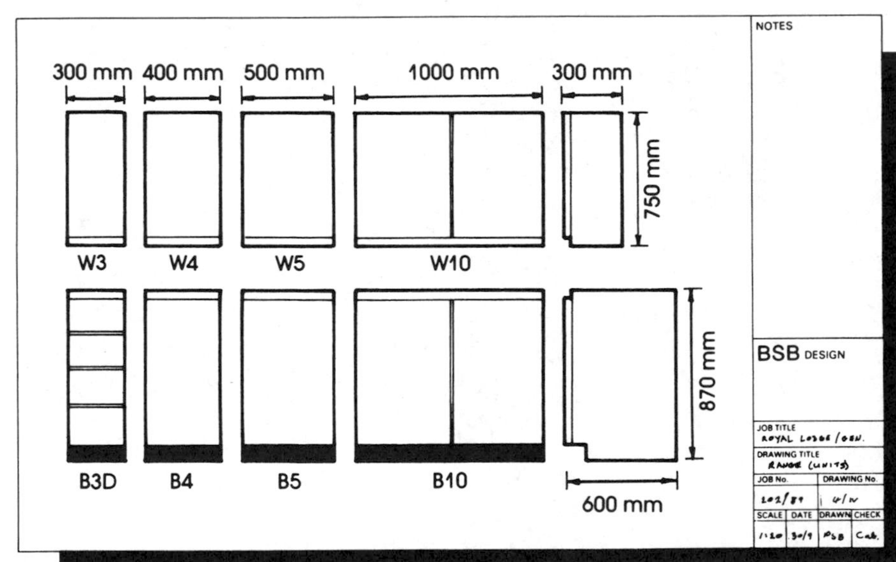

Detail drawings show all the information that is required to manufacture a particular component. The scales most commonly used are 1:10, 1:5 and 1:1.

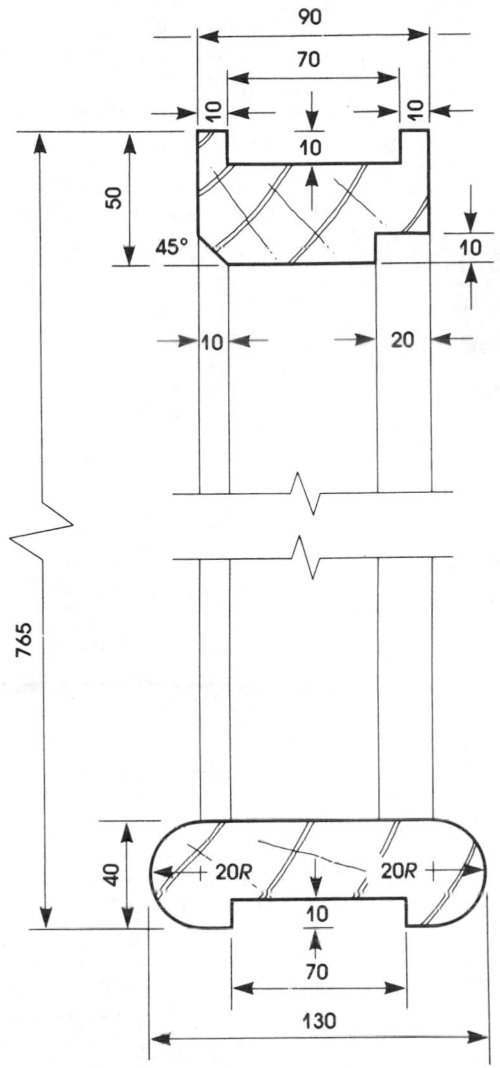

Assembly drawings

Assembly details show the junctions between the various elements and components of a building. The scales most commonly used are 1:20, 1:10 and 1:5.

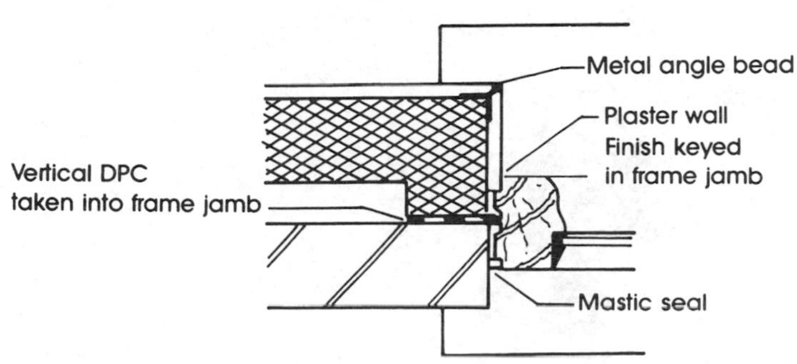

Dimensions are shown on drawings against a lightly drawn line with arrowheads terminating against short cross lines. Actual sizes may be shown individually as separate dimensions or cumulatively along a building as in running dimensions. Running dimensions are to be preferred for setting out rather than separate dimensions, since any inaccuracies or error made in making each separate dimension will have a cumulative effect, throwing each successive position out. Where running dimensions are not shown on a drawing it is best to work them out and indicate against each position. As a check the total of the separate dimensions should equal the final position figure.

To avoid confusing the position of the decimal point, an oblique stroke is often used to separate metres and millimetres. Where the dimension is less than a metre a nought is inserted before the stroke.

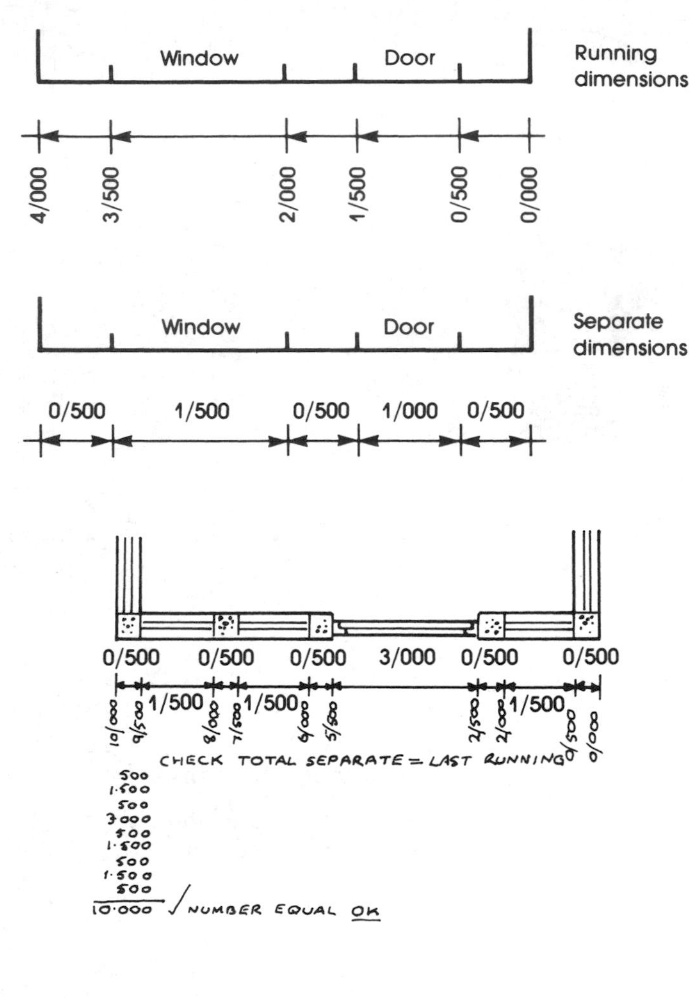

Learning task

Mark up the outline plan to show running dimensions including the check.

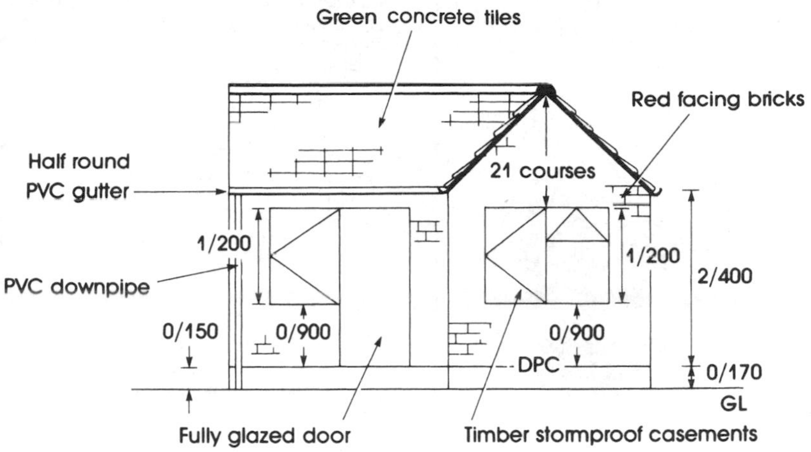

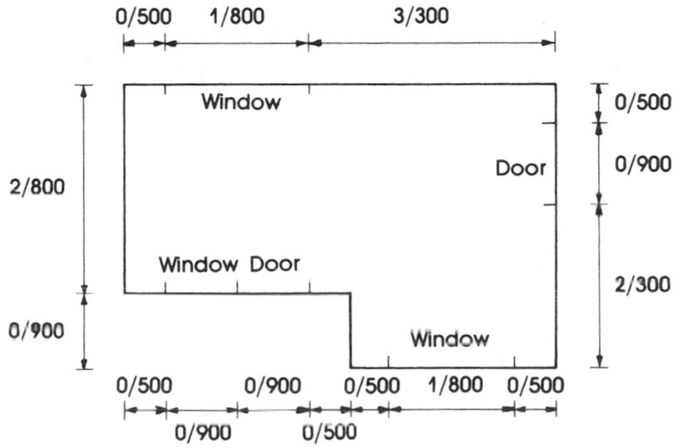

Specification

This is a contract document that supplements the architect's working drawings. It contains precise descriptions of all the essential information and job requirements that will affect the price, but cannot be shown on the drawings. This will include any site restrictions, availability of site services, description of workmanship and materials and any other requirements such as site clearance, making good on completion and who is responsible for approving finished work etc.

STUDY
THE
SPECIFICATION

BBS DESIGN

Specification of the works to be carried out and the materials to be used in the erection and completion of a new house and garage on plot 3, Hilltop Road, Brackendowns, Bedfordshire, for Mr W. Whiteman, to the satisfaction of the architect.

1.00 General conditions

1.01
1.02

1.03
1.04

2.00
2.01
2.02

2.03
2.04
2.05

2.06
2.07
2.08
2.09

10.00 Woodwork

10.01 Timber for carcassing work to be of GS or MGS grade as laid down in BS 4978

10.02 Timber for joinery shall be a species approved by the architect, selected from and conforming to BS 1186.

10.03 Moisture content of all timber at time of fixing to be appropriate to the situation and conditions in which it is used. To this effect all timber and components will be protected from the weather prior to their use.

10.04

10.05

10.06

10.18 Construct the first floor using 50 mm × 200 mm sawn softwood joists at 400 mm centres supported on mild steel hangers.

Provide 75 mm × 200 mm trimmer and trimming around stairwell, securely tusk-tenoned together.

Provide and fix to joists 38 mm × 38 mm sawn softwood herring-bone strutting at 1.8 m maximum intervals.

Provide and fix galvanized restraint straps at 2 m maximum intervals to act as positive ties between the joists and walls.

10.19 Provide and secret fix around the trimmed stairwell opening a 25 mm Brazilian mahogany apron lining, tongued to a matching 25 mm × 100 mm nosing.

10.20 Provide and lay to the whole of the first floor 19 mm × 100 mm prepared softwood tongued-and-grooved floor boarding, each board well cramped up and surface nailed with two 50 mm flooring brads to each joist. The nail heads to be well punched down.

Specification

HOW'S
IT
GOING?

STUDY THE SCHEDULE

Schedule

This is a contract document that is used to record repetitive design information about a range of similar components e.g. doors, ironmongery, finishes and sanitary ware, etc. There will also be a schedule which lists all of the drawings related to the job.

Description	Plot 2				Plot 6				Plot 7				Plot 8				Plot 9				Plot 10				Plot 12			
	Kitchen	Cloaks	Bath	En-suite	Kitchen	Cloaks	Bath	En-suite	Kitchen	Cloaks	Bath	En-suite	Kitchen	Cloaks	Bath	En-suite	Kitchen	Cloaks	Bath	En-suite	Kitchen	Cloaks	Bath	En-suite	Kitchen	Cloaks	Bath	En-suite
ITEM (see range)																												
Inset sink	X				X				X				X				X				X				X			
Waste disposal unit	X								X												X							
Close Couple WC		X	X	X		X	X	X		X	X	X		X	X	X		X	X	X		X	X	X		X	X	X
Bidet				X				X				X				X				X				X				X
Pedestal wash basin			X	X			X	X			X	X			X	X			X	X			X	X			X	X
Wall hung corner basin		X				X				X				X				X				X				X		
Bath			X				X				X				X				X				X				X	
Shower tray				X				X				X				X				X				X				X
STYLE (see range)																												
Anne							X	X														X		X			X	
Sarah		X				X				X				X	X	X										X		
James			X	X							X	X						X	X	X			X					X
Single drainer					X								X				X											
Double drainer	X								X												X				X			
COLOUR																												
Penthouse Red							X	X																X				
Indian ivory						X					X	X				X							X					
Honeysuckle				X	X										X							X				X	X	X
White		X								X				X				X	X	X								
BRASS WORK																												
Chrome plated	X	X	X	X	X	X			X	X	X	X	X	X			X	X	X	X	X	X	X		X	X	X	X
Gold plated							X	X							X	X								X				

Notes

BBS DESIGN

JOB TITLE
Lakeside Estate

DRAWING TITLE
Schedule for sanitary appliances

JOB NO.	DRAWING NO.		
SCALE	DATE	DRAWN	CHECKED

READ THE INSTRUCTIONS AND COMPLETE THE TASK

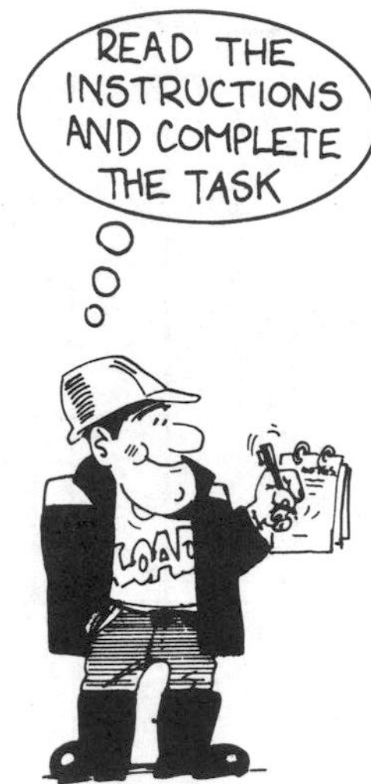

Learning task

Using the schedule for sanitary appliances, estate plan and house plan, complete the order/requisition form for the sanitary appliances required for plot number 6.

These are required on site for installation on the 15 March 2002 at the latest.

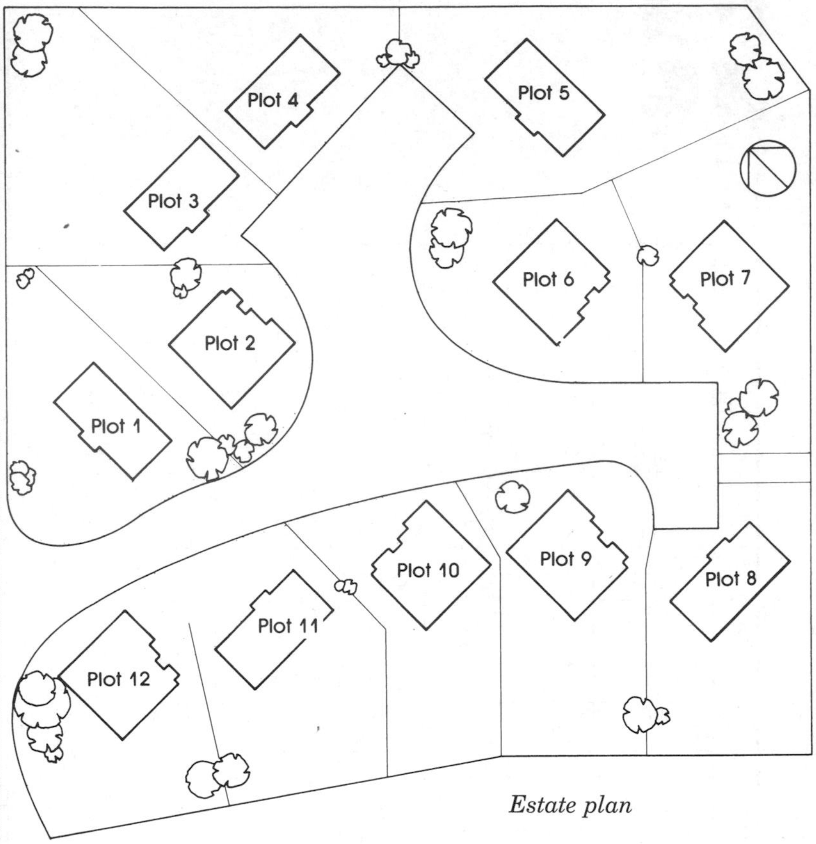

Plot 1
Plot 2
Plot 3
Plot 4
Plot 5
Plot 6
Plot 7
Plot 8
Plot 9
Plot 10
Plot 11
Plot 12

Estate plan

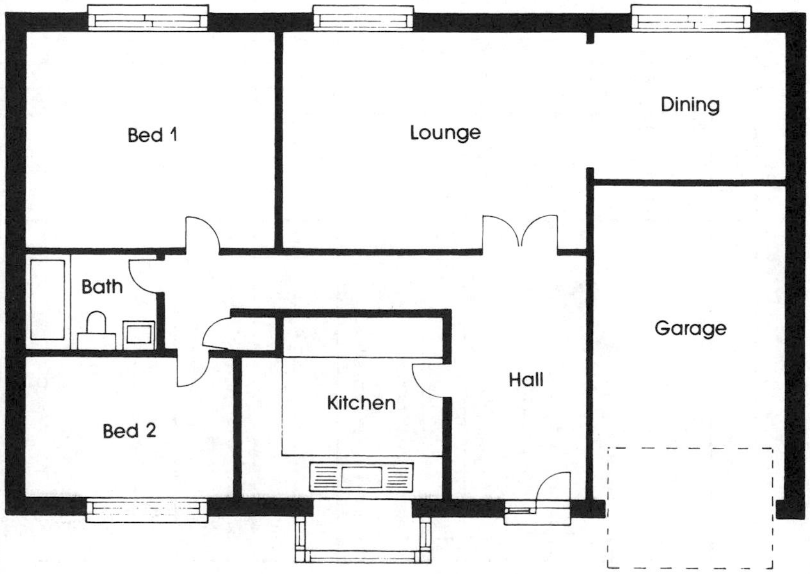

Bed 1
Lounge
Dining
Bath
Garage
Hall
Kitchen
Bed 2

The lakeside bungalow plan

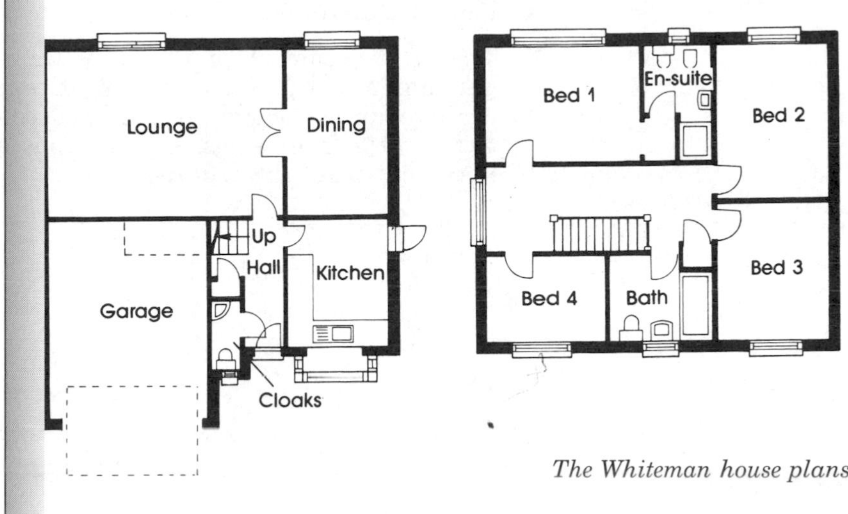

The Whiteman house plans

STUDY AND FILL IN THIS FORM

BBS CONSTRUCTION
ORDER/REQUISITION

Registered office

No. ____________________

Date ____________________

To ____________________ From ____________________

Address ____________________ Site address ____________________

____________________ ____________________

Please supply or order for delivery to the above site the following:

Description	Quantity	Rate	Date required by

Site manager/foreman ____________________

Note Please advise site within 24 hours of request if order cannot be fulfilled by the date required

Bill of quantities (BOQ)

This is a document prepared by the quantity surveyor. It gives a description and measure of quantities of labour, materials and other items required to carry out a building contract. It is based on the architect's working drawings, specifications and schedules and forms part of the **contract documents**.

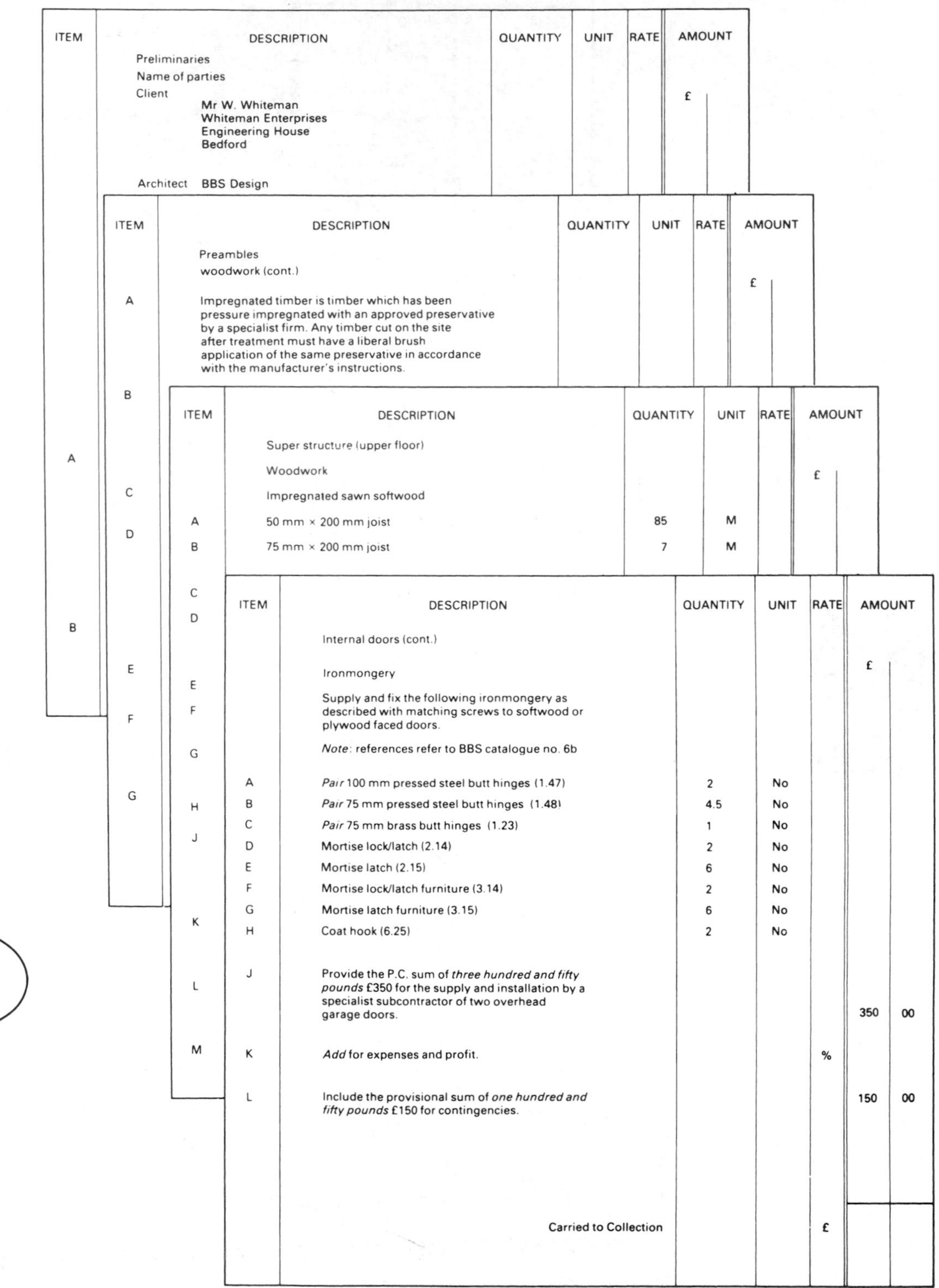

ITEM	DESCRIPTION	QUANTITY	UNIT	RATE	AMOUNT £	
	Preliminaries					
	Name of parties					
	Client					
	Mr W. Whiteman Whiteman Enterprises Engineering House Bedford					
	Architect BBS Design					
A						
B						

ITEM	DESCRIPTION	QUANTITY	UNIT	RATE	AMOUNT £	
	Preambles					
	woodwork (cont.)					
A	Impregnated timber is timber which has been pressure impregnated with an approved preservative by a specialist firm. Any timber cut on the site after treatment must have a liberal brush application of the same preservative in accordance with the manufacturer's instructions.					
B						
C						
D						
E						
F						
G						

ITEM	DESCRIPTION	QUANTITY	UNIT	RATE	AMOUNT £	
	Super structure (upper floor)					
	Woodwork					
	Impregnated sawn softwood					
A	50 mm × 200 mm joist	85	M			
B	75 mm × 200 mm joist	7	M			
C						
D						
E						
F						
G						
H						
J						
K						
L						
M						

ITEM	DESCRIPTION	QUANTITY	UNIT	RATE	AMOUNT £	
	Internal doors (cont.)					
	Ironmongery					
	Supply and fix the following ironmongery as described with matching screws to softwood or plywood faced doors.					
	Note: references refer to BBS catalogue no. 6b					
A	*Pair* 100 mm pressed steel butt hinges (1.47)	2	No			
B	*Pair* 75 mm pressed steel butt hinges (1.48)	4.5	No			
C	*Pair* 75 mm brass butt hinges (1.23)	1	No			
D	Mortise lock/latch (2.14)	2	No			
E	Mortise latch (2.15)	6	No			
F	Mortise lock/latch furniture (3.14)	2	No			
G	Mortise latch furniture (3.15)	6	No			
H	Coat hook (6.25)	2	No			
J	Provide the P.C. sum of *three hundred and fifty pounds* £350 for the supply and installation by a specialist subcontractor of two overhead garage doors.				350	00
K	*Add* for expenses and profit.			%		
L	Include the provisional sum of *one hundred and fifty pounds* £150 for contingencies.				150	00
	Carried to Collection			£		

HOW'S IT GOING?

ALL CONTRACT DOCUMENTS MUST BE STUDIED TO SHOW THE FULL EXTENT OF THE WORK

Contract documents

These various documents together form the legal contract. Normally they consist of the architects' working drawings, specification, schedules, bill of quantities and the conditions of contract. A standard document is normally used but it could be specially prepared.

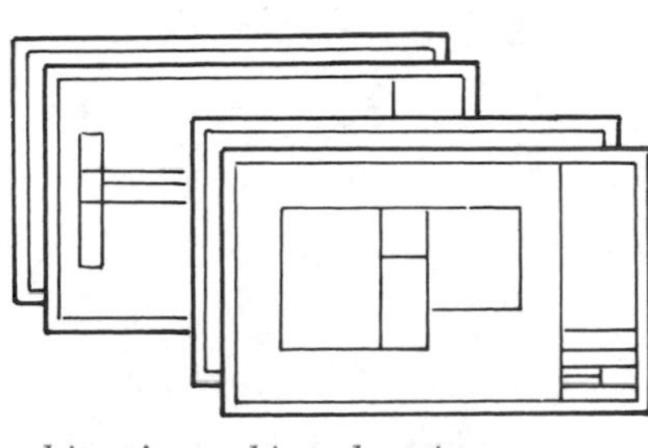

Architect's working drawings

Specification

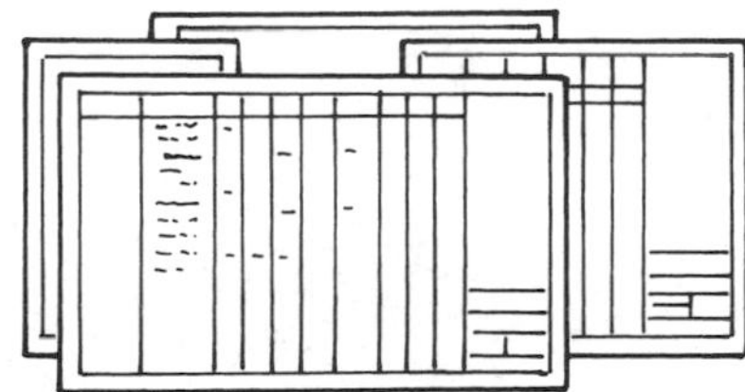

Schedules

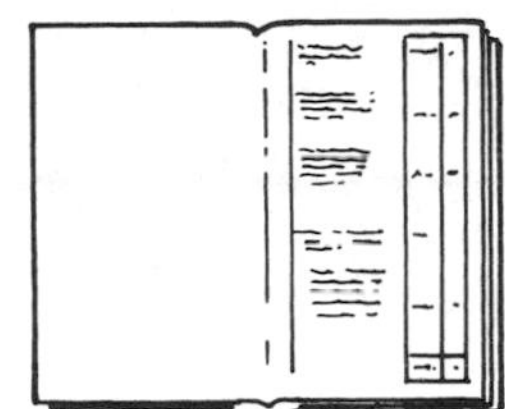

Bill of quantities

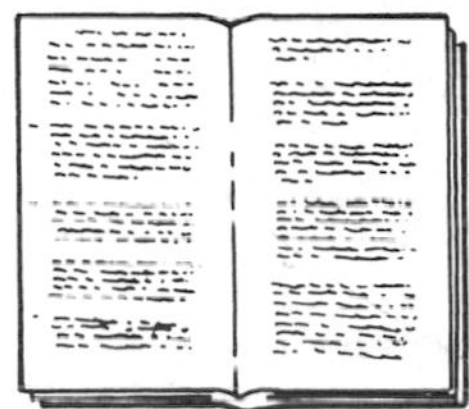

Conditions of contract

Delivery notes and records

When materials and plant are delivered to a site, someone is required to sign the driver's delivery note. A careful check must be made to ensure that all the materials are there and undamaged. Any missing or damaged goods must be clearly indicated on the delivery note and followed up by a letter to the supplier. Many suppliers send out an advice note prior to delivery which states details of the materials and the expected delivery date. This enables the site management to make arrangements for unloading and storage. Delivery records are often completed to provide a record of all the materials received on site. These are normally filled in and sent to the organisation's head office along with the copy of the delivery notes on a weekly basis. This record is used to confirm goods have been delivered before paying suppliers' invoices.

STUDY THE
DELIVERY
NOTE

Learning task

You have supervised the delivery of materials shown on the note below. On checking the delivery, only 48 lengths of 50 × 50 were received and several of the shrink-wrapped hardwood packages had splits in them. Sign the delivery note and make any comments you think applicable.

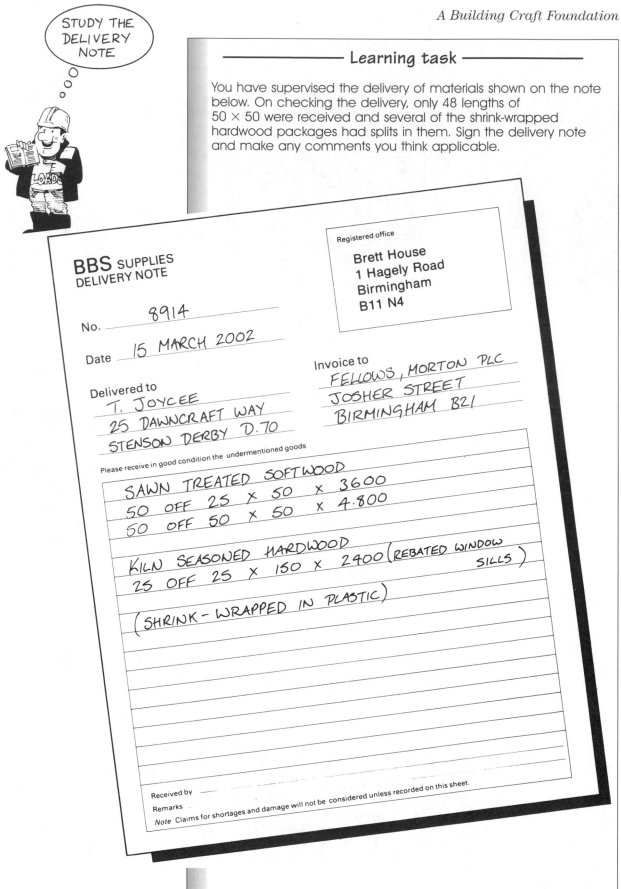

BBS SUPPLIES
DELIVERY NOTE

Registered office
Brett House
1 Hagely Road
Birmingham
B11 N4

No. 8914

Date 15 MARCH 2002

Delivered to
T. JOYCEE
25 DAWNCRAFT WAY
STENSON DERBY D.70

Invoice to
FELLOWS, MORTON PLC
JOSHER STREET
BIRMINGHAM B21

Please receive in good condition the undermentioned goods

SAWN TREATED SOFTWOOD
50 OFF 25 X 50 X 3600
50 OFF 50 X 50 X 4.800

KILN SEASONED HARDWOOD
25 OFF 25 X 150 X 2400 (REBATED WINDOW SILLS)

(SHRINK-WRAPPED IN PLASTIC)

Received by

Remarks

Note Claims for shortages and damage will not be considered unless recorded on this sheet.

Complete the deliveries record for the last order.

STUDY
AND FILL IN
THIS FORM

T. Joycee Construction
DELIVERIES RECORD

Registered office
Ridge House
Norton Road
Cheltenham
GL59 1DB

Week no. P3 WK3 Date MARCH 02
Job title STENSON FIELDS

Delivery note no.	Date	Supplier	Description of delivery	For office use only	
				Rate	Value
241	14/3/02	G. BOGGS	SANITARY WARE		
1535	14/3/02	I. BLUNDER	READY-MIX CONCRETE		
			Total		

Site manager/foreman ______

Note Send weekly to head office with delivery notes

HOW'S
IT
GOING?

TRY AND ANSWER THESE

Questions for you

1. State why scale drawings are used in the construction industry.

2. Explain the purpose of a range drawing.

3. Explain the purpose of a specification.

4. What action should you take if a materials delivery does not match the delivery note?

5. Explain why graphical symbols are used on drawings.

6. State **TWO** main details shown on a site plan.

7. What would 5 mm on a drawing to a scale of 1:20 actually represent full size?

WELL, HOW DID YOU DO?
WORK THROUGH THE SECTION AGAIN IF YOU HAD ANY PROBLEMS

Messages

MAKE SURE
ALL MESSAGES
ARE CLEAR

Much communication within and between organisations takes place by means of forms. The issuer of the form is able to ask precisely the information required and in the desired order. There is normally no opportunity to waffle or give irrelevant details.

When you fill in a form for any reason, remember the following basic rules:

PLEASE PRINT FULL NAME

Please complete in ink
MARCH

BLOCK LETTERS ONLY

- Read the instructions carefully (e.g. do they ask for handwritten, BLOCK CAPITALS, or a black pen, etc.
- Read the questions carefully (do not squash in an answer if there is an opportunity to give that information elsewhere).
- Ensure your writing is legible.
- If your name is Peter Stephen Brett, then your surname, family or last name is Brett; your forename, first name or Christian name is Peter; your other name is Stephen; your initials are PSB. Where a maiden name is asked for this would be a married woman's surname before marriage.
- Complete all dates, times, etc. accurately.
- Delete inappropriate details as asked.
- Do not leave blanks, always write 'not applicable' or 'N/A'.
- Do not forget to sign the form if required. This is normally your initials and surname. This is your signature.
- Do not write where you see these:
  - For official use only
  - For office use only
  - For store use
  - For company use, etc.
- Finally, read through the form again to ensure all sections have been completed correctly.

HOW'S IT GOING?

B CUSTOMER TO COMPLETE (BLOCK CAPITALS PLEASE)

When completing this form ensure the details show clearly on both copies.

Mr/~~Mrs~~/~~Ms~~ P. S. BRETT

Delivery 70 SHALIMAR RD.

Address STEPPING BROOK

NORTHAMPTON

Full Postcode NN7 8WL

Tel: Home (STD 01604) No. 743521

Office (STD 021) No. 345001

I understand that the trusses will be manufactured to the correct sizes based upon the dimensions I have provided and I accept responsibility for the dimensions.

CUSTOMER'S SIGNATURE PSBrett

DATE 15/3/02

Learning task

ORDER FORM

FOR MADE TO MEASURE REPLACEMENT WINDOWS

When completing this form ensure the details show clearly on all copies.
MR/MRS/MS ....
(INITIALS) (SURNAME)
DELIVERY ADDRESS ....
....
.... FULL POSTCODE ....
Please indicate where you would like the goods left, if delivered in your absence.
....
....
TELEPHONE: HOME (STD ....) No ....
OFFICE (STD ....) No ....
I understand that the windows will be manufactured to the correct sizes based upon the dimensions I have provided and I accept responsibility for the dimensions.

CUSTOMER'S SIGNATURE .... DATE ....

STORE USE ONLY

STORE [0| |]

DATE OF ORDER ....
PURCHASE ORDER No. ....
DRL No. ....
ADMIN. CHECKED BY ....
RECEIPT No. ....
TENDER TYPE: CHEQUE/CREDIT/CASH

Blue - Order Copy
White - Store Copy
Green - Goods Inwards Copy
Yellow - Customer Copy

SEE REVERSE SIDE FOR ORDERING INSTRUCTIONS AND GUIDANCE (Enter required details and complete all sections)

SKETCH YOUR WINDOWS SHOWING DESIGNS AND DIMENSIONS HERE. N.B. ALWAYS VIEWED FROM THE OUTSIDE
(See HOW TO ORDER WINDOWS notes for opening lights min/max)
(Please note that our range of Made to Measure Windows vary in specification to our standard stock range.)

Do you require sills? YES ☐ NO ☐
Do you intend to use these windows in conjunction with our range of standard windows? YES ☐ NO ☐
Have you ordered made to measure windows from us before? ☐ If so, please state approx. date of order ....

STUDY AND FILL IN THIS FORM

ASSUME A SITUATION

Memoranda or memos

These are forms of written communication which are used within an organisation. They would not be sent out to customers or suppliers. When you send a memo for any reason, remember the following points:

- Be brief but use formal English
- Deal with one topic only
- May be hand- or typewritten

STUDY THIS MEMO

BBS CONSTRUCTION MEMO

From JOHN PESSAL To IAN CARPENTER

Subject POWER TOOLS Date 16 JANUARY 2001

Message

I HAVE MADE AN ORDER FOR THE POWER TOOLS. THEY SHOULD BE WITH YOU BY THE 20 JANUARY. TRUST THIS IS OKAY. JOHN.

Learning task

Complete the following memo to Christine Baldwin advising her that you will be able to attend the Safety Meeting next Friday.

BBS CONSTRUCTION MEMO

From ______ To ______

Subject ______ Date ______

Message

READ THE INSTRUCTIONS AND COMPLETE THE TASK

Letters

Letters provide a permanent record of communication between organisations and individuals. They can be handwritten, but formal business letters give a better impression of the organisation if they are typed. They should be written using simple concise language. The tone should

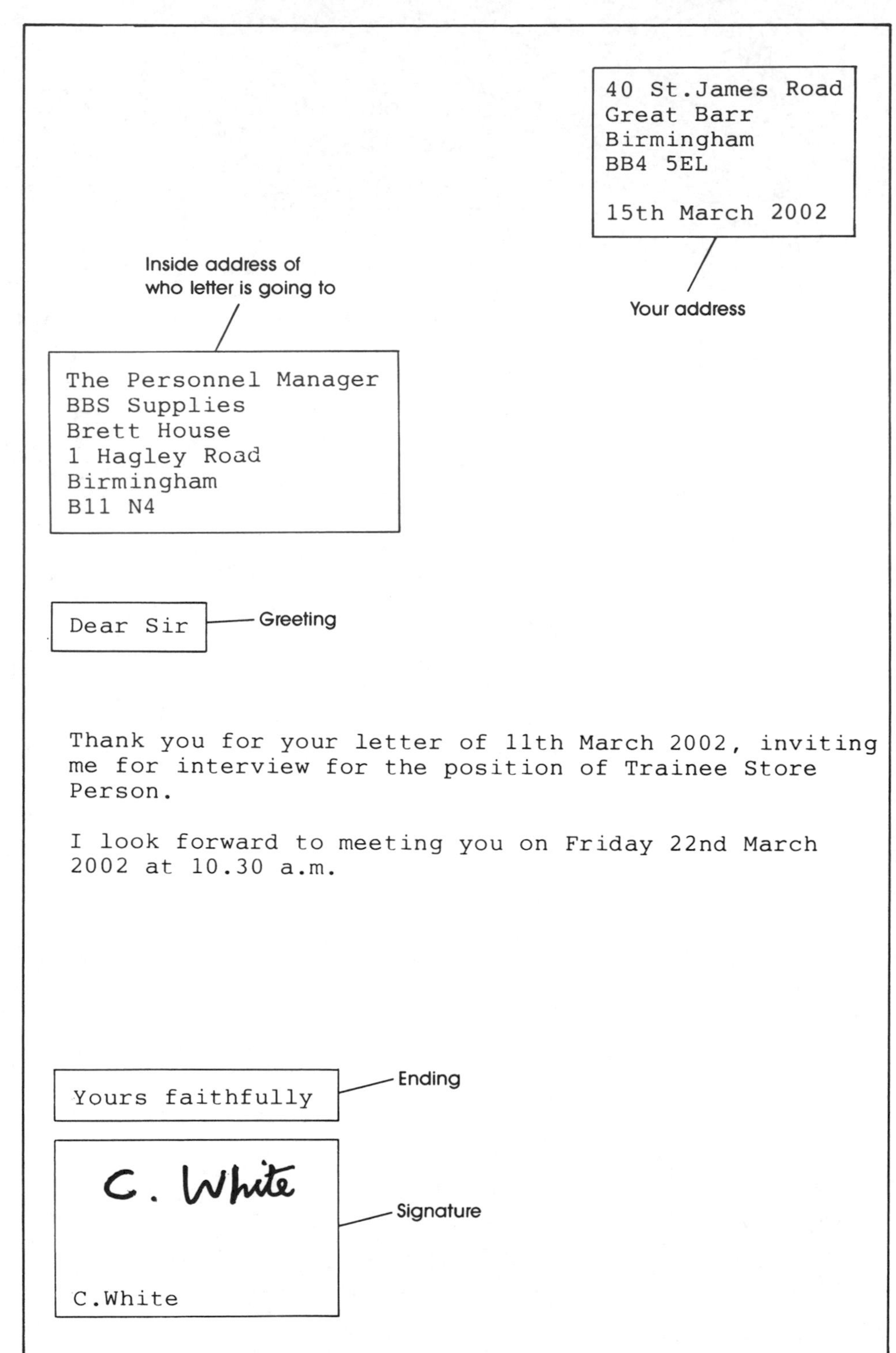
40 St.James Road
Great Barr
Birmingham
BB4 5EL

15th March 2002

The Personnel Manager
BBS Supplies
Brett House
1 Hagley Road
Birmingham
B11 N4

Dear Sir

Thank you for your letter of 11th March 2002, inviting me for interview for the position of Trainee Store Person.

I look forward to meeting you on Friday 22nd March 2002 at 10.30 a.m.

Yours faithfully

C. White

C.White

be polite and business-like, even if it is a letter of complaint. They must be clearly constructed with each fresh point contained in a separate paragraph for easy understanding. When you write a letter for any reason, remember the following basic rules:

- Your own address should be written in full, complete with the postcode.
- Include the inside address. This is the title of the person (plus name if known) and the name and address of the organisation you are writing to. This should be the same as appears on the envelope.
- Write the date in full, e.g. 30 January 2002.
- Greetings. Use 'Dear Sir/Madam' if you are unsure of the sex of the person you are writing to, or 'Dear Sir' or 'Dear Madam' as applicable. Use the person's name if you know it.
- Endings. Use 'Yours faithfully' for all letters unless you have used the person's name in the greeting, in which case use 'Yours sincerely'.
- Signature. Sign below the ending. Your name should be printed below the signature, and status if applicable.

Telephone

Telephones play an important communication role both within an organisation and to customers and suppliers. Its advantage over a written message is the speed with which people are put in touch with one another.

Telephone manner – Remember that you cannot be seen, you will have no facial expressions or other body language to help make yourself understood. The tone, volume and pace of your voice is important. Speak clearly and loud enough to be heard without shouting; sound cheerful, speak at a speed at which the recipient can take down any message, key words or phrases that you are trying to relay.

Making calls – If you initiate a call you are more likely to be in control of the conversation and when you have achieved your objective you will be in the best position to end the call without causing offence. Make notes before you begin. Have times, dates and other necessary information ready. Write down your name and address before making a call if you find it difficult to spell out words 'from your head'. The call may take the following form:

'Good morning' or 'Good afternoon'.

'This is (your name) speaking, of (organisation).'

Give the name of the person you wish to speak to, if a specific individual is required.

State the reason for your call.

Keep the call brief.

Thank the recipient, even if the call did not produce the results required.

Learning task

Imagine that you are an employee of T. Joycee Construction and write a letter of complaint to your materials supplier concerning the delivery you received on 15 March 2002 (See page 106.)

T. Joycee Construction

Ridge House
Norton Road
Cheltenham
GL59 1DB

READ THE INSTRUCTIONS AND COMPLETE THE TASK

Receiving a call – A good telephone manner is as vital as when making a call. The call may take the following form:

> 'Good morning' or 'Good afternoon'; (organisation) (your name) speaking, how can I help you?'

If the call is not for you and the person required is unavailable ask politely if you can take a message.

Telephone messages – It is important that you understand what someone is saying to you on the telephone, and you may make notes of the conversation. However, when the message is not for you it is *essential* that you make written notes during the call, even though you may be seeing the person soon and be able to give the message verbally. Always make sure that the message contains all the necessary details. Any vagueness or omission of details could lead to problems.

Joe has to leave a message for Sid about two telephone calls. If all he writes is:

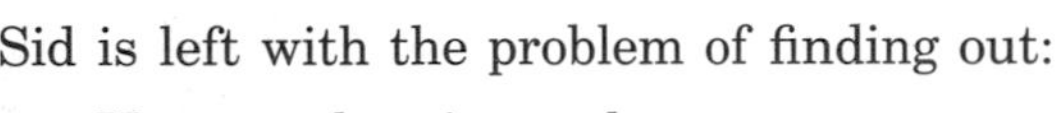

Sid is left with the problem of finding out:

- How much paint each wants.
- Which colour each wants.

A separate message for each in the following form would be much clearer in order to answer the questions Who? Where? When? What? How?

NUMBERS STARTING WITH '07' ARE MOBILE NUMBERS

Telephone Message

Date 15 MARCH Time 0945

Message for SID

Message from (Name) JUDY

(Address) STENSON FIELDS

CONTRACT DERBY

(Telephone) 07814 445 733

Message CAN YOU BRING TWO

5 LITRE CANS OF BLUE GLOSS

PAINT WHEN YOU ATTEND THE

SITE MEETING TOMORROW

MANY THANKS

Message taken by JOE

HOW'S IT GOING?

Extracting information

Throughout your working life at various levels in industry, you will have to make decisions and solve problems. To do this effectively you will have to consult various sources of information. Specialist information may be obtained from Regulations, Standards, manufacturers and Trade Development Association publications, textbooks and periodicals.

Learning task

At 1030 on Monday 14 March you take a telephone call for the general foreman Helen Oakes, from Vic Aston the carpentry sub-contractor, based in Birmingham, telephone number 0121 354 6878.

READ THE INSTRUCTIONS AND COMPLETE THE TASK

I CAN HANG THE DOORS AT STENSON ON THURSDAY, IF YOU HAVE GOT THEM. GIVE US A BUZZ BACK TO CONFIRM. I'VE GOT THE IRONMONGERY.

Use this message to fill in the form.

Telephone Message

Date .... Time ....

Message for ....

Message from (Name) ....

(Address) ....

....

(Telephone) ....

Message ....

....

....

....

....

Message taken by ....

READ THE INSTRUCTIONS AND COMPLETE THE TASK

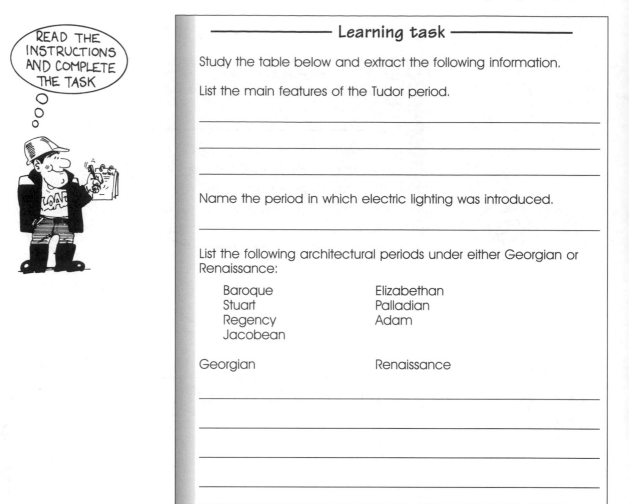

Learning task

Study the table below and extract the following information.

List the main features of the Tudor period.

____

____

____

Name the period in which electric lighting was introduced.

____

List the following architectural periods under either Georgian or Renaissance:

Baroque	Elizabethan
Stuart	Palladian
Regency	Adam
Jacobean	

Georgian	Renaissance

Date	*Period/style*	*Title*	*Features*
1485–1558	Tudor		Red brick, palaces, timber-box frame, reformation and dissolution of the monasteries, decorative brick chimneys
1558–1603	Elizabethan	Renaissance	Classical style, numerous large windows, the Civil War, Cromwell's government, restoration of monarchy, Great Fire of London, window tax
1603–1625	Jacobean		
1625–1702	Stuart		
1695–1720	Baroque	Georgian (1702–1837)	Town planning, squares and terraces of fine houses, sliding sash windows, stucco, brick tax, elaborate interiors
1720–1760	Palladian		
1760–1800	Adam		
1800–1837	Regency		
1837–1901	Victorian	Victorian	Battle of styles, Gothic revival, Crystal Palace, Industrial Revolution, spread of industrial towns, back-to-back housing, cast iron, Portland cement, monumental public buildings, electric lighting, telephone, Public Health Act, tap water and outside lavatories become common
1901–1910	Edwardian	Twentieth century	Introduction of garden cities, first Town Planning Act, two World Wars, jerry-building, ribbon development, new towns, reinforced concrete, steel frame, high-rise construction, system building, council housing, increasing comfort in housing
1910–onwards	Modern		

Personal communications

Yourself and work colleagues

It is necessary, in order for companies to function effectively, that they establish and maintain good working relationships within their organisational structure. This can be achieved by good co-operation and communication between the various sections and individual workers; good working conditions (pay, holidays, status, security, future opportunities and a pleasant safe working environment); and finally by nurturing a good team spirit, where people are motivated, rewarded for their success and allowed to work on their own initiative under supervision for the good of the company as a whole. Most companies have a hierarchical structure, which you have studied earlier in this package. Your working relationships with your immediate colleagues is equally important to the team spirit and overall success. Remember, always plan your work to ensure ease of operation, co-ordination and co-operation with other members of the workforce.

Customers

Remember, the customers pay your wages and they should be treated with respect. You should be polite at all times, even with those that are difficult. Listen carefully to their wishes and pass to a higher authority in the company anything you cannot deal with to the customer's satisfaction.

Remember, when working in their property you are a guest and you should treat everything accordingly. Always treat customers' property with the utmost care. Use dust sheets to protect carpets and furnishings when working internally. Clean up periodically and make a special effort when the job is complete. If any problem occurs contact your supervisor.

Ensure good standards of personal hygiene especially when working in occupied customers' premises. A smelly, dirty workperson will make the customer think the work will be poor. This may cause them to withdraw their offer of employment or may result in further work being given to another company.

- Wash frequently.
- Use deodorant if you have a perspiration problem.
- Wear clean overalls and have them washed at least once a week.
- Take off your muddy boots etc. when working internally on customers' premises.

Questions for you

8. State the purpose of the national code in a telephone number.

____

9. Who should you contact if any problem occurs in a customer's home?

____

TRY AND
ANSWER
THESE

WORD-SQUARE SEARCH

Hidden in the word-square are the following 20 words associated with '*Communications*'. You may find the words written forwards, backwards, up, down or diagonally.

Bill
Scale
Messages
Orthographic
Memo
Component
Assembly
Communication
Drawings
Oblique
Plans
Customer
Symbol
Sketch
Range
Schedules
Specification
Letter
Contract

Draw a ring around the words, or line in using a highlight pen thus:

EXAMPLE

EXAMPLE

P	C	R	Y	L	B	M	E	S	S	A	D	P	S	C	A	L	E
C	L	R	I	S	I	E	S	A	L	T	B	R	Y	O	T	E	F
C	R	A	S	L	L	M	F	D	N	D	H	O	M	M	P	P	G
O	I	C	N	V	L	O	A	E	L	C	P	H	B	M	E	O	H
N	A	D	D	S	Y	A	N	A	T	O	C	I	O	U	H	R	A
T	U	I	L	N	T	O	T	E	G	T	C	B	L	N	E	T	T
R	A	N	G	E	P	A	K	A	O	T	D	I	A	I	L	E	E
A	T	W	V	M	E	S	S	A	G	E	S	T	H	C	M	D	S
C	A	T	O	S	S	S	Y	D	I	E	O	I	T	A	E	B	G
T	C	C	O	C	C	T	R	U	C	T	I	O	N	T	T	C	N
X	O	O	R	T	H	O	G	R	A	P	H	I	C	I	S	A	I
U	B	N	E	L	E	R	T	R	S	L	A	F	A	O	T	I	W
R	L	S	P	V	D	D	A	C	C	I	D	E	N	N	P	E	A
O	I	T	V	C	U	S	T	O	M	E	R	D	I	C	C	A	R
L	Q	R	D	E	L	B	N	H	I	B	I	T	T	I	O	N	D
K	U	U	T	I	E	L	E	T	T	E	R	S	O	D	C	D	V
F	E	C	T	A	S	B	M	L	A	D	E	D	E	R	S	I	E
C	S	P	E	C	I	F	I	C	A	T	I	O	N	D	E	N	T

WELL, HOW
MANY DID YOU
GET?

4 Numerical skills

The use of numbers is a communication skill widely used in the construction industry. Typical questions are:

- How many do we need?
- How long is it?
- What is the area or volume?
- How long will it take?
- How much will it cost?
- How much pay will I get?

We need to revise all we know about numbers and how to use them for expressing answers to the above questions.

The number system

The figures or digits 0, 1, 2, 3, 4, 5, 6, 7, 8 and 9, and their position or place, are used to give the value of a number. We say that 264 is:

- two hundred and sixty four, or
- 200 and 60 and 4, or
- 2 hundreds, 6 tens and 4 units.

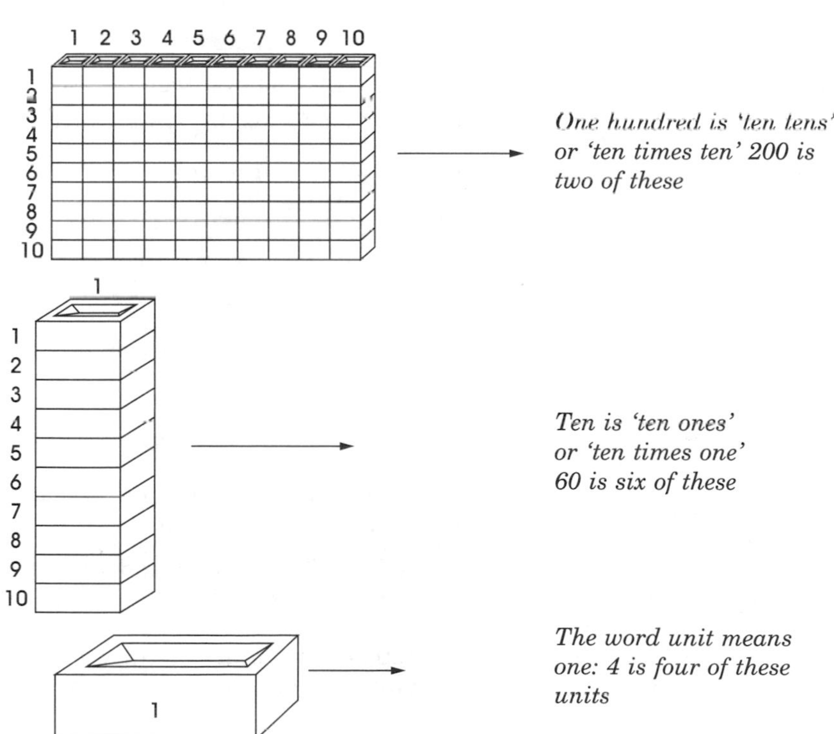

One hundred is 'ten tens' or 'ten times ten' 200 is two of these

Ten is 'ten ones' or 'ten times one' 60 is six of these

The word unit means one: 4 is four of these units

Each figure has a different value when put in a different place e.g.

- In 246 the 6 stands for 6 units.
- In 462 the 6 stands for 6 tens.
- In 624 the 6 stands for 6 hundreds.

Counting in ones or units

Figure	*Word*
0	zero
1	one
2	two
3	three
4	four
5	five
6	six
7	seven
8	eight
9	nine
10	ten
11	eleven
12	twelve
13	thirteen
14	fourteen
15	fifteen
16	sixteen
17	seventeen
18	eighteen
19	nineteen
20	twenty
21	twenty one
22	twenty two
23	twenty three
24	twenty four
25	twenty five

A zero means 'nothing' and is used in the number system to fill an empty space or place, e.g.

- 102 is one hundred and two or one hundred, zero tens and two units. The zero fills the empty space in the place for the missing tens.
- 120 is one hundred and twenty or one hundred, two tens and zero units. The zero keeps a place for the missing units.

Counting in tens

10	ten	=	1 ten
20	twenty	=	2 tens
30	thirty	=	3 tens
40	forty	=	4 tens
50	fifty	=	5 tens
60	sixty	=	6 tens
70	seventy	=	7 tens
80	eighty	=	8 tens
90	ninety	=	9 tens
100	one hundred	=	10 tens

Counting in large numbers

For numbers larger than 999 (nine hundred and ninety nine), **thousands** are used (10 hundreds).

1000	one thousand = ten hundreds
2000	two thousand = twenty hundreds
10 000	ten thousand
20 000	twenty thousand
100 000	one hundred thousand
200 000	two hundred thousand
4240	four thousand, two hundred and forty
21 361	twenty one thousand, three hundred and sixty one
320 636	three hundred and twenty thousand, six hundred and thirty six

Note: when using large numbers containing more than 4 digits, a gap or comma is normally added between each group of three figures starting from the right-hand end (units).

For numbers larger than 999 999, **millions** are used (1000 thousand).

1 000 000	one million = one thousand thousand
10 000 000	ten million
100 000 000	one hundred million
6 071 324	six million, seventy one thousand, three hundred and twenty four
24 650 150	twenty four million, six hundred and fifty thousand, one hundred and fifty
502 081 015	five hundred and two million, eighty one thousand and fifteen

For numbers larger than 999 999 999, **billions** are used (1000 million).

1 000 000 000	one billion = one thousand million
10 000 000 000	ten billion
100 000 000 000	one hundred billion

Number groups

Specific names are sometimes used for certain groups or quantities of numbers:

- 1 dozen = 12
- 1 gross = 144 (1 dozen dozen)
- 1 quire = 25 (sheets of paper)
- 1 ream = 500 (20 quire).

Roman numbers

Roman numbers are rarely used today, apart from some clock faces and some dates. They would be much harder to work with, as there is no zero to keep place value.

Roman number	*Value*
I or i	1
II or ii	2
III or iii	3
IV or iv	4
V or v	5
VI or vi	6
VII or vii	7
VIII or viii	8
IX or ix	9
X or x	10
L	50
C	100
D	500
M	1000

Roman numbers are written by putting the letters side by side:

MCXV	1115
MDCCLVI	1756
MMCCCLXIII	2363

READ THE INSTRUCTIONS AND COMPLETE THE TASK

Learning task

Fill in the missing figures and words in the table.

Number in words	*Number in figures*	*Billions*	*Millions*	*Thousands*	*Hundreds*	*Tens*	*Units*
Sixty four thousand and one	64 001			64			1
	607 098 644						
		1		5	6	4	0
Three hundred and fifty thousand, six hundred and eight							
			9	10	5	0	4
	10 964 001						
Nine million, six hundred and fifty thousand, four hundred and sixty nine							
			2	800	6	4	9
	58 048						
Eighty eight thousand and eight							

TRY AND ANSWER THESE

Questions for you

1. Write the following numbers in both normal figures and roman figures.

(a) Sixteen

___

(b) One hundred and forty five

___

(c) Two thousand, five hundred and six

___

(d) Five hundred and ninety seven

___

2. List the following numbers in size order, starting with the lowest.

(a) 34700 ___

(b) 7043 ___

(c) 60047 ___

(d) 120012 ___

(e) 3404152 ___

(f) MMMDCCLXV ___

3. Write the following numbers in words

(a) 42 304

___

(b) 8 704 312

___

(c) 10 01

___

(d) CCCLXV

___

WELL, HOW DID YOU DO?
WORK THROUGH THE SECTION AGAIN IF YOU HAD ANY PROBLEMS

Types of numbers

Positive and negative numbers

Positive numbers – have a value greater than zero. They may be written either without any sign in front of them or with a plus (+) sign e.g.

10 240 +1147 +523

Negative numbers – have a value less than zero. They are written with a minus (–) sign in front of them, e.g.

–10 –45 –115 –6

Negative numbers may be used to show a temperature below freezing point or on a bank statement to show an overdrawn balance.

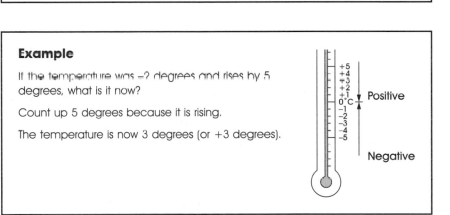

Example

If the temperature is 5 degrees and falls by 6 degrees, what is it now?

Count down 6 degrees because it is falling.

The temperature is now –1 degree.

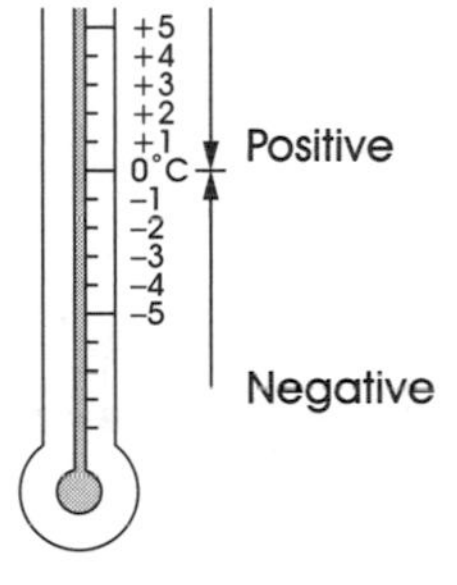

Example

If the temperature was –2 degrees and rises by 5 degrees, what is it now?

Count up 5 degrees because it is rising.

The temperature is now 3 degrees (or +3 degrees).

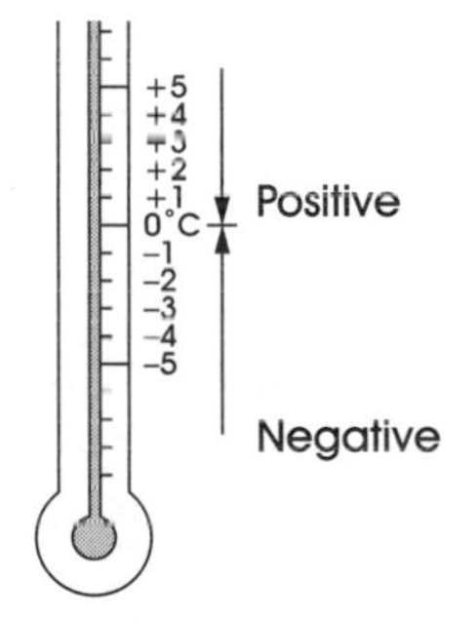

Directed numbers – are numbers which may be positive or negative

Fractions

Fractions are parts of a whole number.

For half a brick we write the fraction:

- ½ a brick, which means one part out of two parts.

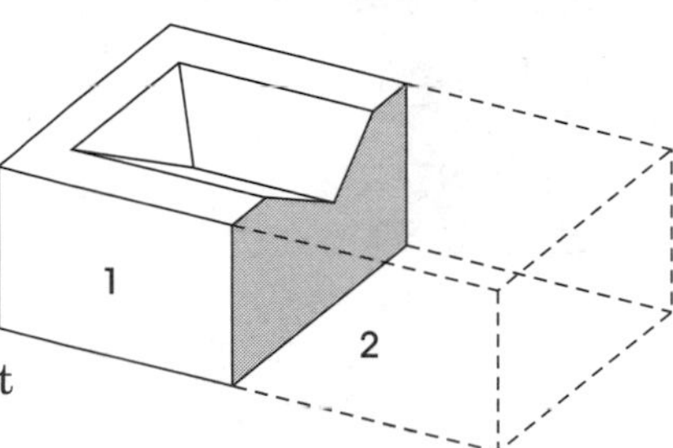

For three quarters of a plank of wood, we write the fraction:

- ¾ of a plank which means 3 parts out of 4 parts.
- The top number of a fractions is the *numerator*, the bottom one is the *denominator*
- A fraction like ½ or ¾ is called a *proper fraction*
- A fraction like $\frac{3}{2}$ or $\frac{4}{3}$ is called an *improper fraction*
- A fraction like 1½ or 2¼ is called a *mixed number*

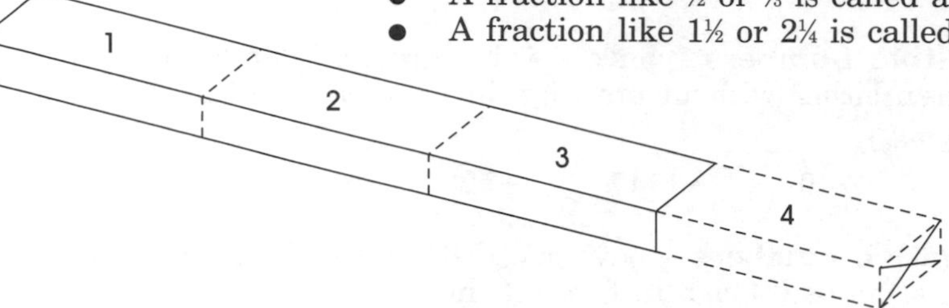

Decimals

Decimals are also used for parts of a whole number. The word 'decimal' is short for 'decimal fraction'. Decimals are just another way of expressing fractions. However, decimal fractions are based on tenths and hundreds, not quarters or thirds, etc.

Note: A decimal point is used to separate the whole number from the decimal fraction.

1.5 tins of paint means $1\frac{5}{10}$ which is the same as 1½ tins.

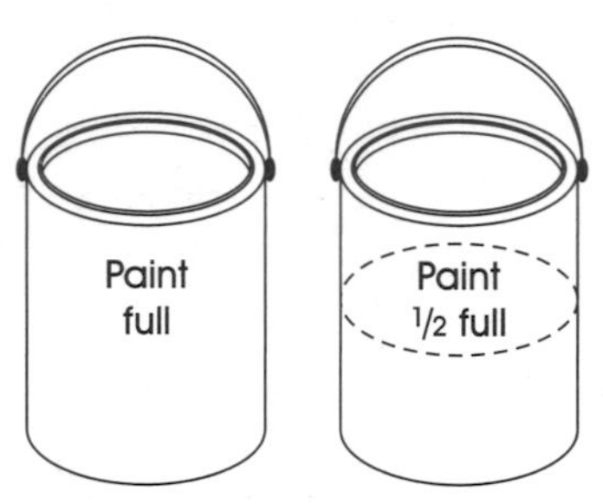

TRY AND ANSWER THESE

Questions for you

4. Underline the whole number in the following numbers:

6.8 3½ 46.46 14¼ 109¾ 1.25

5. Describe in words the following numbers. For example, positive fraction, negative number, etc.

0.5 ________________

–1 ________________

9 ________________

¾ ________________

–½ ________________

–0.75 ________________

Basic rules for numbers

Mathematical operators

The basic signs used when operating with numbers are:

+ plus (add)
– minus (subtract)
× times (multiply)
÷ divide
= equals

These are called operators.

Addition

Adding involves putting, summing, adding, counting things together.

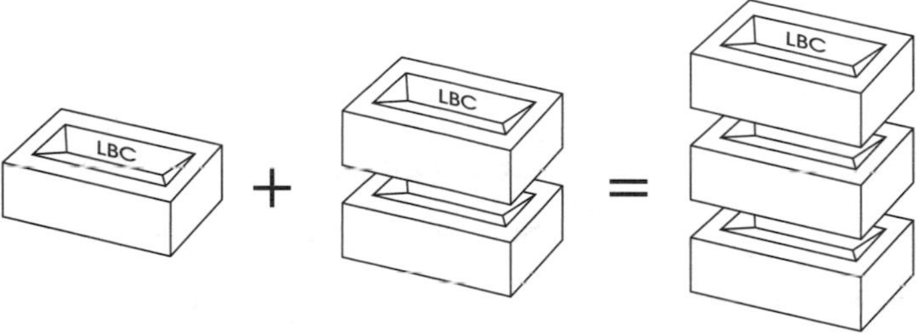

1 brick plus 2 bricks equals 3 bricks

$$1 + 2 = 3$$

When adding it is useful to make sure that the units are lined up underneath each other. The columns can then be easily added starting from the right.

Example

Add 1124 + 287 + 35

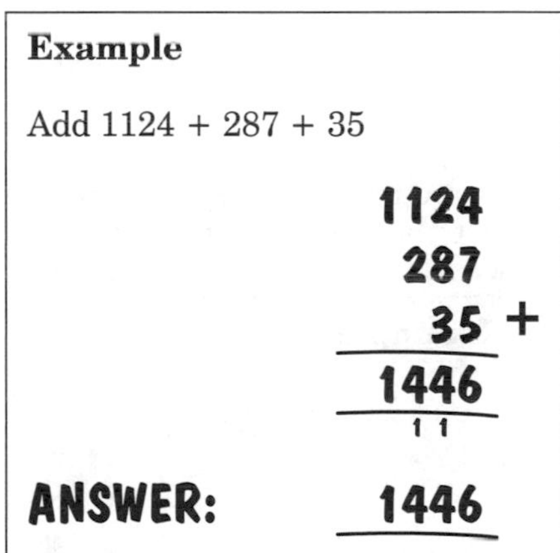

- The right-hand column 4 + 7 + 5 = 16. As this is 1 ten and 6 units, the 6 is put in the units column and the 1 is carried forward to the tens column.
- When adding the tens column the 1 carried forward must be included. 2 + 8 + 3 + 1 = 14. As this is 1 hundred and 4 tens, put 4 in the tens column and carry the 1 forward to the hundreds.
- Continue adding the remaining columns.

The final answer is called the **sum**. The sum of 1124, 287 and 35 is 1446.

Addition of decimals – Numbers containing decimal points must be written down with the points lined up underneath each other when preparing to add.

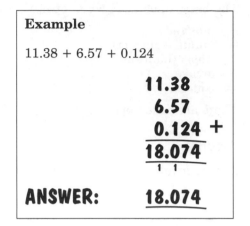

Example

11.38 + 6.57 + 0.124

```
 11.38
  6.57
  0.124 +
 ------
 18.074
  1 1
```

ANSWER: 18.074

Subtraction

Subtraction involves taking things away.

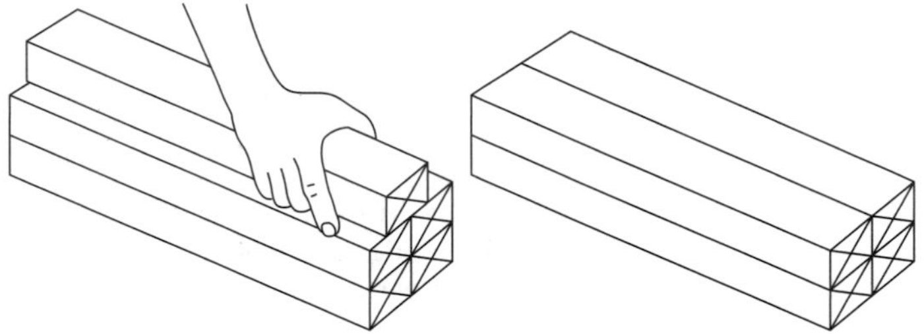

5 pieces of wood, take away (minus) 1 leaves 4

5 – 1 = 4

Again when subtracting, line up the units underneath each other and subtract the columns starting from the right.

Example

Subtract 48 from 694

```
  8 1
  6̸94
   48 –
  ---
  646
```

ANSWER: 646

- When subtracting the units, we say in this example '8 from 4 will not go, so borrow 1 from the tens columns', making 14 units and 8 tens.
- When subtracting the next column, we must remember there is one less at the top.

Subtraction of decimals – Numbers containing decimal points are again written down with the points lined up underneath each other.

Example

16.697 – 8.565

```
 0 1
 16.697
  8.565  –
  8.132
```

ANSWER: 8.132

Multiplication

This is a quick way of adding groups of equal numbers:

$4 + 4 + 4 + 4 + 4 = 20$ is the same as $4 \times 5 = 20$

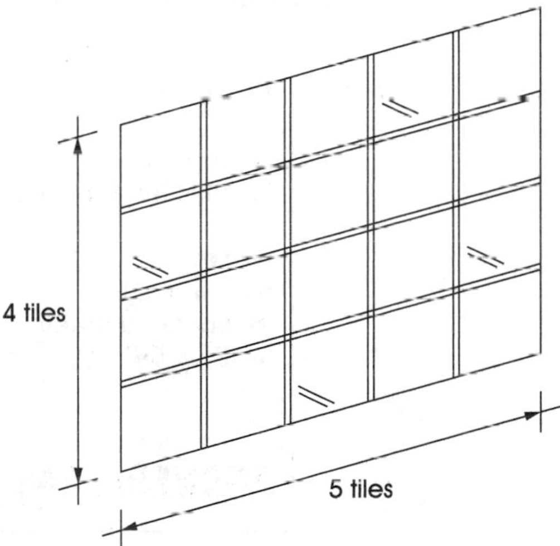

You do not have to count the number of tiles on the wall to tell that there are twenty.

There are 5 lots of 4 tiles. When using the quick way we say 'Four tiles multiplied by five lots equals 20 tiles' or simply 'Four by five equals twenty'. Hence:

$$4 \times 5 = 20$$

20 is called the **product** of 4 and 5. The product of any number up to 10 is shown in a **multiplication table.** In order to multiply you must know your multiplication tables for products at least up to 10 × 10. The example shown in the multiplication table is for the product of 4 × 5.

1	2	3	4	5	6	7	8	9	10
2	4	6	8	10	12	14	16	18	20
3	6	9	12	15	18	21	24	27	30
4	8	12	16	(20)	24	28	32	36	40
5	10	15	20	25	30	35	40	45	50
6	12	18	24	30	36	42	48	54	60
7	14	21	28	35	42	49	56	63	70
8	16	24	32	40	48	56	64	72	80
9	18	27	36	45	54	63	72	81	90
10	20	30	40	50	60	70	80	90	100

Multiplication table

To find 4 × 5, look at the intersection of row 4 and column 5. There you find the product of 4 and 5 which is 20.

Multiplication of large numbers –

Example

A larger wall is 8 tiles high and 56 tiles long. How many tiles?

Note:
8 × 56 means
'8 lots of 56 make'
or '8 by 56'

```
 56
  8 ×
----
448
----
 4
```

ANSWER: 448 tiles

- Start with the units. We say in this example '8 × 6 is 48'. (Use the multiplication table).
- Put the unit figure down (8) and carry the tens forward (4).
- Then continue to multiply 5 × 8 which is 40 plus the carried 4 makes 44.
- The product of 8 × 56 tiles is 448 tiles.

Example

Multiply 146 × 92

```
  146
   92
  292 ×
13140
13432
```

ANSWER: 13432

- In this example, first multiply 146 × 2, to give 292.
- Then multiply 146 × 90. Do this by putting a zero down in the units and multiply 146 by 9.
- 9 times 146 is 1314.
- Add the two rows to get 13432.

Multiplication of decimals – When multiplying, any decimal points can be ignored until the two sets of numbers have been multiplied together.

Example

11.6 × 4.5

Two places to the right of the decimal point to start with.

Count 2 places to left for point in answer

```
 116
  45 ×
 580
4640
5220
```

ANSWER: 52.20

The position of the decimal point can be located by the following rule:

The number of figures to the right of the decimal point in the answer will always equal the total number of figures to the right of the decimal point to start with.

Division

Division involves sharing or dividing things into equal parts. It is the opposite of multiplication.

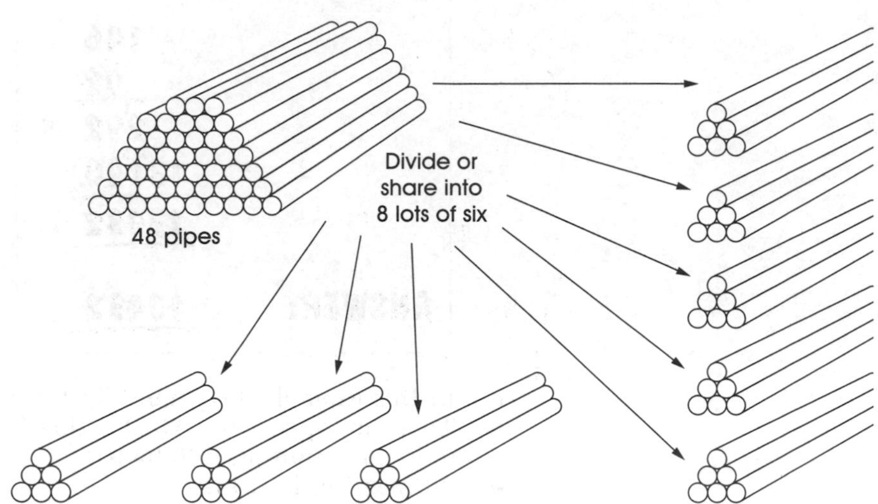
48 pipes
Divide or share into 8 lots of six

Division using tables – Multiplication tables can be used backwards for division.

1	2	3	4	5	6	7	8	9	10
2	4	6	8	10	12	14	16	18	20
3	6	9	12	15	18	21	24	27	30
4	8	12	16	20	24	28	32	36	40
5	10	15	20	25	30	35	40	45	50
6	12	18	24	30	36	42	48	54	60
7	14	21	28	35	42	49	56	63	70
8	16	24	32	40	48	56	64	72	80
9	18	27	36	45	54	63	72	81	90
10	20	30	40	50	60	70	80	90	100

Multiplication table

- Use the multiplication table. Look for 48 in the 6th row.
- Read up the column to get 8.
- Hence 48 shares into 8 equal lots of 6.
- We say '48 divided by (÷) 8 equals (=) 6'.

Division by calculation – Divide 384 by 6. This may be written down as 384 ÷ 6.

The 6 is called the ***divisor,*** 384 is called the ***dividend*** and the answer is called the ***quotient.***

Note how the calculation is laid out in the example. For division we work from the left, not from the right, and the answer is written above, not below.

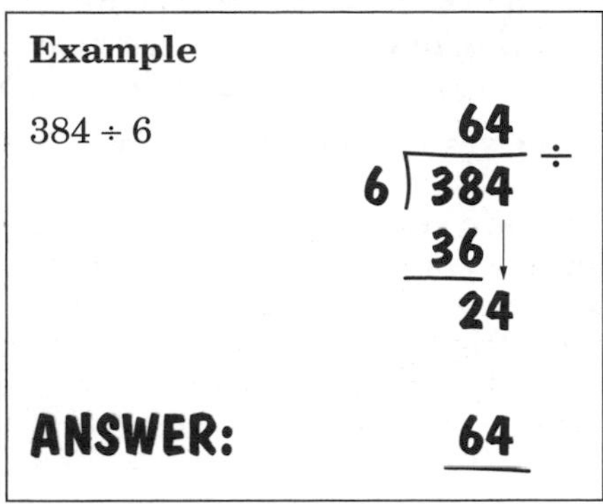

- First we try to find how many times 6 shares into 3.
- We say 6 into 3 will not go. So there is nothing to enter in the first position of the answer line.
- Then we try 6 into 38 (from table) and the nearest number below 38 in the 6th row is 36.
- Read up the column to get 6.
- Put the 6 above the 8 in the second position of the answer line.
- Subtract 36 from 38 to leave 2.
- 'Bring down' the 4 to make 24.
- Divide 24 by 6 (from table) equals 4. We say '6 into 24 goes 4 times'.
- Put the 4 next to the 6 in the third position of the answer line.

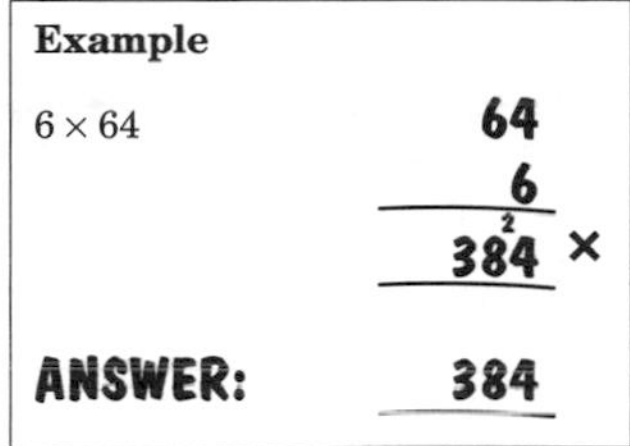

This gives the answer, so that 384 ÷ 6 = 64.

Your division can be checked by multiplying back. This check proves the working out is correct: 6 × 64 = 384.

Division with a remainder – When there is a remainder (left over at the end of the division) the operation can be continued by inserting a decimal point and 'bringing down' zeros. In the example:

Example

37 ÷ 4

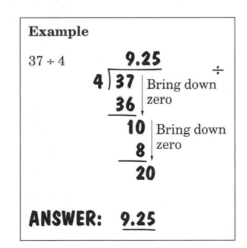

```
      9.25
  4 ) 37      Bring down zero
      36
      10      Bring down zero
       8
      20
```

ANSWER: 9.25

- We say '4 into 3 will not go'. (Nothing is entered in the first position of the answer line.)
- We say '4 into 37 goes 9 remainder 1'. (9 is entered into the second position of the answer line.)

- Insert decimal point next to 9 and bring down zero next to the remainder 1 making 10;
- We say '4 into 10 goes 2 remainder 2'. (2 is entered into the answer line after the decimal point.)
- Bring down another zero next to the remainder 2 making 20.
- We say '4 into 20 goes 5 and nothing left over'.

This gives the answer, so that, 37 ÷ 4 = 9.25.

Division of decimal numbers – when dividing numbers with decimal points, we need first to make the divisor a whole number. We do this by moving its decimal point a number of places to the right until it is a whole number, but to compensate we must also move the decimal point in the dividend by the same number of places. (We may have to add zeros to do this.)

Example

164 ÷ 0.2

164.6 ÷ 0.2

1646 ÷ 2

Move both the decimal points until the division is a whole number

$$2\overline{)1646}\quad 823$$

÷

ANSWER: 823

Operating with fractions

Fractions are best converted into decimals before proceeding with addition, subtraction, multiplication or division operations. Conversion is done by dividing the bottom number into the top number.

Example

Express $^7/_8$ as a decimal

$$8\overline{)7.0}\quad 0.875$$

64

60

56

40

ANSWER: 0.875

Combined operations

You will need to understand the following types of mathematical statements:

- $12 - (6 + 4)$
- $3 \times (4 - 1)$ usually written as $3(4 - 1)$
- $3 \times (12 \div 3)$ usually written as $3(12 \div 3)$
- $(40 \div 2) + (4 \times 6)$

Rules for combined operations –

You must work out the operation contained in the brackets ( ) first before proceeding.

You must then do multiplication and division before addition and subtraction.

A useful made-up word 'BODMAS' can help you remember the order in which calculations should be undertaken:

Brackets, then the Order is Division, Multiplication, Addition and then Subtraction

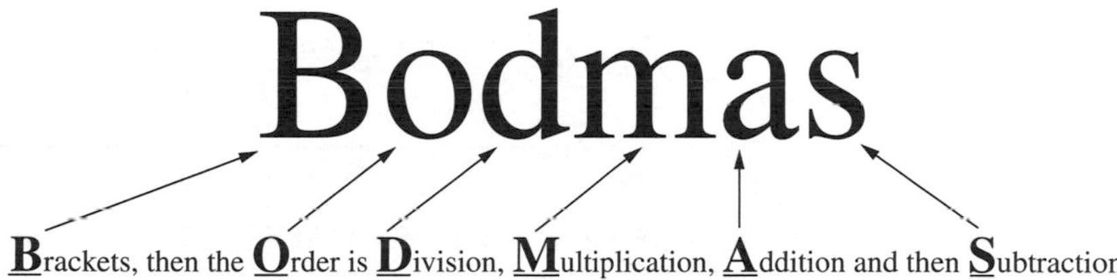
Bodmas
Brackets, then the Order is Division, Multiplication, Addition and then Subtraction

- So, $12 - (6 + 4)$ gives $12 - 10 = 2$
- $3(4 - 1)$ gives $3 \times 3 = 9$

Examples

$3(12 \div 3)$

$$3(12 \div 3)$$
$$= 3(4)$$
$$= 3 \times 4$$
$$= \underline{12}$$

$40 \div 2 + (4 \times 6)$

$$40 \div 2 + (4 \times 6)$$
$$= 40 \div 2 + (24)$$
$$= 20 + 24$$
$$= \underline{48}$$

TRY AND
ANSWER
THESE

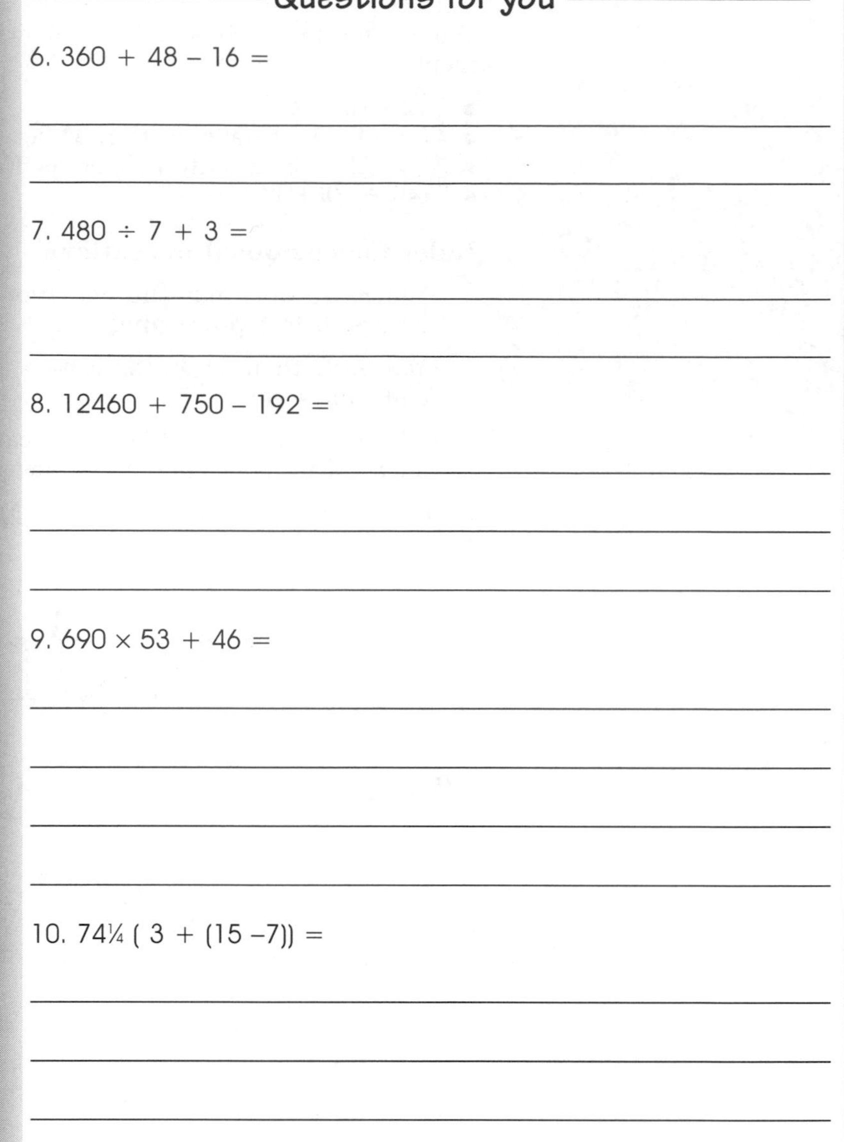

Questions for you

6. 360 + 48 – 16 =

7. 480 ÷ 7 + 3 =

8. 12460 + 750 – 192 =

9. 690 × 53 + 46 =

10. 74¼ (3 + (15 –7)) =

WELL, HOW
MANY DID YOU
GET?

Rough checks

Approximate answers – common causes of incorrect answers to calculation problems are incomplete working out and incorrectly placed decimal points. Rough checks of the expected size of an answer and the position of the decimal point would overcome this problem. These rough checks can be carried out quickly using approximate numbers.

- 4.65 × 2.05 ÷ 3.85
- For a rough check say 5 × 2 ÷ 4 = 2.5
- The actual correct answer is 2.476

The rough checks and the correct answer are of similar size. This confirms that the answer is 2.476 and not 0.2476 or 24.76 etc.

Rough checks will be nearer to the correct answer if, when choosing approximate numbers, some are increased and some are decreased. In cases where the rough check and the correct answer are not of the same size the calculation should be reworked to find the cause of the error.

Electronic calculations

ONCE YOU
HAVE AN
UNDERSTANDING OF
NUMBERS A CALCULATOR
MAY BE USED

All basic electronic calculators look similar. They include the following main key functions:

Numbered keys	[0] [1] [2] [3] [4] [5] [6] [7] [8] [9]
Operation keys	[+] [−] [÷] [×]
Equals key	[=]
Decimal point key	[·]
Square root key	[√]
Percentage key	[%]
Clear last entry key	[CE]
Clear all entries key	[AC] or [ON/C]

The operation of your calculator will vary depending on the model; therefore consult the booklet supplied with it before use. After some practice you should be able to operate your calculator quickly and accurately.

In the following example, first press the AC key to clear all numbers in the display. Then press the keys listed in the left-hand column in turn. Check that the correct numbers appear in the display.

Example

$55.335 \times 2.1 \div 3.52$

Key	*Display*
[AC] or [ON/C]	0.
[5] [5] [·] [3] [3] [5]	55.335
[×]	55.335
[2] [·] [1]	2.1
[=]	116.2035
[÷]	
[3] [·] [5] [2]	3.52
[=]	33.012357

Answer 33.012357

Note: For most modern calculators, the first operation of the = key may be omitted since the next key automatically proceeds with the calculation to give the desired result.

The process of approximating answers covered before, should be carried out even when using an electronic calculator, as wrong answers are often the result of miskeying; even a slight hesitation on a key can cause a number to be entered twice.

TRY AND ANSWER THESE

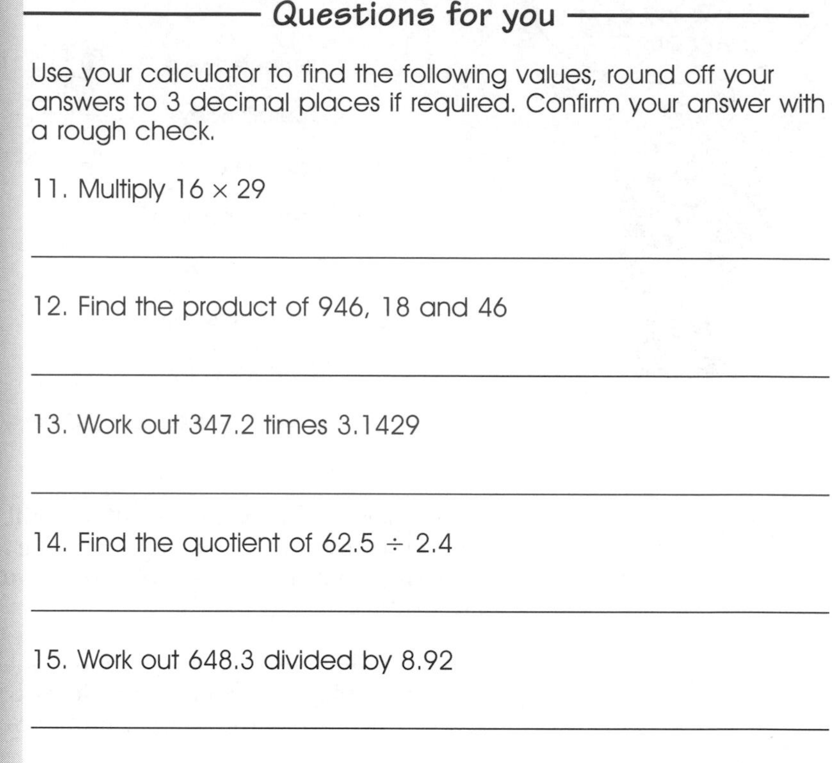

Questions for you

Use your calculator to find the following values, round off your answers to 3 decimal places if required. Confirm your answer with a rough check.

11. Multiply 16 × 29

___

12. Find the product of 946, 18 and 46

___

13. Work out 347.2 times 3.1429

___

14. Find the quotient of 62.5 ÷ 2.4

___

15. Work out 648.3 divided by 8.92

___

WELL, HOW DID YOU DO?
WORK THROUGH THE SECTION AGAIN IF YOU HAD ANY PROBLEMS

Rounding numbers

Number of decimal places – For most purposes, calculations which show three decimal figures are considered sufficiently accurate. These can therefore be rounded off to three decimal places. This however entails looking at the fourth decimal figure; if it is a five or above add one to the third decimal figure. Where it is below five ignore it. For example:

33.012357 becomes 33.012

2.747642 becomes 2.748

Number of significant figures – On occasions a number may have far too many figures before and after the decimal place for practical purposes. This is overcome by expressing it to 2, 3 or 4 significant figures.

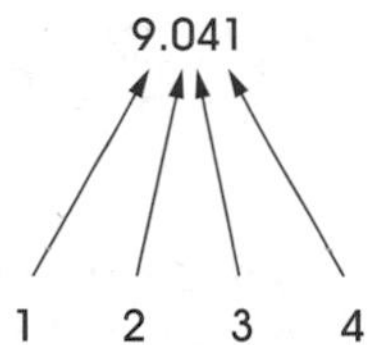
9.041
1
2
3
4

The term 'significant figures' is normally abbreviated to S.F. or sig. fig. This illustration shows a decimal number expressed to 4 S.F.

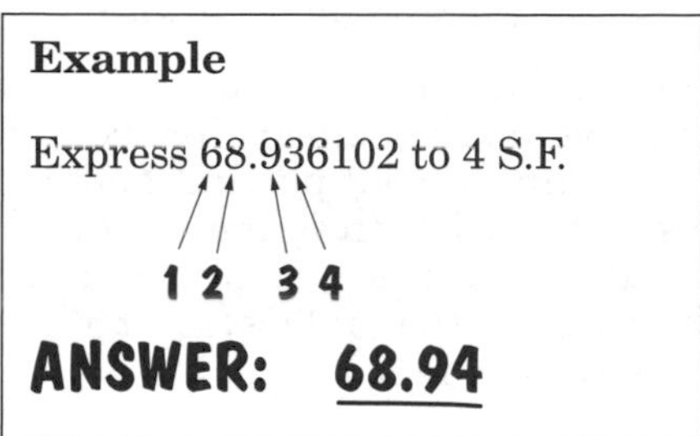
Example
Express 68.936102 to 4 S.F.
1 2 3 4
ANSWER: 68.94

STUDY THIS TABLE

In the example, we have applied the same rule as for rounding numbers to a number of decimal places. If the next figure after the last S.F. is 5 or more then round up the last S.F. by 1. More examples below:

Number	*To 3 S.F.*	*To 2 S. F.*	*To 1 S.F.*
6.308	6.31	6.3	6
5368	5370	5400	5000
0.051308	0.0513	0.051	0.05
0.1409	0.141	0.14	0.1

After rounding up to a number of significant figures, you must take care not to change the place value. For example, 4582 is 4600 to 2 S.F., so zeros are added to the end in order to maintain the place value.

However, 4.582 = 4.6 to 2 S.F. not 4.6000, since trailing zeros are not added after the decimal point.

TRY AND ANSWER THESE

Questions for you

16. Complete the following table:

Number	*To 3 S.F.*	*To 2 S. F.*	*To 1 S.F.*
5.874			
9643			
0.048739			
0.14973			

Units of measurement

Metrification of measurements in the construction industry is almost total. Previously we used imperial units of measurement. However, you may still come across them occasionally, so a knowledge of both is required. To avoid the possibility of mistakes all imperial measurements should be converted to metric ones. All calculations and further work can then be undertaken in metric units of measurement.

The metric units we use are also known as SI units (Systeme International d'Unites). For most quantities of measurement there is a base unit, a multiple unit and a sub-multiple unit. For example the base unit of length is the metre (m). Its multiple the kilometre (km) is a thousand times larger: m × 1000 = km. The submultiple unit of the metre is the millimetre (mm), which is a thousand times smaller: m ÷ 1000 = mm.

It is an easy process to change from one unit to another in the metric system. The decimal point is moved 3 places to the left when changing to a larger unit and three places to the right when changing to a smaller unit.

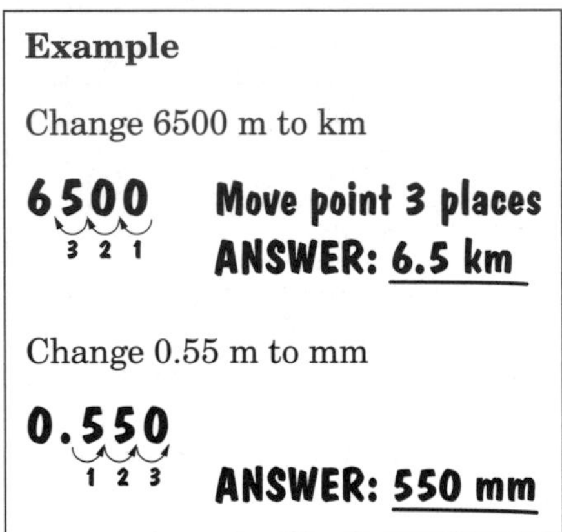
Example

Change 6500 m to km

6500 (3 2 1) Move point 3 places

ANSWER: 6.5 km

Change 0.55 m to mm

0.550 (1 2 3)

ANSWER: 550 mm

The units you will most frequently come across in the construction industry are covered in the following examples.

Length

Length is a measure of how long something is from end to end. The length of this screw is 75 mm.

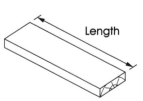

Length

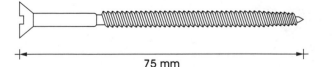

75 mm

Long lengths – such as the distance between two places on a map, are measured in kilometres (km). Miles are used in imperial measure:

- 1 mile = 1760 yards
- 1 mile = 1609 km
- 1 km = 0.62 miles

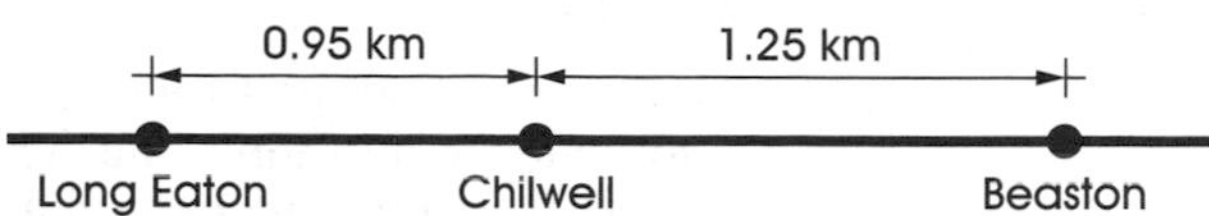

Intermediate lengths – such as a piece of timber, is measured in metres (m). Feet (foot) and yards are used in imperial measurement:

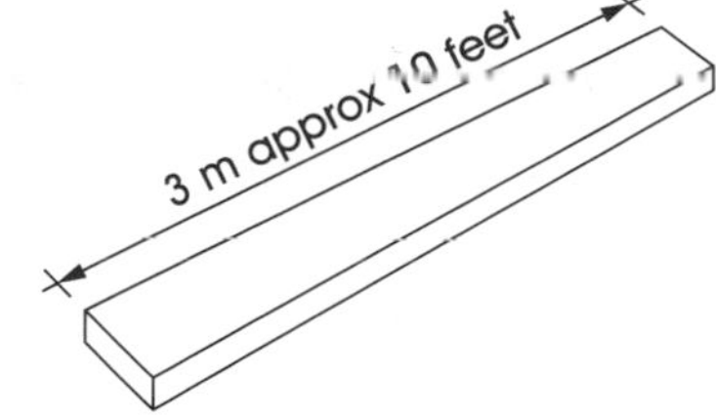

- 3 feet = 1 yard
- 1 yard = 0.9144 metres
- 1 foot = 0.3048 metres
- 1 metre = 1.0936 yards
- 1 metre = 3.281 feet

Small lengths – such as a brick, are measured in millimetres. Inches are used in imperial measure:

215 mm

= Approx 9 inches

- 12 inches = 1 foot
- 1 inch = 25.4 millimetres.

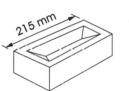

Sometimes centimetres (cm) are encountered as an intermediate measure of length:

- 1 metre = 100 centimetres
- 1 centimetre = 10 millimetres

However the centimetre is not used in the construction industry. Measurement should always be given in metres or millimetres, e.g. 55 cm is written as either 0.55 m or 550 mm.

Area

Area is concerned with the extent or measure of a surface. A patio 3 m × 4 m has an area of 12 square metres. Two linear measurements multiplied together give area (square measure). Both square metres and square millimetres are used for area. These are written as m^2 or mm^2. For example 12 square metres = 12 m^2.

- 1 m^2 = 1.196 square yards
- 1 m^2 = 10.764 square feet
- 1 mm^2 = 0.00155 square inches.

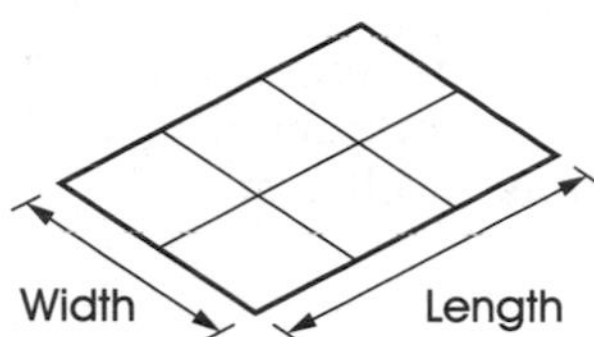

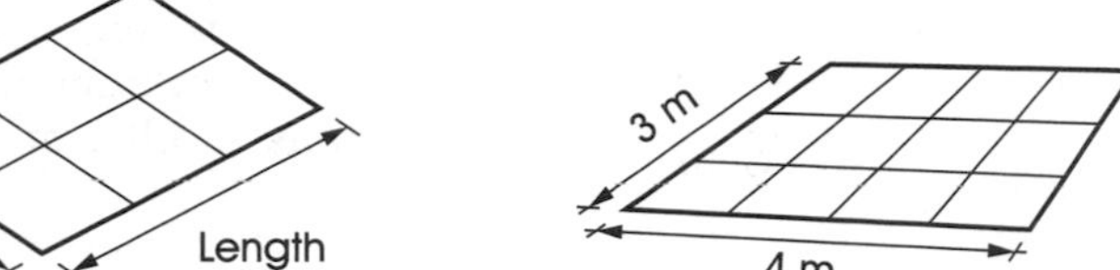

Width × Length = Area

Large areas – are measured in metric hectares or imperial acres.

- 1 hectare = 10 000 m^2
- 1 hectare = 2.471 acres
- 1 acre = 4840 square yards

Volume

Volume is concerned with the space taken up by a solid object. A room 4 m wide by 5 m long and 3 m high has a volume of 60 cubic metres. The three linear measurements multiplied together give the volume (cubic measure). Cubic metres and cubic millimetres are used for volume. These are written as m^3 or mm^3, i.e. 60 cubic metres = 60 m^3.

- 1 m^3 = 1.31 cubic yards
- 1 m^3 = 35.335 cubic feet
- 1 mm^3 = 0.0000611 cubic inches

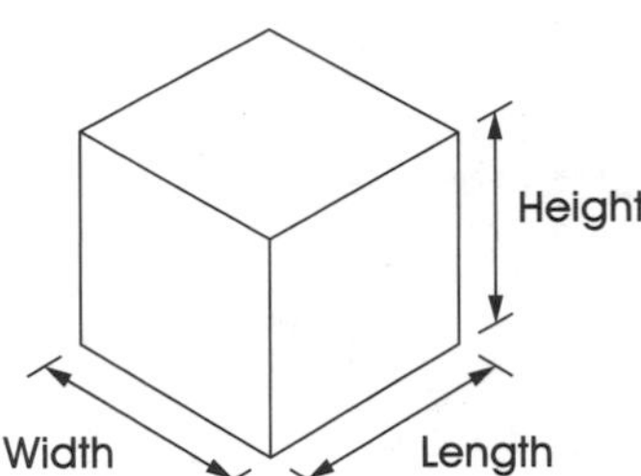

Width × Length × Height = Volume

Width × Length × Height = Volume

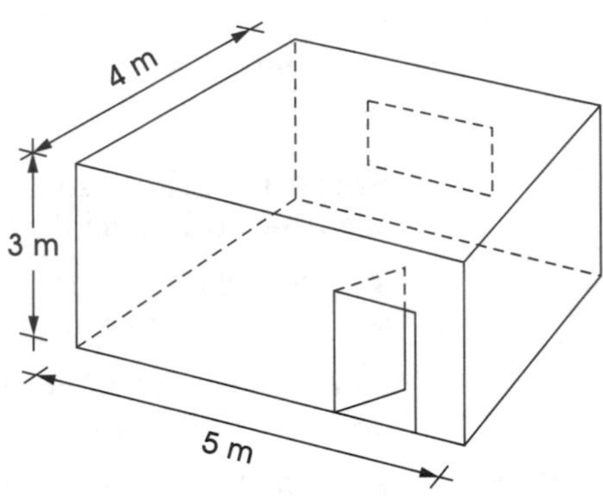

Volume = 4 × 5 × 3
= 60 m^3

Capacity

Capacity is concerned with the amount of space taken up by liquids or the amount of liquid that can fit into a given volume. Litre is the unit of capacity (to avoid confusion with the number 1 'litre' is best written in full), the millilitre (ml) is the sub-multiple. 1 litre = 1000 ml. Gallons, pints and fluid ounces are the imperial units of capacity.

- 1 gallon = 8 pints
- 1 pint = 20 fluid ounces
- 1 litre = 1.76 pints
- 1 litre = 0.22 gallons
- 1 ml = 0.035 fluid ounces
- 1 ml = 0.0017 pints.

Capacity is linked with volume. A carton 100 mm × 100 mm × 100 mm contains a litre of water. It would take 1000 of these cartons to fill 1 m^3.

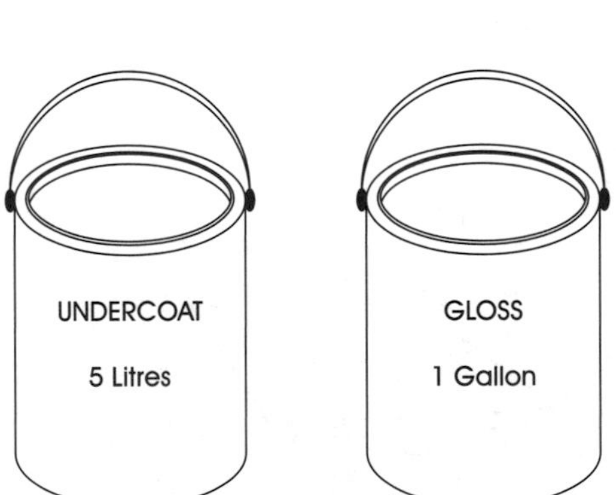

5 litres is more than 1 gallon
4.5 litres = approx 1 gallon

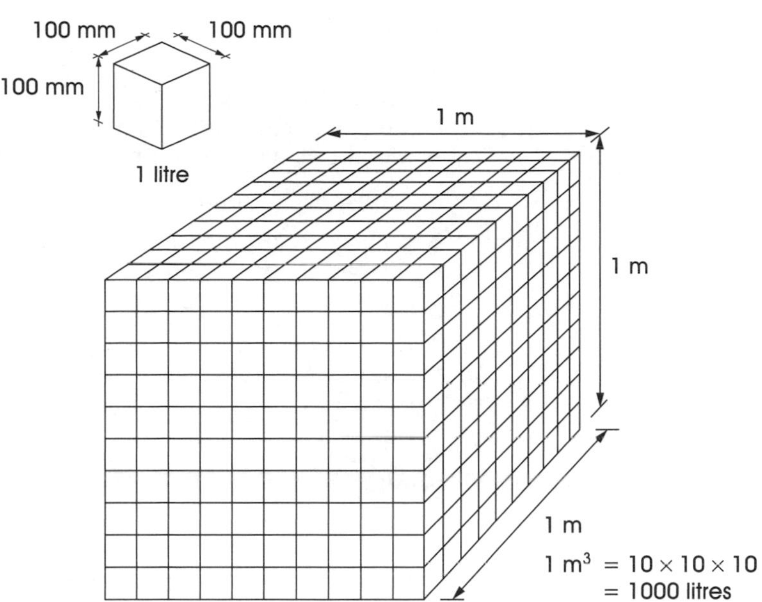

1200 cc engine = 1.2 litre engine

Although not used in construction, 1 ml = 1 cubic centimetre (cc). Thus a car with a 1200 cc engine is 1200 ml or 1.2 litres in size.

Mass

Mass is concerned with the weight of an object. The metric unit of mass is the kilogram (kg). Its multiple is the tonne (to avoid confusion with the imperial ton no abbreviation is used). The sub-multiple unit is the gram (g), for very small objects the milligram (mg) is used. Imperial units are tons (ton), hundredweights (cwt), stones (st), pounds (lb) and ounces (oz).

- 1 tonne = 1000 kg
- 1 kg = 1000 g
- 1 g = 1000 mg
- 1 ton = 2240 lb
- 112 lb = 1 cwt
- 14 lb = 1 st
- 1 lb = 16 oz
- 1 tonne = 2205 lb
- 1 kg = 2.2 lb
- 30 g = 1 oz

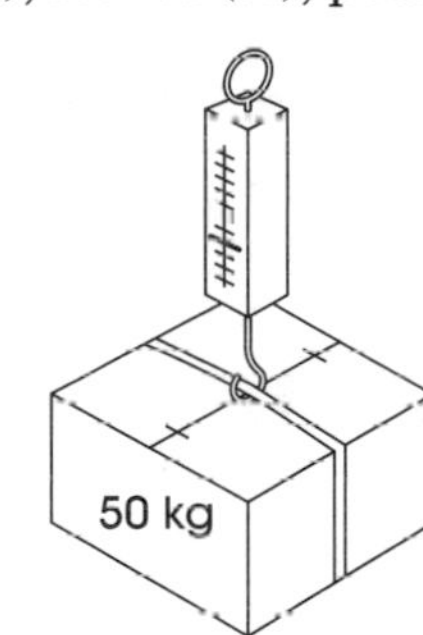

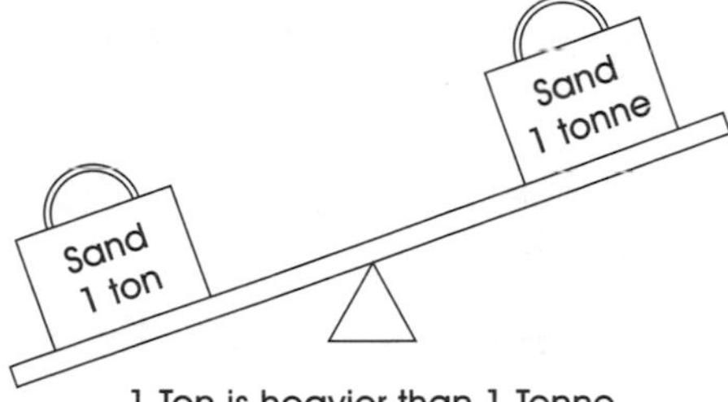

1 Ton is heavier than 1 Tonne

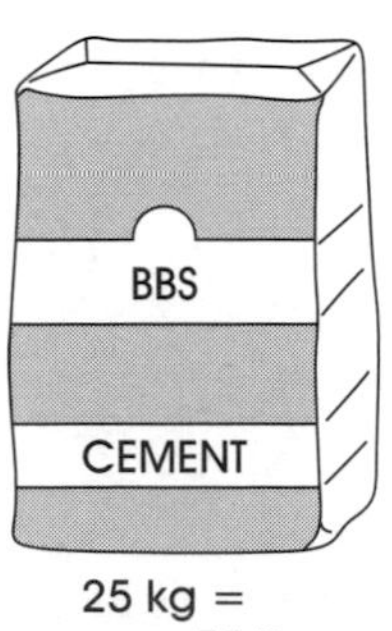

25 kg = approx 55 lb

Density

This is a measure of a material's consistency. A more solid material is generally considered to be more dense, but a true measure of density can only be calculated by measuring its weight and volume. The ratio of weight (mass) to volume is a measure of density, thus:

Mass ÷ Volume = Density

- The density of water is 1000 kg/m^3
- The density of softwood is typically 450 kg/m^3
- The density of concrete is typically 2400 kg/m^3

Note that to compare density of different materials, the ratio of weight to volume has to be converted to a standard unit of measurement – in this case kg/m^3.

Specific gravity – The density of materials is normally expressed in smaller units than kg/m^3 by dividing values by 1000. This is referred to as the specific gravity (sg) and the specific gravity of water is the value 1 sg. Materials heavier than water have an sg greater than 1 sg and sink; materials lighter than water have an sg smaller than 1 sg and float.

Density ÷ 1000 = Specific gravity or relative density

- The specific gravity of water is 1 sg
- The specific gravity of softwood is typically 0.45 sg
- The specific gravity of concrete is typically 2.4 sg

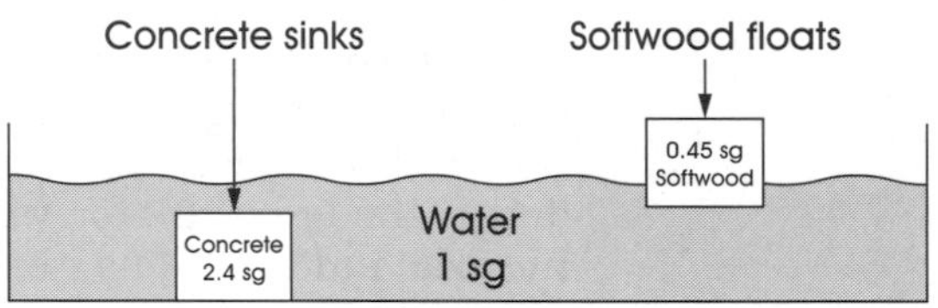

Example

Determine the density of a sample which has a mass of 25 kg and a volume of 0.125 m^3. Will it sink or float?

25 ÷ 0.125 = 200

ANSWER:
Density = 200 kg/m^3

200 ÷ 1000 = 0.2

Specific gravity = 0.2 sg

ANSWER: It will float

Imperial to metric conversion

An approximation between imperial and metric measurements can be made using the comparisons given.

Examples

Convert 350 feet to metres

350 × 0.3048 = 106.68
ANS: 350 feet = 106.68 m

Convert 12 yards to metres

12 × 0.9144 = 10.9728
ANS: 12 yards = 10.973 m

Convert 1 gallon to litres

1 ÷ 0.22 = 4.545
ANS: 1 gallon = 4.545 litres

Convert 10 pounds to kilograms

10 ÷ 2.2 = 4.545
ANS: 10 pounds = 4.545 kg

TRY AND ANSWER THESE

Questions for you

17. Change to millimetres:

(a) 1.2 m ____________________

(b) 0.95 m ____________________

(c) 24.6 m ____________________

(d) 0.070 m ____________________

18. Express in metres:

(a) 1264 mm ____________________

(b) 920 mm ____________________

(c) 21 950 mm____________________

(d) 68 mm ____________________

19. Convert 6 feet 6 inches into metres

20. Change to kilograms:

(a) 1.2 tonne

(b) 5500 g

(c) 0.5 tonne

(d) 250 g

21. Convert 10 lb into kilograms

22. You are asked to purchase 2 dozen 3 inch screws. The supplier only has metric screws in the following sizes: 50 mm, 62 mm, 75 mm, 82 mm and 100 mm. How many and what size will you buy?

23. You are informed that the last time a factory was painted out it took 10½ gallons. How many 5-litre tins of paint are required this time?

24. You are told over the phone to pick up a 4 foot × 2 foot paving slab. Which is the nearest size metric slab?

(a) 450 × 900 m

(b) 600 × 900 mm

(c) 600 × 1200 mm

(d) 900 × 1200 mm

WELL, HOW DID YOU DO?
WORK THROUGH THE SECTION AGAIN IF YOU HAD ANY PROBLEMS

Time

This is a measure of the continued progress of existence, i.e. the past, the present and the future. The second is the main unit of time:

The time shown is:
3.55 or we can say 5 to 4 o'clock

- 60 seconds (s) = 1 minute (min)
- 60 minutes = 1 hour (h)
- 24 hours = 1 day
- 7 day = 1 week (wk)
- 31, 30, 29 or 28 days = 1 month (mth)

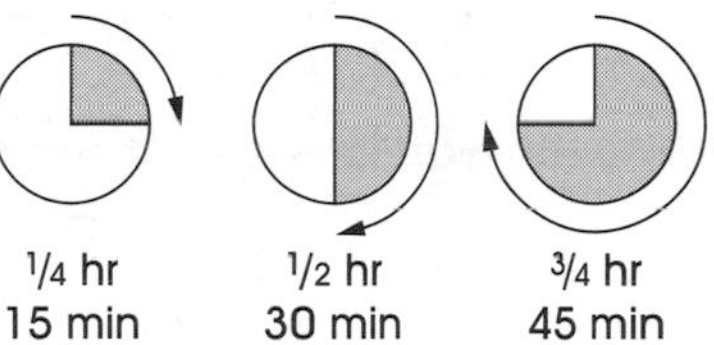

(Remember: 30 days for September, April, June and November. All the rest have 31 excepting February alone, which has 28 and in a leap year 29.)

- 13 weeks = 1 quarter
- 26 weeks = 1 half year
- 12 months = 1 year
- 365 days = 1 year
- 366 days = 1 leap year (every 4 years to include 29 February)
- 52 weeks = 1 year

In the winter months we use GMT (Greenwich Mean Time). In the summer our clocks are put forward by 1 hour, called British Summer Time.

Times in other parts of the world will not be the same due to time zone differences. The time of day is normally expressed using a 12-hour clock, with the morning and afternoon being distinguished by the use of AM or PM.

- AM is in the morning (ante meridian)
- PM is in the afternoon (post meridian)
- midday or noon is usually written as 12 AM
- midnight is usually written 12 PM

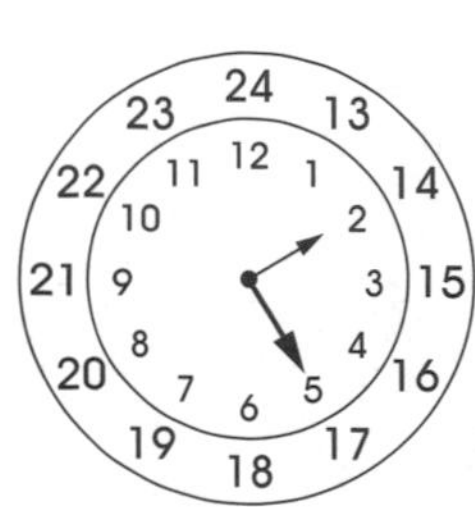

24 hour clock

Example
Convert 2.25 PM to the 24-hour clock

$$\begin{array}{r} 2.25 \\ 12 \\ \hline 1425 \end{array} + \quad \textbf{ANSWER: The time is 1425}$$

The 24-hour clock – Timetables are normally based on the 24-hour clock. This runs from 0000 hrs at midnight, to 2400 again at midnight.

- 8.30 AM is 0830 hours; 8.30 PM is 2030 hours
- Noon is 1200 hours; midnight is 0000 or 2400 hours
- 6.45 AM is 0645 hours; 6.45 PM is 1845 hours
- (Add 12 hours to the AM hours and drop the point to convert to PM.)

READ THE INSTRUCTIONS AND COMPLETE THE TASK

Learning task

Use the portion of a bus timetable to answer the following:

Long Eaton	*Chilwell*	*Beaston*	*Nottingham*
0700	0715	0730	0745
0730	→	→	0800
0800	0815	0830	0845
0830	0845	0900	0915
1000	→	1020	1040
1230	1245	1300	1315
1400	1415	→	1435

1. Which bus is the fastest between Long Eaton and Nottingham?

____

2. You arrive at Chilwell at 7.15 AM. How long is it before the next bus arrives?

____

3. How long does it take the fastest bus to travel between Chilwell and Nottingham?

____

4. You wish to arrive in Nottingham by 9 AM. Which bus from Long Eaton will you catch?

____

WELL, HOW MANY DID YOU GET?

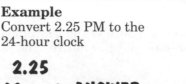

Money

£5.00 £1.00 10p

£1.00 = 100p

Pounds (£) and pence (p) are the UK monetary units:

£1 (pound) = 100p (pence)

£6.50 = 6 pounds 50 pence and is often read as 'six pounds fifty' and £9.07 is read as '9 pounds 7' or '9 pounds 7 pence'. It is important to remember the zeros as they keep the place for the missing units and tens.

Calculations with money can be simply undertaken using the previously covered rules of numbers.

Example

Add £10.65 and £32.40

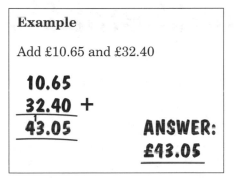

```
 10.65
 32.40 +
 43.05
```

ANSWER: £43.05

Example

You give a £10 note to pay for a £6.47 lock. What is the change?

£10 – £6.47

```
 10.00
  6.47 –
  3.53
```

ANSWER: There is £3.53 change

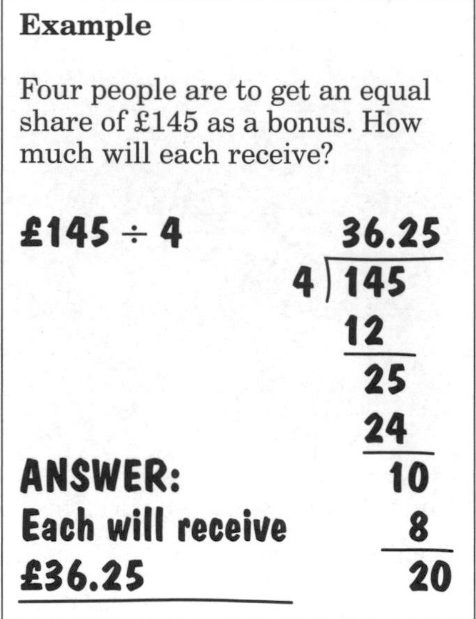

Example

Four people are to get an equal share of £145 as a bonus. How much will each receive?

£145 ÷ 4

```
   36.25
4 )145
   12
    25
    24
     10
      8
     20
```

ANSWER:
Each will receive
£36.25

Applied mathematics

Wages and salaries

People that work for an employer are paid either a wage or a salary. *Wages* are based on an hourly rate and normally paid weekly.

A basic week consists of a set number of hours typically up to 40 hours. The amount paid per hour is known as the **basic rate.** After completing the basic week, additional work is paid at an **overtime rate**, time and a half or double time.

Suppose you are paid £7.50 an hour for the first 8 hours worked in a day and time and a half after that. How much will you earn for working a 10 hour day?

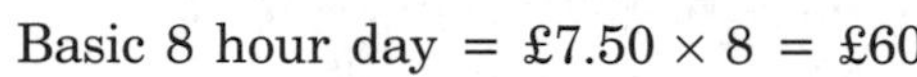

Basic 8 hour day = £7.50 × 8 = £60

2 hours overtime = £7.50 × 2 × 1.5 = £22.50

Total days pay = £82.50

A **bonus** is sometimes paid by employers for doing more than a set amount of work in a week.

Hourly paid workers may have to fill in a **time sheet**, to show the hours they have worked.

Salaries are based on a fixed annual amount which is normally paid monthly.

A project manager is paid a salary of £28 500 pa.

Monthly salary = £28 500 ÷ 12 = £2375

BBS Recruitment Solutions: WEEKLY TIME SHEET | Name: J. JAMES | Works No. 57

Day	Date	Start Time	Lunch	Finish Time	Total basic hours	Total overtime hours
Monday	15/3	8.00 am	½ hr	4.30 pm	8	—
Tuesday	16/3	8.00 am	½ hr	6.30 pm	8	2
Wednesday	17/3	8.00 am	½ hr	6.00 pm	8	1 ½
Thursday	18/3	8.00 am	½ hr	6.30 pm	8	2
Friday	19/3	8.00 am	½ hr	4.30 pm	8	—
Saturday						
Sunday						
Signature: J JAMES					40	5 ½

Overtime is not normally paid to salaried persons. However they may receive an annual bonus or other performance-related payment. This is often based on the annual profits made by their employer.

Commission is paid to sales people. They might receive a basic wage or salary, plus a percentage of the value of the goods sold.

Deductions are made from wages or salaries for income tax, national insurance and pension fund payments etc.

- The wage or salary before deductions is known as the *gross amount*.
- The wage or salary after deductions is known as the *net amount* or 'take-home pay'

PAY-ADVICE	: BSS.plc.SALARIED.PAYROLL	TAX PERIOD	: 07
PAY-DATE	: 26/10/2002	N.I. NUMBER	: DB/06/78/01/B
EMPLOYEE REF	: NOTT/MANF/NADM/005695	N.I. TABLE	: F
EMPLOYEE NAME	: A. C. WHITEMAN	TAX CODE	: 453L
PAY METHOD	: B.A.C.S.	BASE RATE	: 18706 P.A.
OCCUPATION	: JOINERY MANAGER		

--------------ALLOWANCES--------------		--------------DEDUCTIONS--------------		---------------YTD TOTALS---------------		
BASIC SALARY GROSS	1558.83	PAYE TAX	234.19	PAYE TAX		1638.95
		NAT. INS.	99.19	NAT. INS.		694.33
		SOCIAL CLUB	0.86	TAXABLE PAY		10696.49
		EES NI REBATE	−1.06	SAVE	FSC	0.00
		DEFINED CONTRI	30.76	LOAN BALANCE B		0.00
				PENSION	FSC	0.00
				AVC	FSC	0.00
				DEFINED C	FSC	215.32
TOTAL ALLOWANCES	1558.83	TOTAL DEDS	363.94	**NET PAY**		1,194.89

Ratio and proportions

Ratios and proportions are ways of comparing or stating the relationship between two similar or related quantities.

A bricklaying mortar mix may be described as 1:6 ('one to six'). This means the ratio or proportion of the mix is 1 part cement to 6 parts sand (or aggregate).

You could use bags, buckets or barrows as the unit of measure, providing the proportions are kept the same they will make a suitable mortar mix. All that changes is the volume of mortar produced.

To share a quantity in a given ratio:

- add up the total number of parts or shares
- work out what one part is worth
- work out what the other parts are worth.

For example, if £72 is to be shared by two people in a ratio of 5:3, what will each receive?

Number of shares = 5 + 3 = 8
One share = 72 ÷ 8 = £9
Five shares = 5 × 9 = £45
Three shares = 3 × 9 = £27

Answer: They will receive £45 and £27 respectively.

Example

A 1:3 ('one in three') pitched roof has a span of 3.6 m: what is its rise? This means for every 3 m span the roof will rise 1 m.

Rise = Span ÷ 3

$$3\overline{)3.6} = 1.2$$

ANSWER: The rise is 1.2 m

Percentages

This is a standard way of representing a portion or part of a total quantity. Percentage (%) means 'per hundred' or 'per cent'.

In the diagram, what part or portion of the total do these 4 squares make? We can write this in four ways:

- **Proportion:** 4 in 80 of the squares have been filled in, or
- **Ratio:** $^4/_{80}$ of the squares have been filled in, or
- **Percentage:** 5% of the squares have been filled in, or
- **Decimal:** 0.05 of the squares have been filled in.

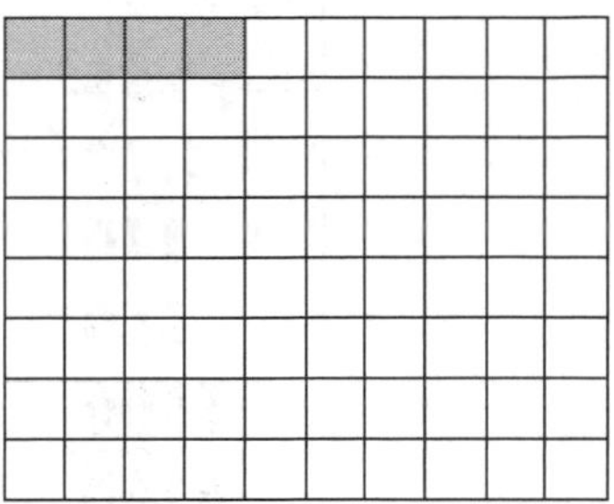

Converting to a percentage – If we need to convert a number to a percentage:

- First divide the number by the total, i.e. $^4/_{80}$ = 0.05
- Then multiply by 100 to find the percentage, i.e. 0.05 × 100 = 5%.

Converting from a percentage – If we need to convert a percentage to a number:

- First divide the number by 100. i.e. 5% ÷ 100 = 0.05
- Then multiply by the total quantity, i.e. 0.05 × 80 = 4

Hence 10% becomes 0.1; if the total number is 250, then 10% of 250 = 25

Using percentages – There are three circumstances where percentages are used:

1. *Where a straightforward percentage of a number is required*

Turn the percentage into decimal and multiply the total by it.

Example

Find 12% of a total of 55 bags of cement.

```
  55
0.12
----
 110
 550
----
 660
```

ANSWER: 6.6 bags

2. *Where a number plus a certain percentage (increase) is required*

Turn the percentage into decimal, place a one in front of it to include the original quantity and multiply by it.

Example

Find 55 plus 12%
(55 + 12% = 55 + 6.6 (see above example) = 61.6 bags.
Alternatively, multiply 55 by 1.12:)

```
   55
 1.12
 ----
  110
  550
 5500    ANSWER:
 ----
 6160    61.6 bags
```

3. *Where a number minus a certain percentage (decrease) is required.*

Take away percentage from 100, convert to decimal and then multiply by it.

Example

Find 55 minus 12%

```
   55
 0.88
 ----
  440
 4400
 ----
 4840   ANSWER: 48.4
```

Statistics

We can make collections of numbers (statistics) speak to us in words and pictures. The presentation of statistics help us to make sense of groups of numbers.

Averages

An average is the *mean value* of several numbers or quantities. It is found by adding the quantities and dividing by the number of quantities.

Average of the numbers = Sum of numbers ÷ Number of numbers

Example: To find the average of these 6 numbers: 2, 4, 6, 7, 9, 15 we add them together and divide by 6:

$$\begin{aligned} \text{Average} &= (2+4+6+7+9+15) \div 6 \\ &= 43 \div 6 \\ &= 7\tfrac{1}{6} \\ &= 7.167 \end{aligned}$$

Example

Find the average mark obtained for a number of college assessments: 48%, 27%, 49%, 75%, 84%, 44%, 65%.

$$\begin{aligned} \text{Ave} &= \frac{48+27+49+75+84+44+65}{7} \\ &= 392 \div 7 \\ &= 56 \end{aligned}$$

ANSWER: Average mark is 56%

Mean, median and mode

These are all types of averages used in statistics:

- *Mean* is the true average we mainly refer to. It equals the sum of values divided by the number of values in the range, group or set of numbers.
- *Median* is the middle value when the numbers are put in order of size.
- *Mode* is the value that occurs most frequently.

Example: The number of people late for work over a ten-day period was: 4, 2, 0, 1, 2, 3, 3, 6, 4, 2.

$$\begin{aligned} \textit{Mean} &= (4+2+0+1+2+3+3+6+4+2) \div 10 \\ &= 27 \div 10 \\ &= 2.7 \text{ people late per day} \end{aligned}$$

Median

- First put in size order: 0, 1, 2, 2, 2, 3, 3, 4, 4, 6
- Then cross off numbers from either end to find the middle value(s)
- The median is the middle value or the mean of the two middle values 2 and 3
- Median = 2.5 people late per day

Mode = 2 people late per day, because it occurs the most times.

Range of numbers – When using *mode* the range should be stated, to show how much the information is spread. In the above example:

Range = highest value – lowest value
= 6 – 0
= 6

Graphs

Graphs and charts are used to give a pictorial representation of numerical data.

Line graphs – are used to show the relationship between two or more quantities.

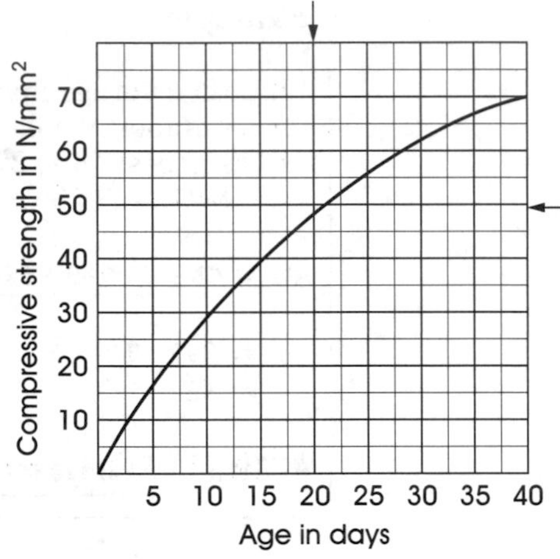

It can be seen from the graph that concrete is expected to have a compressive strength of nearly 50 N/mm^2 after 20 days.

Bar charts – These use parallel bars (horizontal or vertical) to compare things or show change over time. The length of each bar being proportional to the quantity represented. Hence, by collecting data on work productivity, we could use our statistics to show how Friday has the lowest productivity of the week.

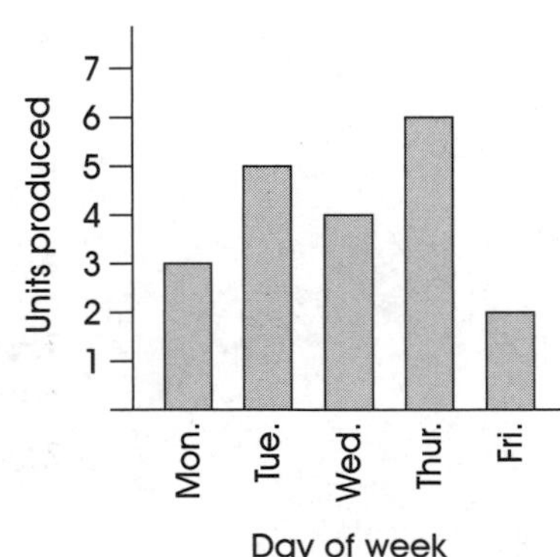

Grouped bar chart – This shows more than one set of data on the same chart and helps us to picture a large range of statistics. Hence we could compare the costs of wages and materials against profits over a number of years in the same chart.

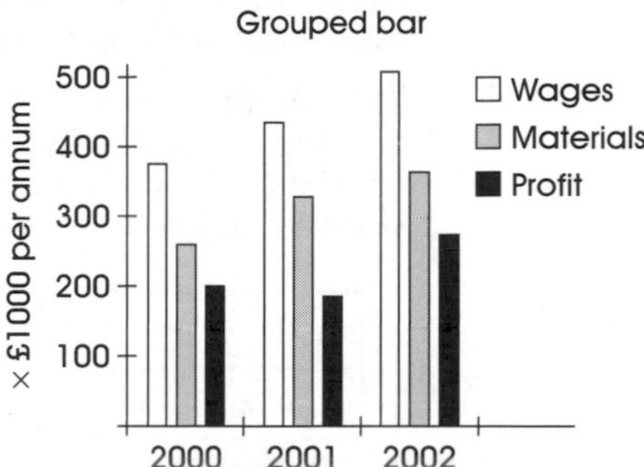

Stacked bar chart – This shows the size of items in sub-groups or sub-sets of numbers, drawn in proportion to each other. Hence we could picture the number or percentage of different trades employed on-site at any one time.

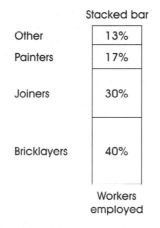

Stacked bar

Other	13%
Painters	17%
Joiners	30%
Bricklayers	40%

Workers employed

Pie charts – These also show the proportional size of items that make up a set of data, pictured in the shape of a circle or 'pie'.

Hence we could show pictorially how a company spends more of its earnings on wages than on its material costs.

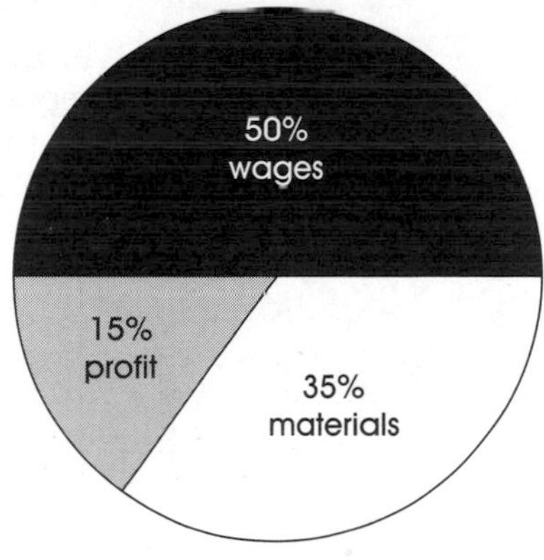

Distribution of earnings

Tally charts – These are used for collecting data. They show the frequency of occurrence of identical numbers or values in a range. A tally is just a group of lines or strokes, which are written down in fives for ease of counting. The fifth line forms a 'gate'.

Hence we could use a tally chart to show the number (frequency) and size (value) of critical measurements on factory-made components.

The following measurements were taken during a quality control check of 1.8 m panel size lengths in a joinery shop. A tally chart was filled in by measuring the panel widths to 5 significant figures. The chart provides a quick observation of the accuracy of the manufacturing process.

1799.5	1799.5	1800	1800.5	1800.5
1800	1800	1800	1799	1799.5
1800.5	1800	1800	1799.5	1801
1799	1800	1800	1800.5	1800
1799.5	1799.5	1799.5	1800	1800
1800.5	1799	1800	1800	1800.5
1799.5	1800	1800	1800	1800.5
1799.5	1799.5	1799.5	1800	1800
1800	1799	1800	1800	1800.5
1799	1800	1800	1800.5	1799.5

Measurement	Tally	Frequency
1799	卌	5
1799.5	卌 卌 ǁ	12
1800	卌 卌 卌 卌 ǀǀǀ	23
1800.5	卌 ǀǀǀǀ	9
1801	ǀ	1
Total	*50*	50

Histograms – A histogram or frequency diagram may be drawn to represent the frequency of occurrence determined in a tally chart. These look similar to bar charts except that there are no gaps between the bars.

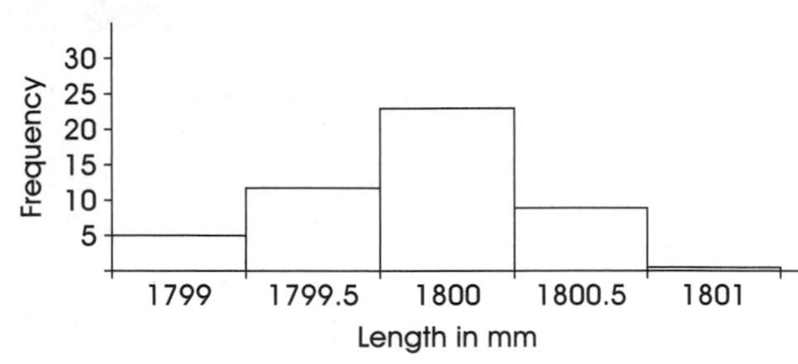

Powers and roots of numbers

Powers

A simple way of writing repeated multiplications of the same number is shown below:

$10 \times 10 = 100$ or 10^2

$10 \times 10 \times 10 = 1000$ or 10^3

$10 \times 10 \times 10 \times 10 = 10\,000$ or 10^4

and so on.

The small raised number is called the power or index. Numbers raised to the power 2 are usually called square numbers. We say 10^2 is '10 squared', and 10^3 is '10 cubed' but we say 10^4 is '10 to the power 4', etc.

Large numbers – can therefore be written in a standard shorthand form by the use of an index or power:

$30\,000\,000 = 3 \times 10\,000\,000$ or 3×10^7

$6\,600\,000 = 6.6 \times 1000\,000$ or 6.6×10^6

$990 = 9.9 \times 100$ or 9.9×10^2

Note that the index number is the number of places that the decimal point has to be moved to the right if the number is written in full.

Small numbers – This standard form can also be used for numbers less than one, by using a negative power or index:

$0.036 = 3.6 \times 10^{-2}$ or 36×10^{-3}

$0.0099 = 9.9 \times 10^{-3}$ or 99×10^{-4}

$0.00012 = 1.2 \times 10^{-4}$ or 12×10^{-5}

The negative index is the number of places that the decimal point will have to be moved to the left if the number is written in full. For example:

$99.0 \times 10^{-4} = 0.0099$

Roots

It is sometimes necessary to find a particular root of a number. Finding a root of a number is the opposite process of finding the power of a number. Hence, the **square root** is a number multiplied by itself once to give the number in question:

- The square of 5 is 5^2 or $5 \times 5 = 25$.
- Therefore the square root of 25 is 5.

The common way of writing this is to use the square root sign $\surd$

$\sqrt{25} = 5$

The **cube root** is a number multiplied by itself twice to give the number in question.

- The cube of 5 is 5^3 or $5 \times 5 \times 5 = 125$.
- Therefore, the cube root of 125 is 5.

The common way of writing this is to use the root sign and an index of 3, i.e. $\sqrt[3]{}$

$\sqrt[3]{125} = 5$

From this we can see why there is a connection between powers and roots as opposite processes.

$10^2 = 100$ $\quad \sqrt{100} = 10$

$10^3 = 1000$ $\quad \sqrt[3]{1000} = 10$

Estimating roots – Roots can be found by estimation.

For example, $\sqrt{58.2}$ lies between the whole number squares of 49 (7×7) and 64 (8×8). So the $\sqrt{58.2}$ will be a decimal number between 7 and 8. And 64 is closer to 58.2 than 49 is so the root will be closer to 8 than 7.

- Try $7.6 \times 7.6 = 57.76$ this is close but too small
- Try $7.7 \times 7.7 = 59.29$ this is close but too big

57.76 is closer to 58.2 than 59.29 so the root will be closer to 7.6 than 7.7.

- Try $7.62 \times 7.62 = 58.0644$ this is close but too small
- Try $7.63 \times 7.63 = 58.2169$ this close but slightly too big

However 7.63 is a closer estimate than 7.62 and also correct to 1 decimal place.

Thus $\sqrt{58.2} = 7.63$ to 1 decimal place.

Calculating roots – Roots are more easily found using a calculator with a √ function key.

Example

To find the square root of 58.2 using a calculator

Keys to press	*Display*
[5] [8] [·] [2]	58.2
[√]	7.6288924

So rounded to 3 decimal places $\sqrt{58.2} = 7.629$ (very close to our estimate)

Angles and lines

Angles

An angle is the amount of space between two intersecting straight lines. The angle does not depend on the length of the lines or the distance from the intersection that it is measured. The size of an angle is measured in degrees (°).

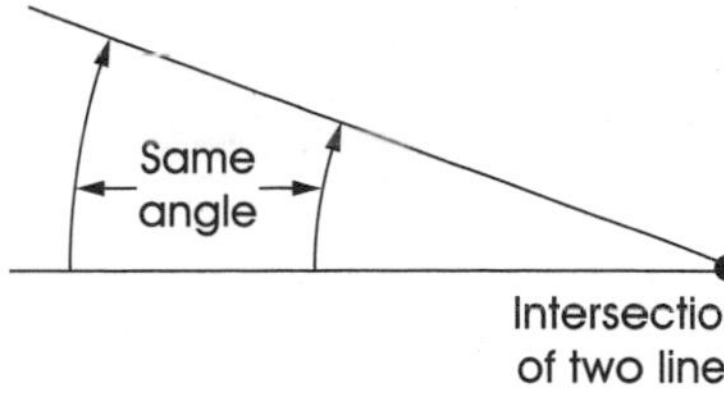

Angles are also a measure of turning.

Standing on a spot looking in one direction, if you were to turn completely round to look again in the same direction. You would have turned through 360° (degrees).

- A three-quarter turn is 270°.
- A half turn is 180°.
- A quarter turn is 90°. This is called a right angle.

Simple names and angle properties

- A **right angle** is 90°. They are shown on drawings by either a small square in the corners, or an arc with arrow heads and the figures 90°.

- An **acute angle** is between 0° and 90°.

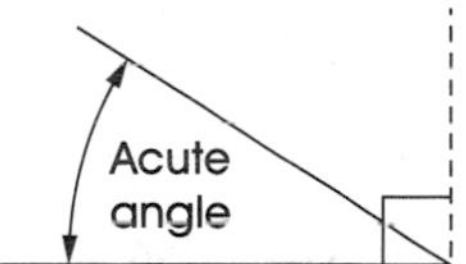

- An **obtuse angle** is between 90° and 180°.

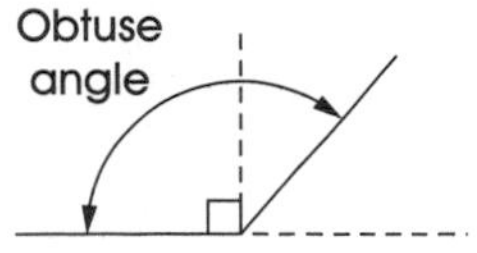

- A **reflex angle** is between 180° and 360°.

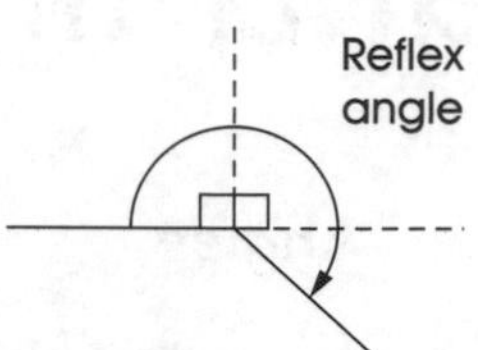

The addition of angles

Angles on a **straight line** always add up to 180°:

A + B + C = 180°

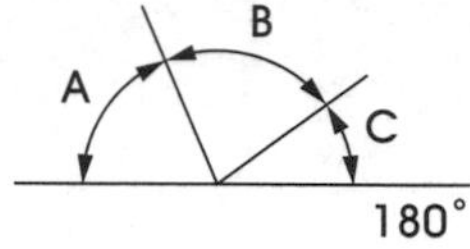

Angles in a **triangle** (three-sided figure) always add up to 180°:

A + B + C = 180°

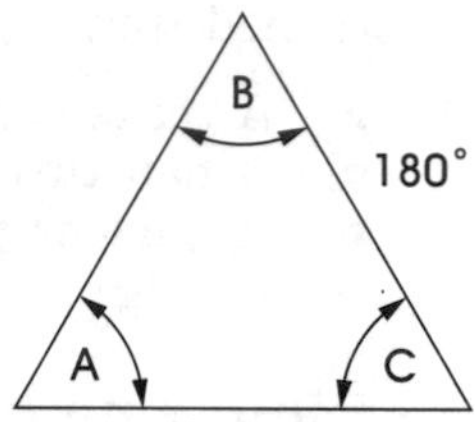

Angles at a **point** always add up to 360°:

A + B + C = 360°

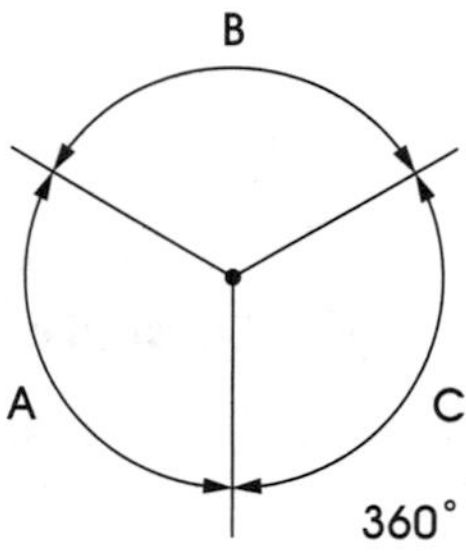

Angles in a **quadrilateral** (four-sided figure) always add up to 360°:

A + B + C + D = 360°

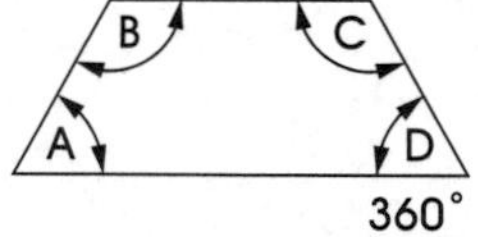

Angles in a **polygon** (more than four sides) always add up to 360°:

A + B + C + D + E = 360°

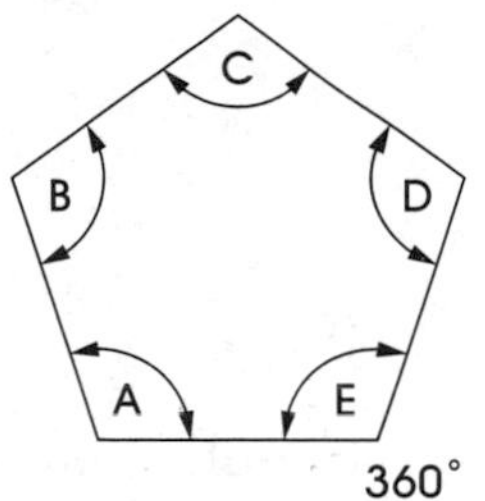

Vertically opposite angles are always equal:

A = B; C = D

A + C = B + D = 180°

A + C + B + D = 360°

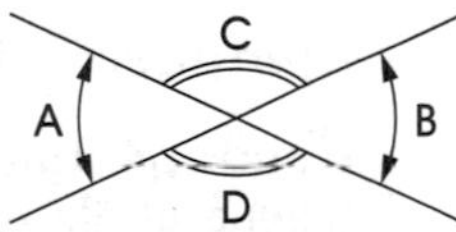

Lines and angles

Parallel lines – are always the same distance apart: they never meet.

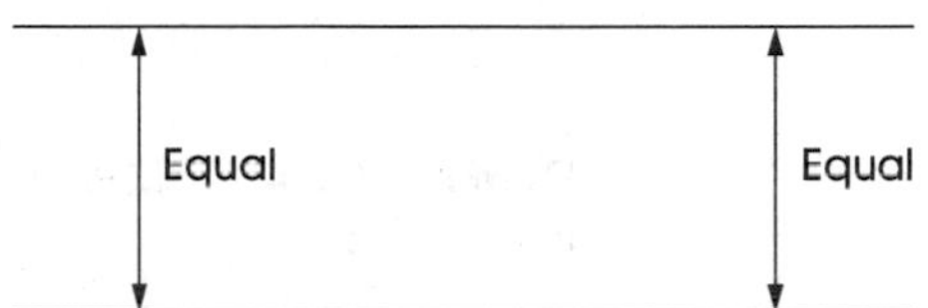

They are shown on a drawing using either a single or double pair of arrow heads.

A line drawn across a pair of parallel lines is called a **traverse.**

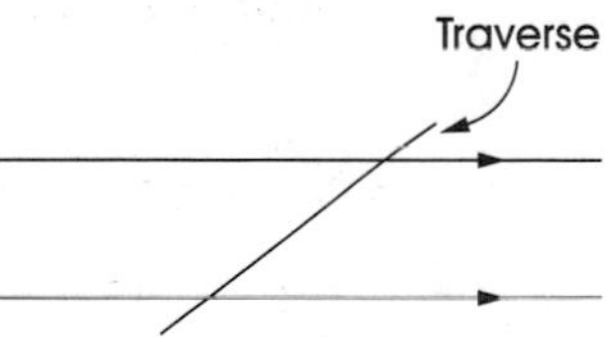

Alternate angles are always equal.

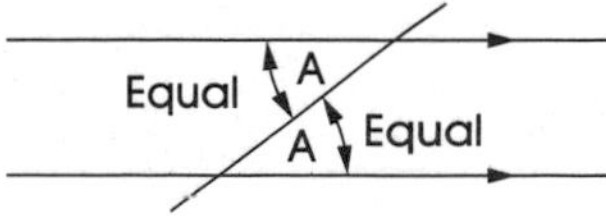

A = A

Corresponding angles are always equal.

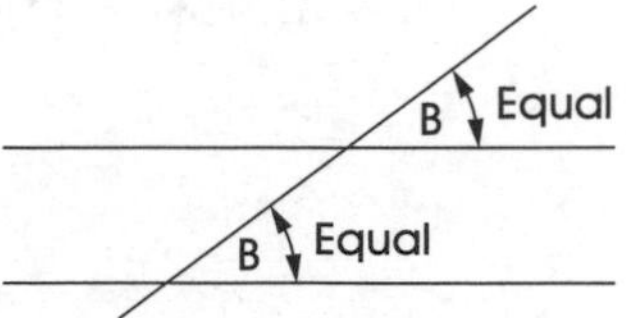

B = B

Supplementary angles always add up to 180°.

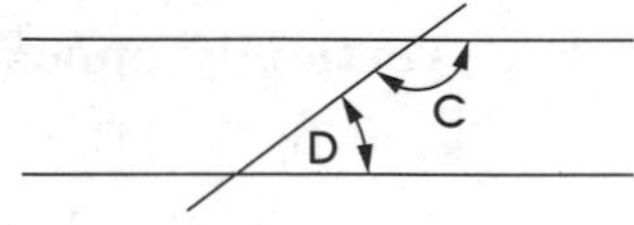

C + D = 180°

Measuring the size of angles

Reading angles – If asked to find the angle between AB and CB (called angle ABC), it means you have to find the angle of the middle letter, B.

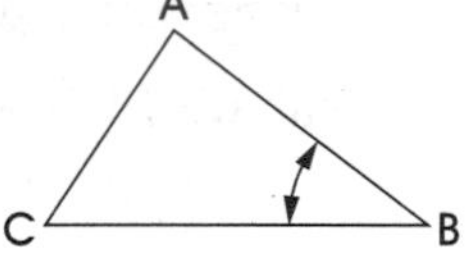

Protractors – Measuring angles can be done with the aid of a protractor.

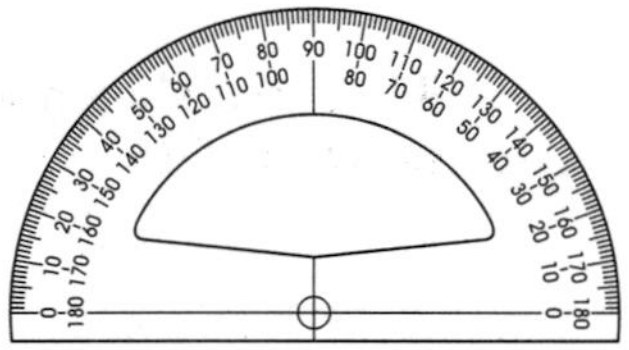

- Place the protractor with its centre point over the intersection of the lines AB and CB and its baseline over one of the lines.
- Read off the angle at the edge of the protractor. It is 40° or 140° and is obviously less than 90°, so it must be 40° in this case.

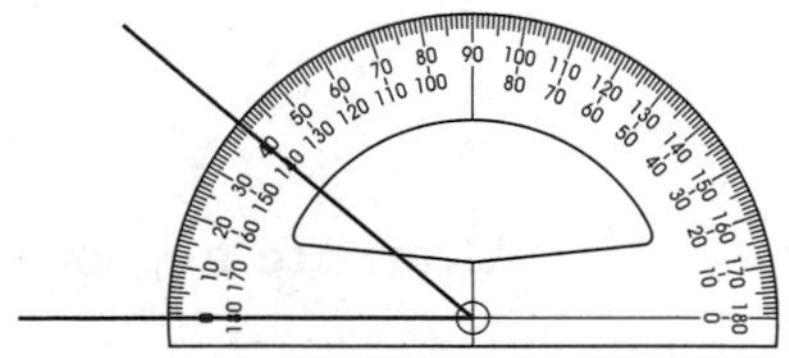

Shapes and solids

Shapes

A plane – is a flat surface. Its shape is formed by lines which are known as the sides of the shape giving the length and breadth. The total distance all the way round the sides is called the perimeter. Plane shapes are classified by their number of sides.

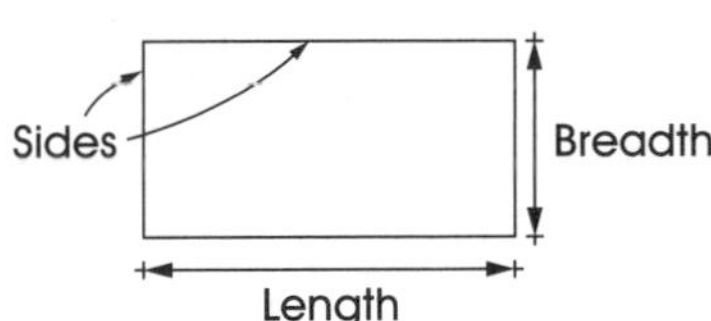

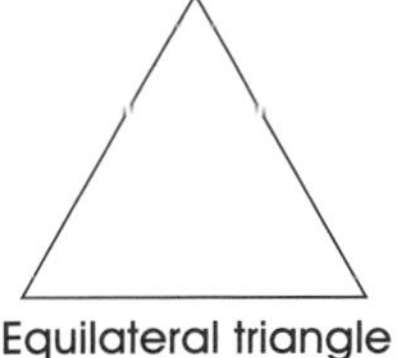
Equilateral triangle

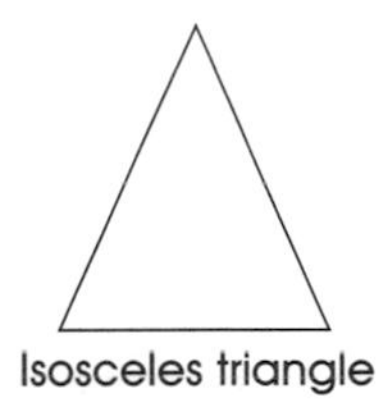
Isosceles triangle

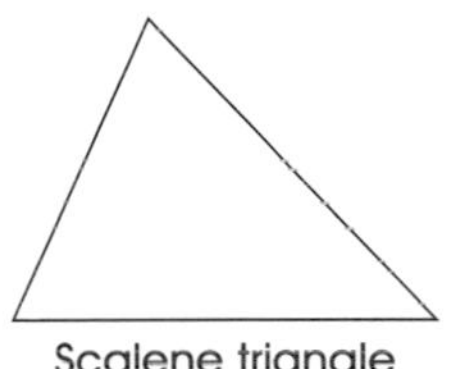
Scalene triangle

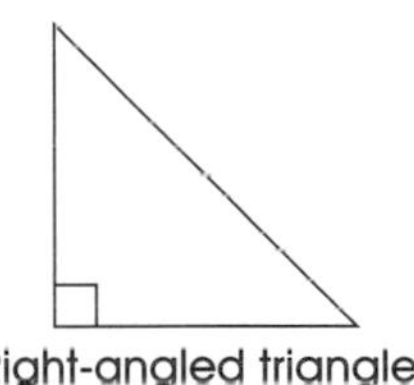
Right-angled triangle

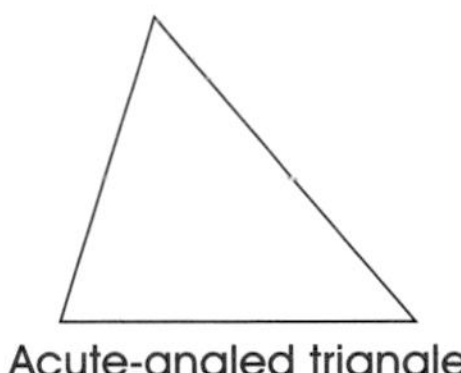
Acute-angled triangle

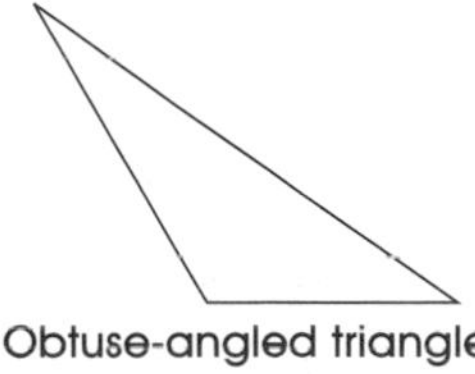
Obtuse-angled triangle

Triangle –A plane shape which is bounded by three straight lines.

- The vertex is the angle opposite the base.
- The altitude is the vertical height from the base to the vertex.

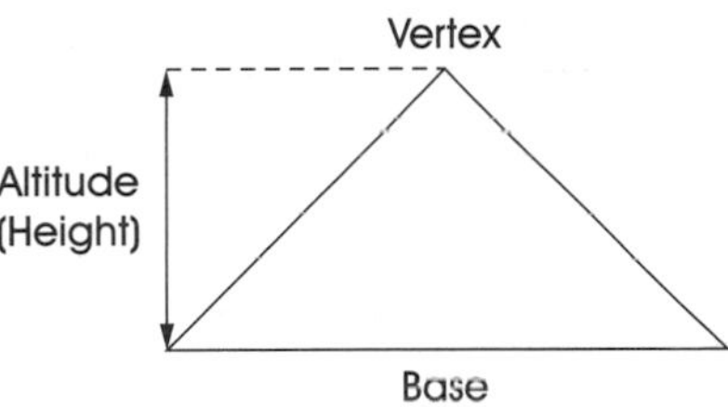

Triangles are classified by either the length of their sides or by the size of their angles.

- An *equilateral triangle* has three equal length sides and three equal angles.
- An *isosceles triangle* has two sides of equal length.
- A *scalene triangle* has sides which are all unequal in length.
- A *right-angled triangle* has one 90° angle.
- In an *acute-angled triangle* all angles are less than 90°.
- In an *obtuse-angled triangle* one of the angles is between 90° and 180°.

Quadrilateral – A plane shape which is bounded by four straight lines. A straight line joining opposite angles is called a diagonal, it divides the figure into two triangles.

- A *square* has all four sides of equal length and all angles are right angles.
- A *rectangle* has opposite sides of equal length and four right angles.
- A *rhombus* has four equal sides, opposite sides being parallel, but none of the angles is a right angle.
- A *parallelogram* has opposite sides which are parallel and equal in length, but none of its angles is a right angle.
- A *trapezium* has two parallel sides.
- A *trapezoid* has no parallel sides.

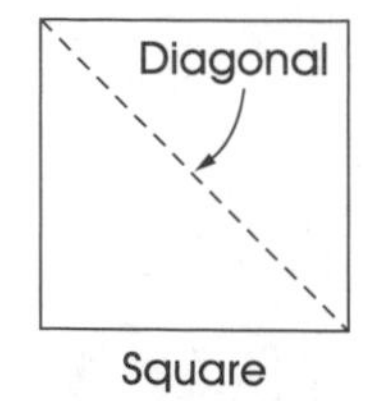
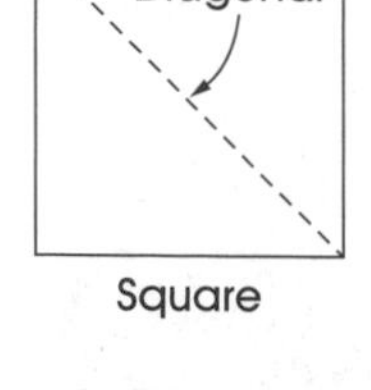

Square

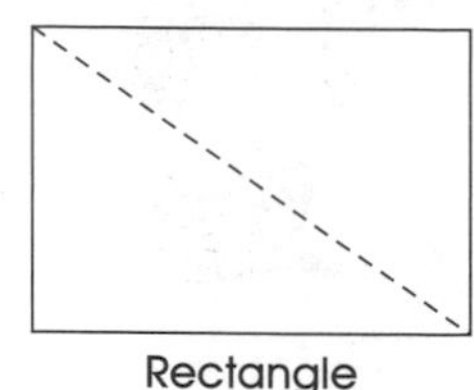
Rectangle

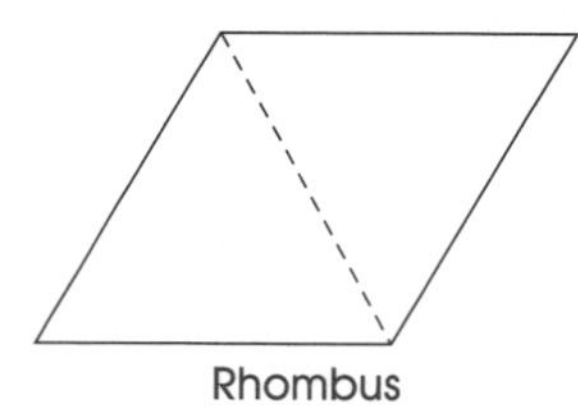
Rhombus

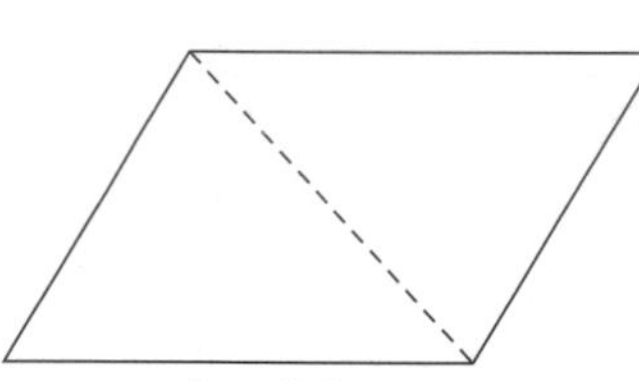
Parallelogram

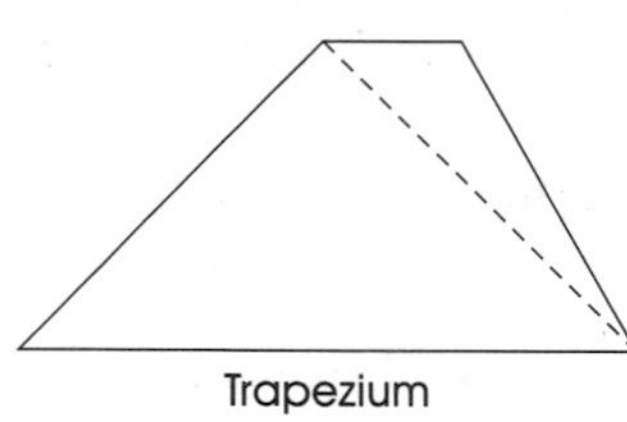
Trapezium

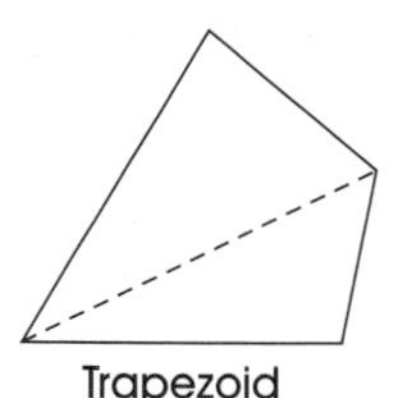
Trapezoid

Polygon – a plane shape which is bounded by more than four straight lines. Polygons may be classified as either:

- *Regular polygon* – these have sides of the same length and equal angles.
- *Irregular polygon* – these have sides of differing length and unequal angles.

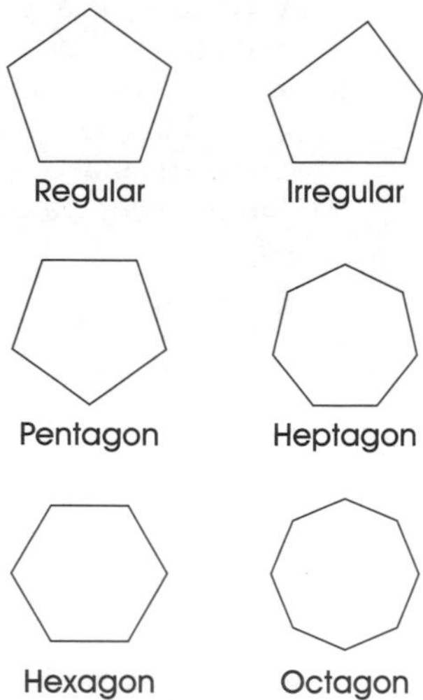

Both regular and irregular polygons are further classified by the number of sides which they consist of:

- a *pentagon* consists of five sides.
- a *hexagon* consists of six sides.
- a *heptagon* consists of seven sides.
- an *octagon* consists of eight sides.

Circles – a circle is a plane shape, bounded by a continuous, curved line, which at every point is an equal distance from the centre. The main elements of a circle are:

- *Circumference* – the curved outer line (perimeter) of the circle.
- *Diameter* – a straight line which passes through the centre and is terminated at both ends by the circumference.
- *Radius* – the distance from the centre to the circumference. The radius is always half the length of the diameter.
- *Chord* – a straight line which touches the circumference at two points but does not pass through the centre.
- *Arc* – any section of the circumference

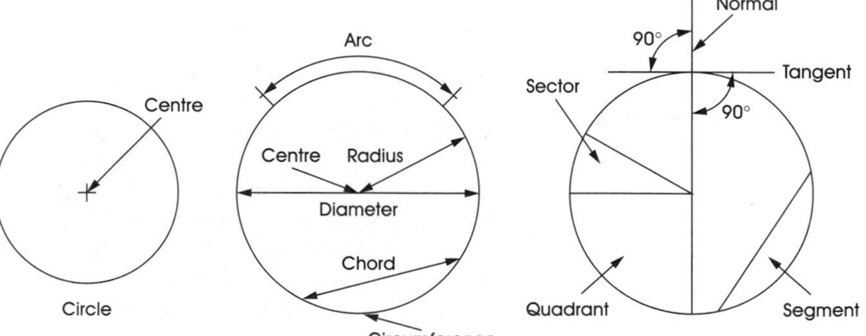

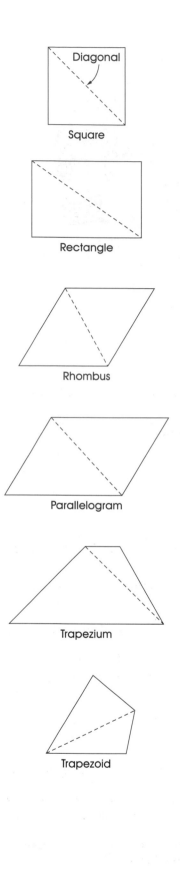

- *Normal* – any straight line which starts at the centre and extends beyond the circumference.
- *Tangent* – a straight line which touches the circumference at right angles to the normal.
- *Sector* – the portion of a circle contained between two radii and an arc. (Radii is the plural of radius.)
- *Quadrant* – a sector whose area is equal to a quarter of the circle.
- *Segment* – the portion of a circle contained between an arc and a chord.

Annulus – The area of a plane shape that is bounded by two circles, each sharing the same centre but having different radii.

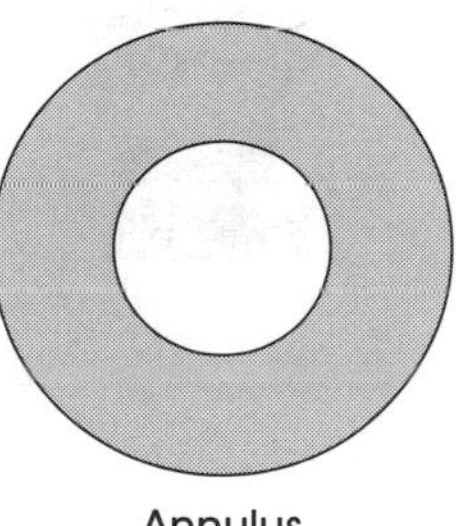

Annulus

Ellipse – A plane shape bounded by a continuous curved line drawn round two points called foci. The longest diameter is known as the major axis and the shortest diameter is the minor axis.

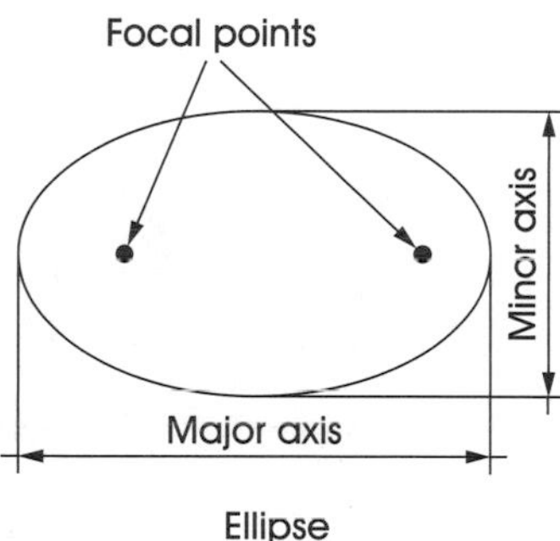

Ellipse

Solids

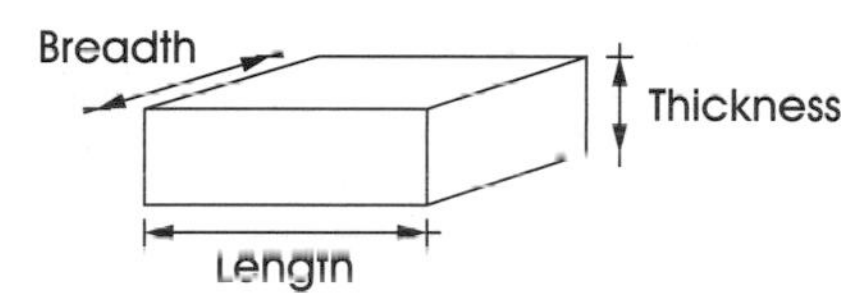

Solid figures are three-dimensional; they have length, breadth and thickness.

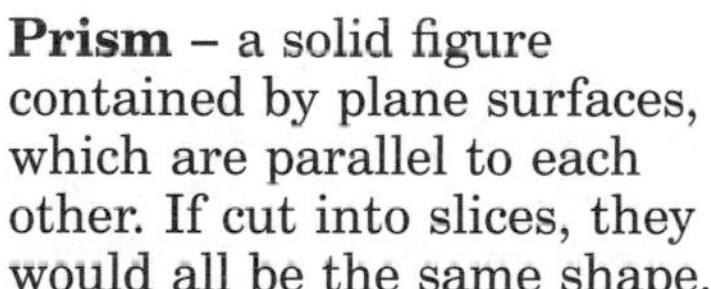

Prism – a solid figure contained by plane surfaces, which are parallel to each other. If cut into slices, they would all be the same shape.

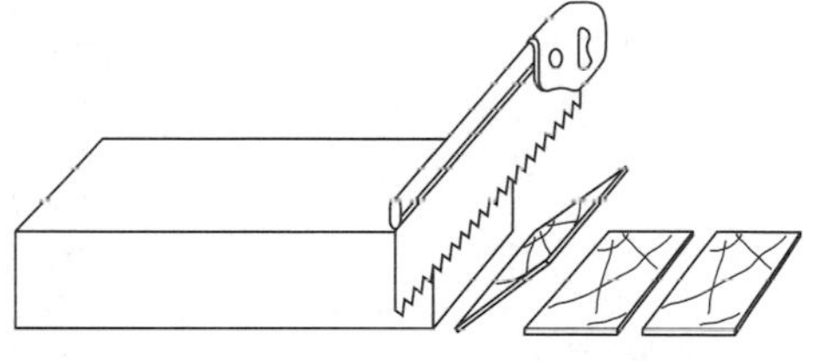

All prisms are named according to the shape of their ends:

- *Cube* – all sides are equal in length and each face is a square.
- *Cuboid or rectangular prism* – each face is a rectangle and opposite faces are the same size. Bricks and boxes are examples of cuboids.
- *Triangular prism* – the ends are triangles and other faces are rectangles.
- *Octagonal prism* – the ends are octagons and other faces are rectangles.

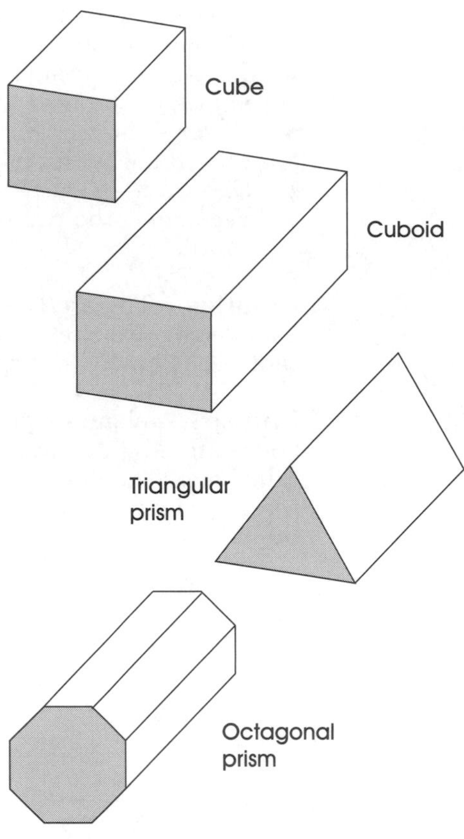

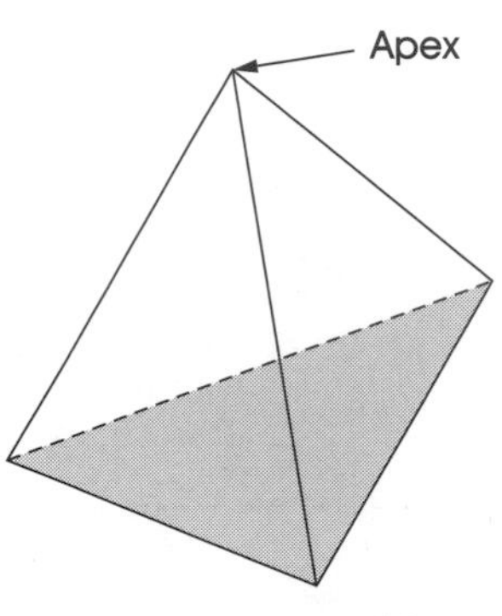

Triangular pyramid

Pyramid – a solid figure contained by a base and triangular sloping sides. The sides meet at a point called the apex. All pyramids are named according to their base shape.

- *Triangular pyramid*
- *Square pyramid*
- *Hexagonal pyramid*

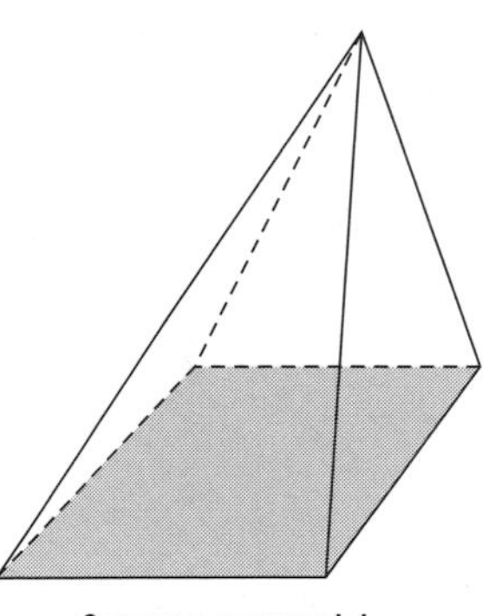
Square pyramid

Circular solids –

- *Cylinder* – a circular prism, the ends are circular in shape. Most tins are cylindrical.
- *Cone* – a circular pyramid, the base is circular in shape.
- *Sphere* – a solid figure, where all sections are circular in shape. Most balls are spherical.

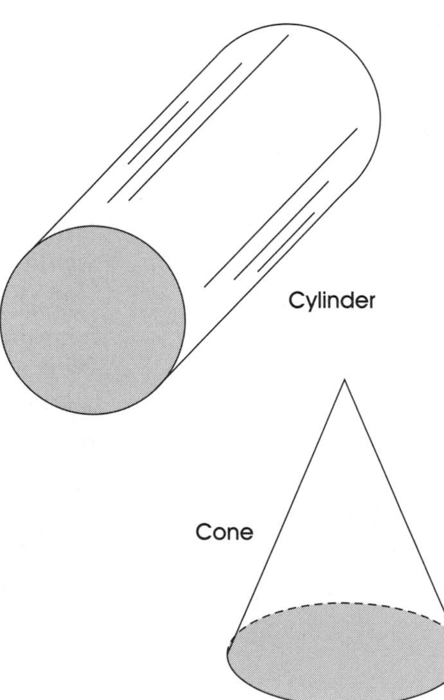

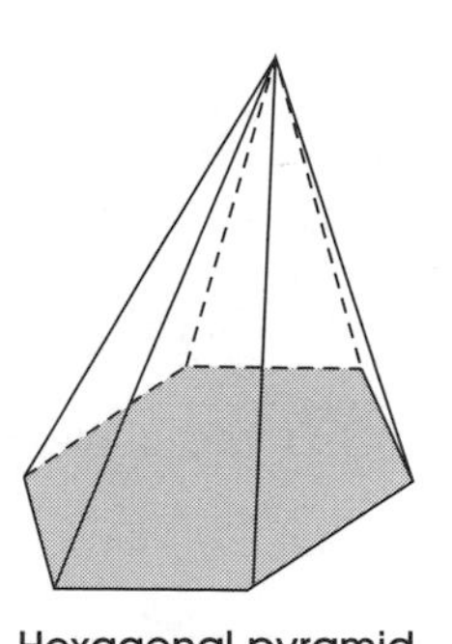
Hexagonal pyramid

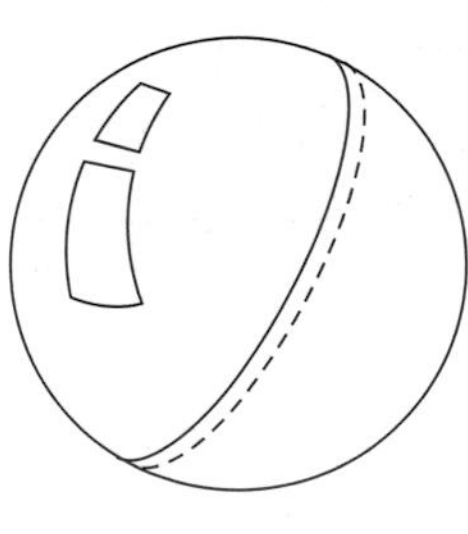
Sphere

Nets of solids – The net of a solid is the two-dimensional (2-D) shape which can be folded to make the three-dimensional (3-D) shape.

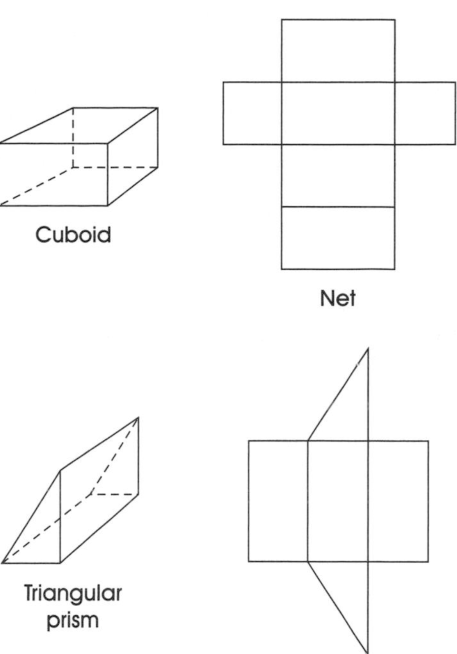

Questions for you

25. Convert 3.45 PM to the 24 hour clock.

26. Jane gets paid £7.55 per hour

(a) How much will she receive for a 40-hour week?

(b) How much is this per annum?

27. A concrete mix is expressed as 1:3:6 being the proportion of cement, fine aggregate and course aggregate. How much of each is required for 2.4 m^3?

28. Jason earns 75% of Julie's salary. How much does Jason earn if Julie earns £18 250 per annum?

29. Find the mean, median, mode and range of the following set of data:

8, 7, 8, 6.5, 10, 8, 7, 7.5, 9, 9, 8, 7.5, 9, 10, 8, 8.5, 9.5, 12.

30. How many degrees are there in two right angles?

31. How many degrees are there in a circle?

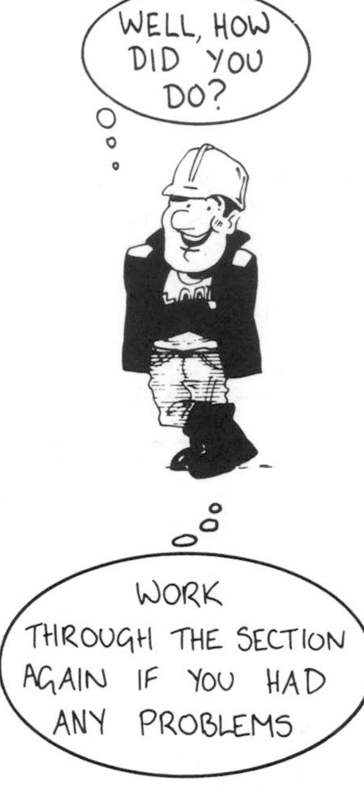

32. Angle ABC is 35 degrees. Which of the three angles does this refer to?

33. How many degrees does each angle measure in an equilateral triangle?

34. What is the name of a prism where all sides are equal in length and each face is a square?

35. What shape is a brick?

36. The pie chart shows the proportion of costs on a building job. Find the cost of the brickwork, if the total cost was £8762.50.

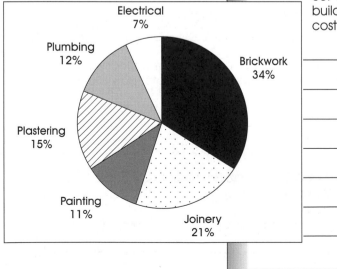

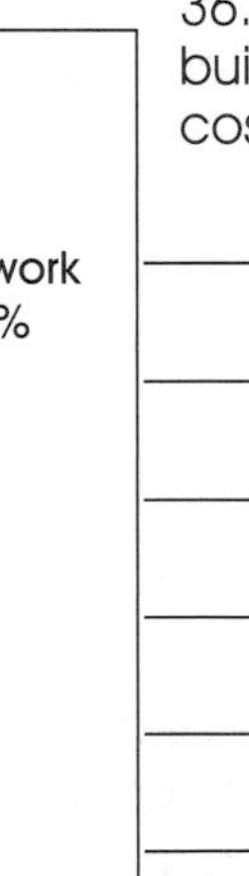

Formulae

Formulae are normally stated in algebraic terms. Algebra uses letters and symbols instead of numbers to simplify statements and enable general rules and relationships to be worked out.

Transposition of formulae

When solving a problem, sometimes formulae have to be rearranged in order to change the subject of the formulae before the calculation is carried out. Basically anything can be moved from one side of the equals sign to the other by changing its symbol. This means that on crossing the equals sign:

- plus changes to minus
- multiplication changes to division
- powers change to roots.

This is also true vice versa.

Alternatively, we can cross-multiply. This means that on crossing the equals sign anything on the top line moves to the bottom line, and conversely anything on the bottom line moves to the top line.

Triangles

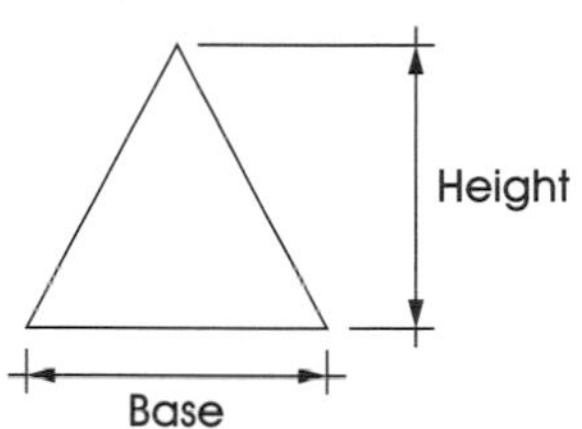

Area – The area of a triangle can be found by using the formula:

Area = Base × Height ÷ 2

If we say:

- Area = A
- Base = B
- Height = H

Then:

$A = B \times H \div 2$

Using algebra, we can abbreviate by missing out the multiplication sign (×) and expressing a division in its fractional form.

$A = \frac{BH}{2}$ means the same as

$A = B \times H \div 2$

Height – If the area and base of a triangle are known but we wanted to find out its height, the formula could be transposed to make the height the subject.

Area = Base × Height ÷ 2

$A = \frac{BH}{2}$

Example

Find the area of a triangle having a base of 2.5 m and a height of 4.8 m.

Area = Base × Height ÷ 2

$$A = \frac{B \times H}{2}$$

$$= \frac{2.5 \times 4.8}{2}$$

$$= \frac{12}{2}$$

$$= 6$$

ANSWER: Area = 6 m²

- Then $\frac{A}{B} = \frac{H}{2}$

  (B moves from above to below on crossing the = sign)
- And $\frac{2A}{B} = H$

  (2 moves from below to above on crossing the = sign)

Therefore Height = 2 × Area ÷ Base

> **Example**
>
> Find the height of a triangle having an area of 4.5 m and a base of 1.5 m
>
> $$\text{Area} = \frac{\text{Base} \times \text{Height}}{2}$$
> $$A = \frac{BH}{2}$$
> $$\frac{2A}{B} = H$$
> $$\frac{2 \times 4.5}{1.5} = H$$
> $$6 = H$$
>
> **ANSWER: Height = 6 m**

Rectangles

Perimeter – The perimeter of a rectangle can be found by using the formulae:

Perimeter = 2 × (Length + Breadth)

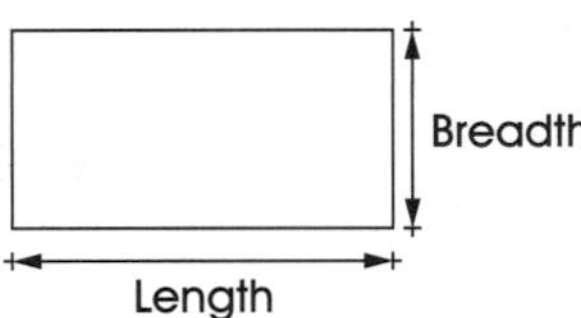

This can be abbreviated to:

P = 2(L + B)

Plus and minus (+ and –) signs cannot be abbreviated and must always be shown in the formulae. For the correct order of working you must use the 'Bodmas' rule. To obtain the correct answer L must be added to B before multiplying by 2.

> **Example**
>
> Find the perimeter of a rectangle having a length of 3.6 m and a breadth of 2.2 m
>
> $$\text{Perimeter} = 2 \times (\text{Length} + \text{Breadth})$$
> $$= 2\,(L + B)$$
> $$= 2\,(3.6 + 2.2)$$
> $$= 2\,(5.8)$$
> $$= 11.6$$
>
> **ANSWER: Perimeter = 11.6 m**

Circles

Perimeter and diameter – The formula for the perimeter or circumference of a circle is:

Circumference = π × Diameter

$C = \pi \times D$

π (spoken as pi) is the number of times that the diameter will divide into the circumference. It is the same for any circle and is taken to be 3.142.

Suppose we were given the circumference and asked to find the diameter. Since $C = \pi \times D$, then $C \div \pi = D$ (π is moved across = and changes from × to ÷). Hence:

Diameter = Circumference ÷ π

> **Example**
>
> Find the diameter of a circle having a circumference of 7.855 m
>
> **Circumference = π × Diameter**
>
> **$C = \pi \times D$**
>
> **$C \div \pi = D$**
>
> **$= 7.855 \div 3.142$**
>
> **$= 2.5$**
>
> **ANSWER: Diameter = 2.5 m**

Formulae for areas and perimeters

Values of **common shapes** can be found by using the following formulae.

- The perimeter of a figure is the distance or length around its boundary, *linear measurement*, given in metres run m.
- The area of a figure is the extent of its surface, *square measurement*, given in square metres (m^2).

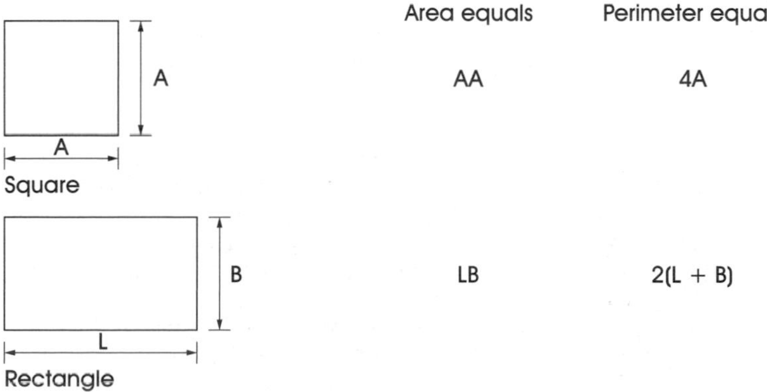

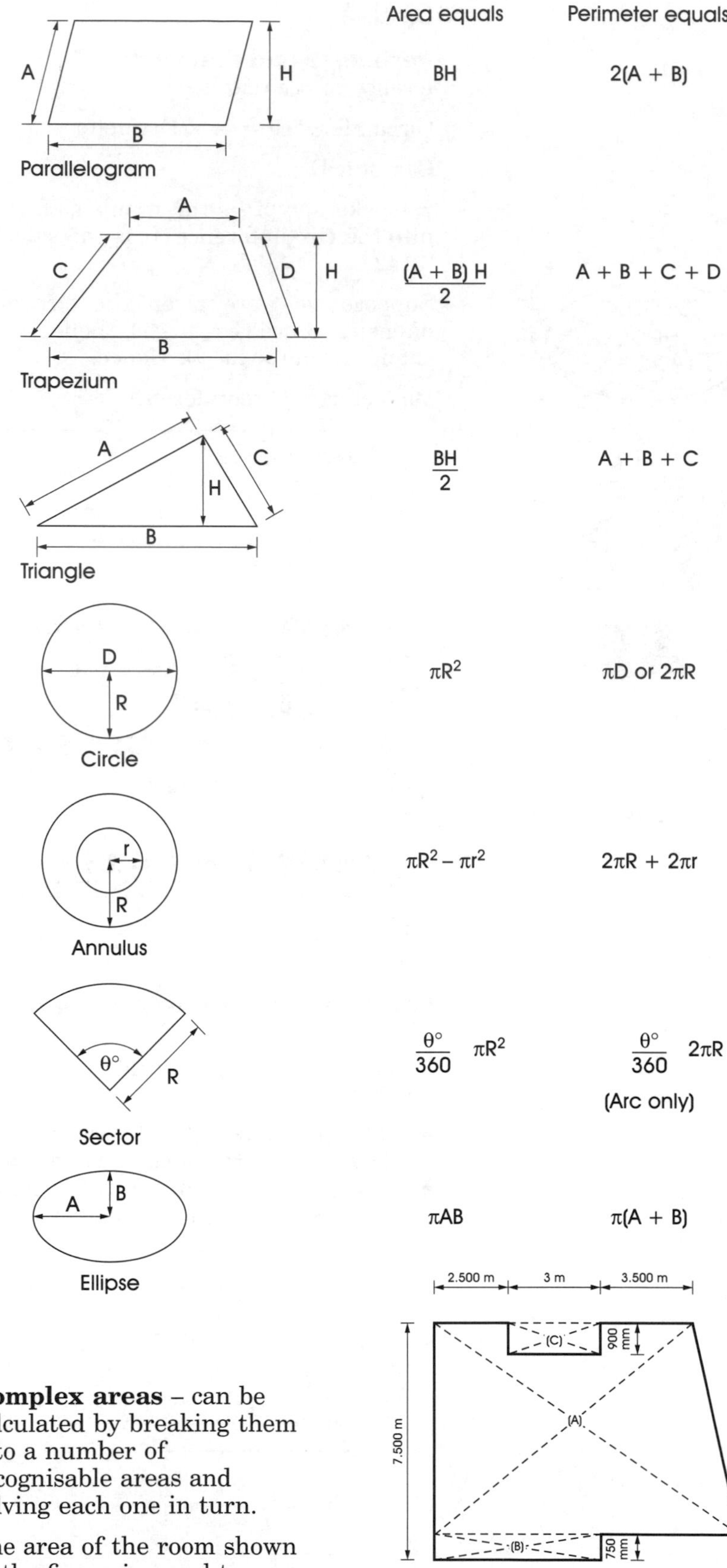

Shape	Area equals	Perimeter equals
Parallelogram	BH	2(A + B)
Trapezium	$\frac{(A + B) H}{2}$	A + B + C + D
Triangle	$\frac{BH}{2}$	A + B + C
Circle	πR^2	πD or $2\pi R$
Annulus	$\pi R^2 - \pi r^2$	$2\pi R + 2\pi r$
Sector	$\frac{\theta°}{360} \pi R^2$	$\frac{\theta°}{360} 2\pi R$ (Arc only)
Ellipse	πAB	$\pi(A + B)$

Complex areas – can be calculated by breaking them into a number of recognisable areas and solving each one in turn.

The area of the room shown in the figure is equal to area A plus area B minus area C.

Example

Find the area of the room above

$$\text{Area A} = \left(\frac{9 + 10.5}{2}\right) \times 6.75$$
$$= 65.813 \text{ m}^2$$
$$\text{Area B} = 0.75 \times 5.5$$
$$= 4.125 \text{ m}^2$$
$$\text{Area C} = 0.9 \times 3$$
$$= 2.7 \text{ m}^2$$
$$\text{Total area } A + B - C$$
$$65.813 + 4.125 - 2.7$$
$$67.238 \text{ m}^2$$

ANSWER: Area = 67.238 m²

Converting to the same units – We can only multiply like terms. Where metres and millimetres are contained in the same problem, first convert the millimetres into a decimal part of a metre by dividing by 1000. (Move the imaginary decimal point behind the number three places forward.) Alternatively, convert all units to millimetres.

Example

Convert 50 mm to metres

.050
3 2 1

ANSWER:
50 mm = 0.05 m

Volume

The volume of an object can be defined as the space it takes up, *cubic measurement*, given in cubic metres (m^3).

Many solids have a uniform cross-section and parallel edges. The volume of these can be found by multiplying their base area by their height:

Volume = Base area × Height

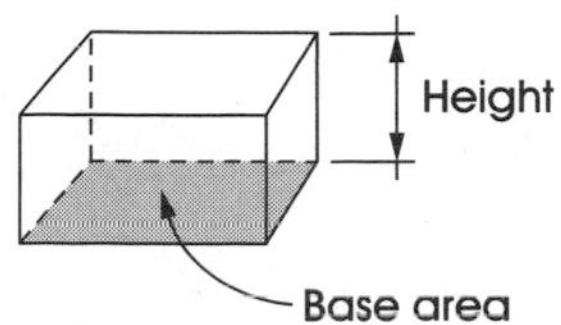

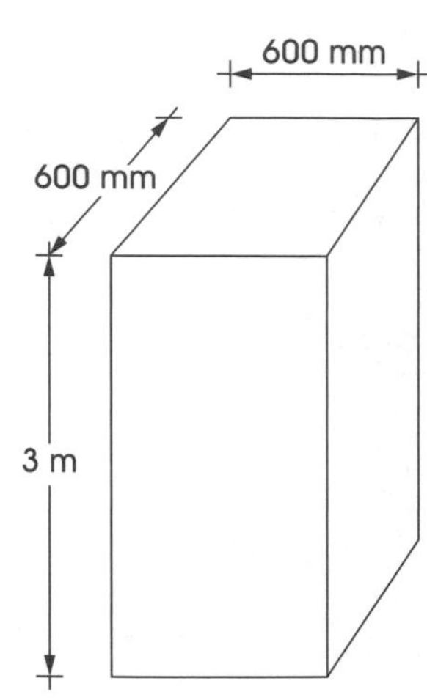

Example

Find the volume of concrete required for a 600 mm square, 3 m high column

Volume = Base area × Height
= 0.6 × 0.6 × 3
= 1.08 m³

ANSWER: Volume = 1.08 m³

Example

A house contains forty-eight 50 mm × 225 mm softwood joists, 4.5 m long. How many cubic metres of timber are required?

Volume of 1 joist = 0.05 × 0.225 × 4.5
= 0.050625 m³

Total volume = 0.050625 × 48
= 2.43 m³

ANSWER:
2.43 m³ of timber are required

Surface area

The lateral surface area of a solid with a uniform cross-section is found by multiplying its base perimeter by its height.

Lateral surface area = Base perimeter × Height

For a **cylinder**, the lateral surface area = π × Diameter × Height.

The *total surface area* can be determined by adding the areas of ends or base to the lateral surface area.

The total surface area of a cylinder = π × Diameter × Height + (π × Radius × Radius × 2) i.e. (Lateral area) × (Area of ends):

Surface area = $\pi DH + (2\pi R^2)$

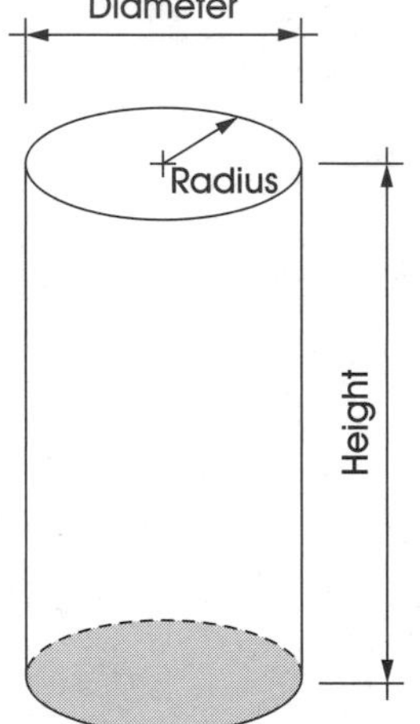

Example

Find the total surface area of a cylinder 1.2 m diameter and 2.4 m in height.

Total surface area
= $\pi D \times H + (\pi R^2 \times 2)$
= (3.142 × 1.2 × 2.4) + (3.142 × 0.6 × 0.6 × 2)
= 9.04896 + 2.26224
= 11.3112 m²
say 11.31m² to 2 decimal places

ANSWER:
Total surface area = 11.31m²

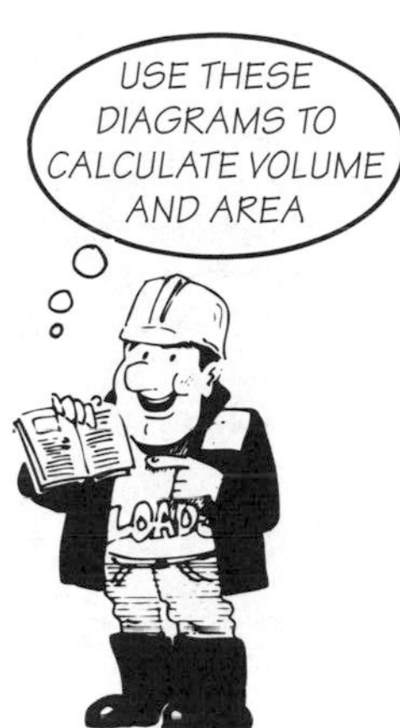

USE THESE DIAGRAMS TO CALCULATE VOLUME AND AREA
LOADS

Formulae for volumes and surface areas

The following formulae can be used for calculating the volume and lateral surface area of frequently used **common solids**. It can be seen that the volume of any pyramid or cone will always be equal to one-third of its equivalent prism or cylinder.

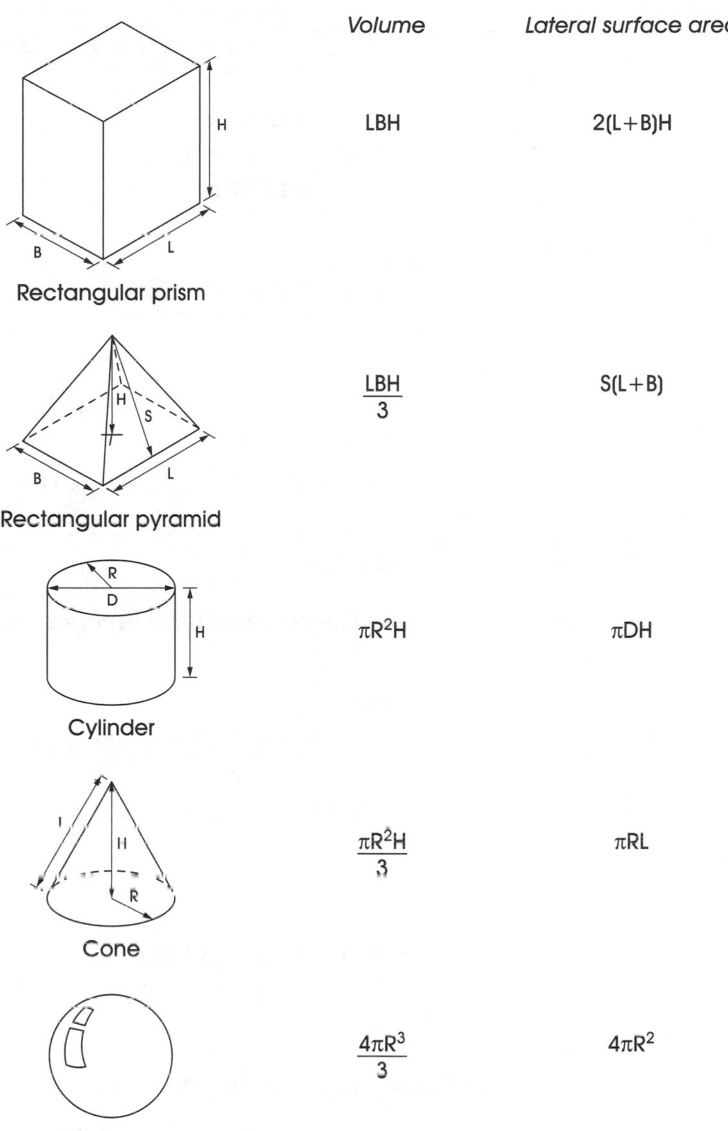

	Volume	*Lateral surface area*
Rectangular prism	LBH	2(L+B)H
Rectangular pyramid	$\frac{LBH}{3}$	S(L+B)
Cylinder	$\pi R^2 H$	πDH
Cone	$\frac{\pi R^2 H}{3}$	πRL
Sphere	$\frac{4\pi R^3}{3}$	$4\pi R^2$

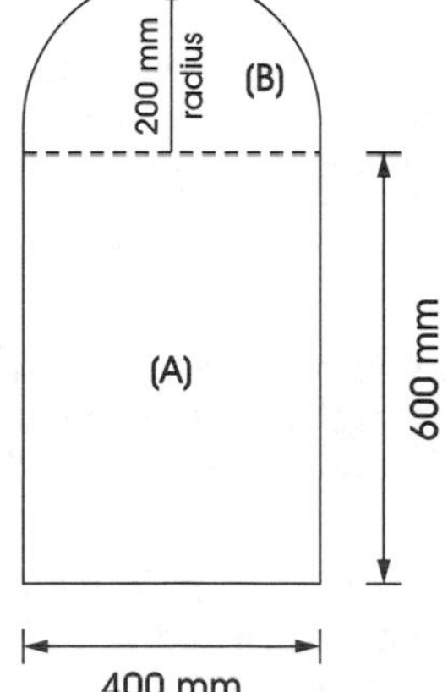

Complex volumes – are found by breaking them up into a number of recognisable volumes and solving for each one in turn. This is the same as the method used when solving for complex areas.

Suppose you were asked to find the volume of concrete required for a 2.4 m high column having the plan or horizontal cross section shown.

The column can be considered as a rectangular prism (A) and half a cylinder (B).

Example

Find the volume of the column shown.

$$\text{Volume A} = 0.4 \times 0.6 \times 2.4 = 0.576\ m^3$$

$$\text{Volume B} = \frac{3.142 \times 0.2 \times 0.2 \times 2.4}{2} = 0.151\ m^3$$

$$\text{Total vol.} = 0.576 + 0.151 = 0.727\ m^3$$

ANSWER:
0.727 m³ of concrete is required

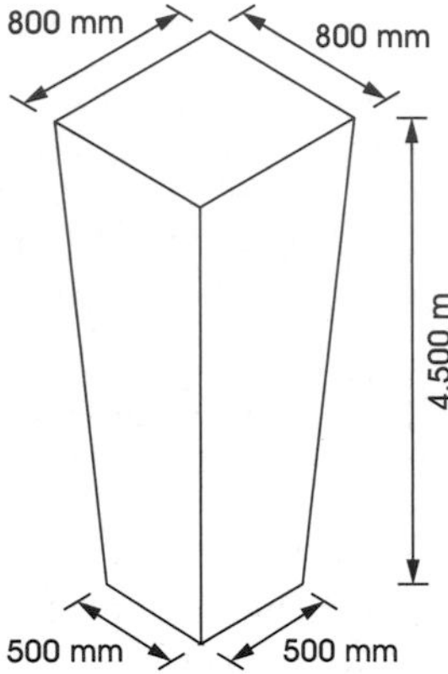

Where a solid tapers, its volume can be found by multiplying its average cross-section by its height.

Volume = Average cross-section × Height

Example

Find the volume of the tapered column shown.

Volume

$$= \frac{(0.8 \times 0.8) + (0.5 \times 0.5)}{2} \times 4.5$$

$$= \frac{0.64 + 0.25}{2} \times 4.5$$

$$= 2.003\ m^3$$

ANSWER: Volume = 2.003 m³

Pythagoras' theorem

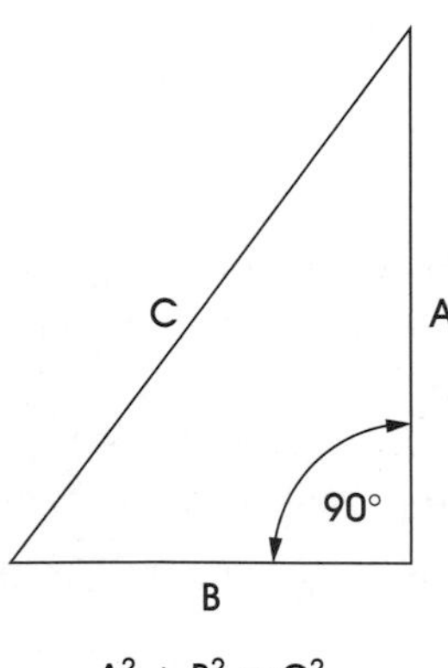

$A^2 + B^2 = C^2$

The lengths of the sides in a right-angled triangle can be found using Pythagoras' theorem. According to this theorem, in any right-angled triangle the square of the length of the longest side is equal to the sum of the squares of the other two sides.

$C^2 = A^2 + B^2$

A simple version of Pythagoras' theorem is known as the *3:4:5 rule.* This is often used for setting out and checking right angles, since a triangle whose sides equal 3 units, 4 units and 5 units must be a right-angled triangle because $5^2 = 3^2 + 4^2$.

If we know the lengths of two sides of a right-angled triangle we can use Pythagoras' theorem to find the length of the third side. In fact this theorem forms the basis of pitched roof calculations.

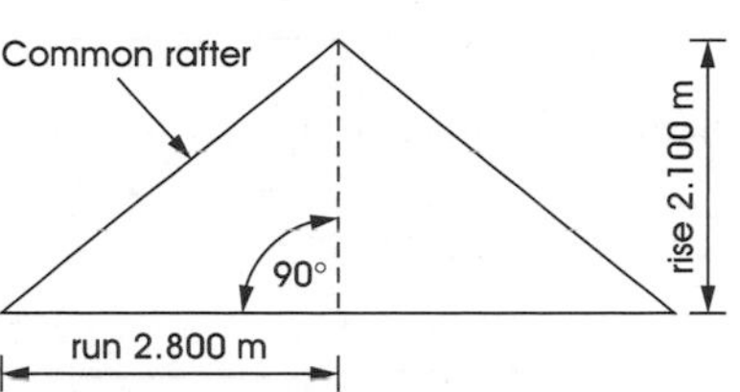

Example

Calculate the length of the common rafter shown.

A = Rise 2.1 m
B = Run 2.8 m
C = Common rafter

$$A^2 + B^2 = C^2$$
$$2.1^2 + 2.8^2 = C^2$$
$$= 4.41 + 7.84 = 12.25$$

Therefore:

$$C = \sqrt{12.25} = 3.5 \text{ m}$$

ANSWER:
The common rafter is 3.5 m long

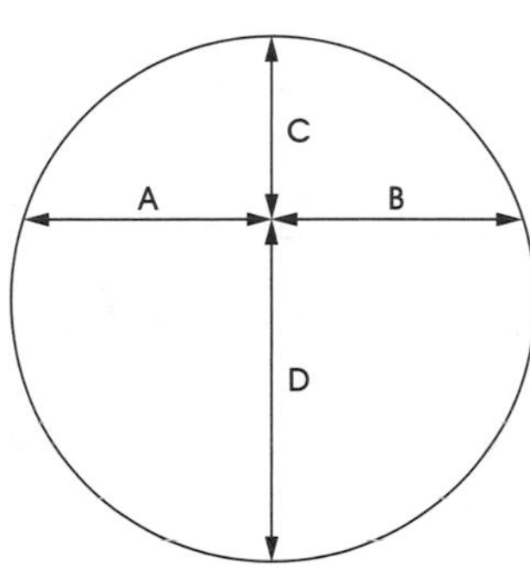

A × B = C × D

Intersecting chords rule

Where two chords intersect in a circle, the product (result of multiplication) of the two parts of one chord will always be equal to the product of the two parts of the other chord. It is very useful for finding radius lengths.

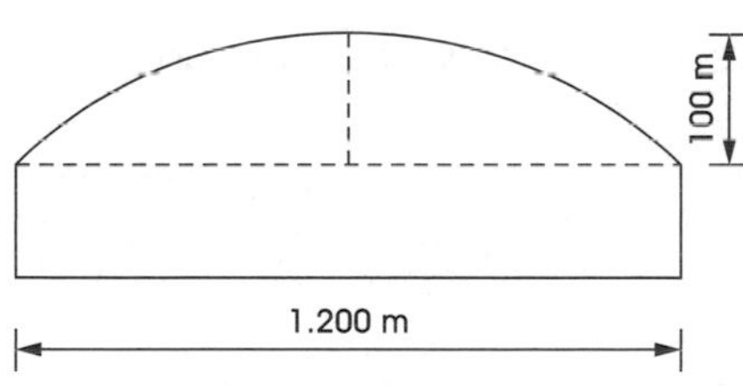

Example

Suppose you were asked to set-out a turning piece having a span of 1.2 m and a rise of 100 m. What is the radius?

A = 0.6 m
B = 0.6 m
C = 0.1 m
D = ?

$$A \times B = C \times D$$
$$\frac{A \times B}{C} = D$$
$$\frac{0.6 \times 0.6}{0.1} = 3.6$$
$$\text{Radius} = \frac{C + D}{2}$$
$$= \frac{0.1 + 3.6}{2}$$
$$= 1.85 \text{ m}$$

ANSWER: The radius is 1.85 m

TRY AND
ANSWER
THESE

Questions for you

37. What is the relationship between the diameter and circumference of a circle?

38. A rectangle has a length of 6.39 m and a breadth of 2.15 m.

(a) What is the area?

(b) What is the perimeter?

39. What is the volume of a room if it measures 4.5 m × 3.65 m × 2.4 m?

40. What is the total surface area of a cylindrical prism if the radius is 600 mm and the height is 1.9 m?

41. The longest side in a right-angled triangle measures 5 m; one of the other sides is 3 m. What is the length of the remaining side?

42. Calculate the area and perimeter of the figure illustrated.

5.292 m
90°
900 mm
3.925 m
Semi circle
90°
0.9 m
6095 mm

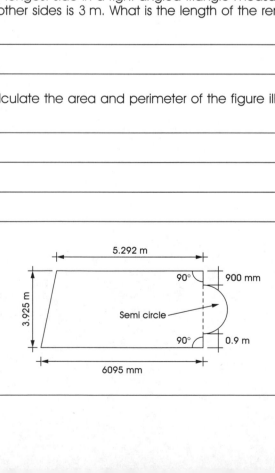

WELL, HOW
DID YOU
DO?
WORK
THROUGH THE SECTION
AGAIN IF YOU HAD
ANY PROBLEMS

Measuring and costing materials

Flooring

In order to determine the amount of floor covering materials required for an area, multiply its width by its length.

Floorboards – To calculate the metres run of floorboards required to cover a floor area of say 4.65 m^2, if the floorboards have a covering width of 137 mm.:

Metres run required = Area ÷ Width of board
= 4.65 ÷ 0.137
= 33.94 m,
say 34 m run

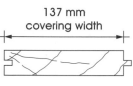

137 mm
covering width

It is standard practice to order an additional amount of flooring to allow for cutting and **wastage**. This is often between 10% and 15%

If 34 m run of floor boarding is required to cover an area, calculate the amount to be ordered including an additional 12% for cutting and wastage.

Amount to be ordered = 34 × 1.12
= 38.08, say 38 m run

In order to determine the number of sheets of plywood or chipboard required to cover a room either:

- Divide area of room by area of sheet, or
- Divide width of room by width of sheet and divide length of room by length of sheet. Convert these numbers to the nearest whole or half and multiply them together.

To calculate the number of 600 mm × 2400 mm chipboard sheets required to cover a floor area of 2.02 m × 3.6 m.

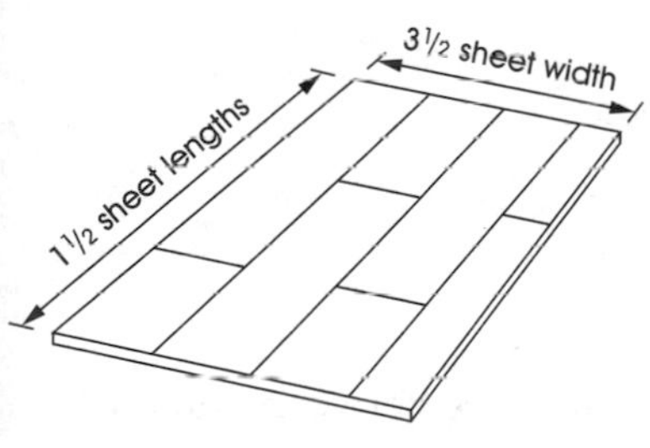

Number of sheets required = Area of room ÷ Area of sheet

Area of room = 2.05 × 3.6 m
=7.38 m^2

Area of sheet = 0.6 × 2.4
=1.44 m^2

Number of sheets = 7.38 ÷ 1.44
= 5.125, say 6 sheets

Alternatively:

Number of sheet widths in room width = 2.05 ÷ 0.6
= 3.417, say 3.5

Number of sheet lengths in room length = 3.6 ÷ 2.4
= 1.5

Total number of sheets = 3.5 × 1.5
=5.25, say 6 sheets

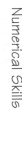

Joists

To determine the number of joists required and their centres for a particular area the following procedure can be used:

- Measure the distance between adjacent walls, say 3150 mm
- The first and last joist would be positioned 50 mm away from the walls. The centres of 50 mm breadth joists would be 75 mm away from the wall. The total distance between end joists centres would be 3000 mm. See (a).
- Divide the distance between end joists centres by the specified joist spacing say 400 mm. This gives the number of spaces between joists. Where a whole number is not achieved round up to the nearest whole number above. There will always be one more joist than the number of spaces so add another one to this figure to determine the number of joists. See (b).

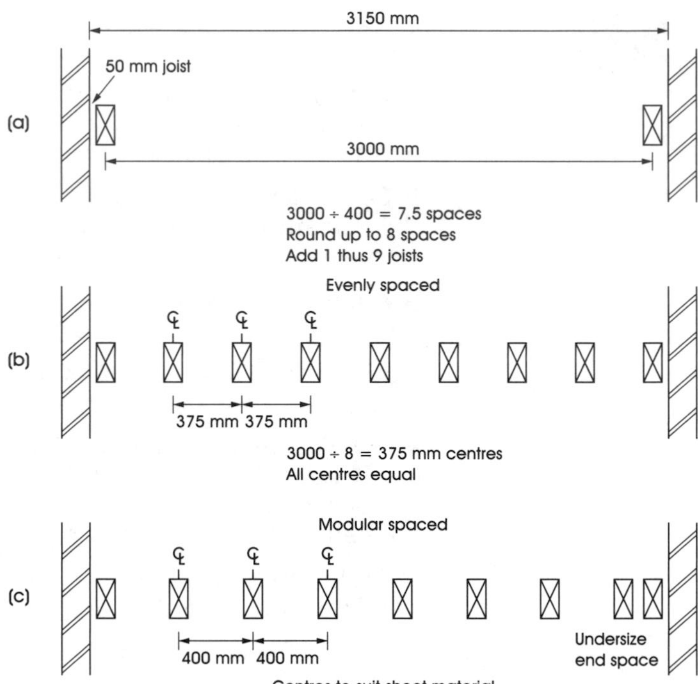

- Where T&G boarding is used as a floor covering the joist centres may be spaced out evenly, i.e. divide the distance between end joist centres by the number of spaces.
- Where sheet material is used as a joist covering to form a floor, ceiling or roof surface, the joist centres are normally maintained at a 400 mm or 600 mm module spacing to coincide with sheet sizes. This would leave an undersized spacing between the last two joists. See (c).

Number of joists =

(Distance between end joist centres $\times$ Joist spacing) + 1

$= (3000 \div 400) + 1$

$= 8.5$, say 9

Stud partitions

To determine the number of studs required for a particular partition the following procedure can be used:

- Measure the distance between the adjacent walls of the room or area which the partition is to divide, say 3400 mm.
- Divide the distance between the walls by the specified spacing, say 600 mm. This gives the number of spaces between the studs. Use the whole number above. There will always be one more stud than the number of spaces so add one to this figure to determine the number of studs. Stud centres must be maintained to suit sheet material sizes leaving an undersized space between the last two studs.
- The lengths of head and sole plates are simply the distance between the two walls.
- Each line of noggins will require a length of timber equal to the distance between the walls.

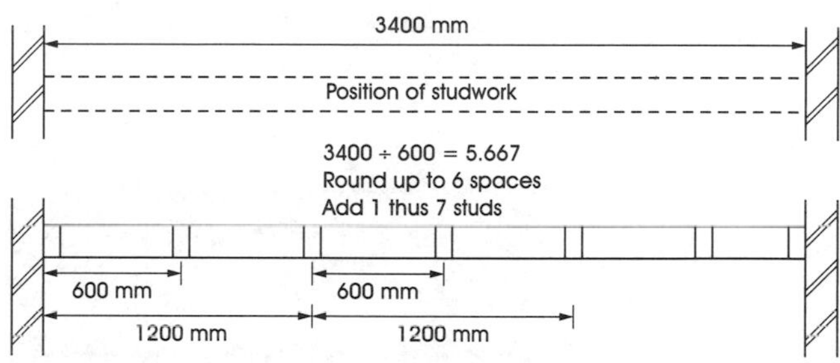

Stud spacing to suit 1200 mm wide 12.5 mm thick plasterboard

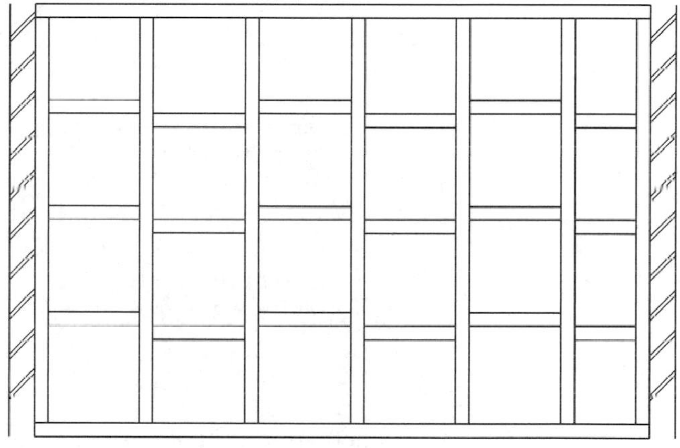

The total length of timber required for a partition, can be determined by the following method:

7 studs at 2.4 mm, 7 × 2.4 = 16.8m

Head and sole plates at 3.4 m, 2 × 3.4 = 6.8 m

3 lines of noggins at 3.4 m, 3 × 3.4 = 10.2 m

Total metres run required, 16.8 + 6.8 + 10.2 = 33.8 m, say 34 m

Rafters

The number of rafters required for a pitched roof can be determined using the same method as used for floor joists. For example, divide the distance between the end rafter centres by the rafter spacing. Round up and add one. Double the number of rafters to allow for both sides of the roof.

Say distance between end rafter centres is 12 m and spacing is 400 mm:

Number of rafters
= Distance between end rafter centres ÷ Rafter spacing + 1
= 12 ÷ 0.4 + 1
= 30 + 1
= 31 rafters

Therefore total number of rafters required for both sides of the roof is 62.

Where overhanging verges are required an additional rafter must be allowed at each end to form gable ladders, which provide a fixing for the barge board and soffit.

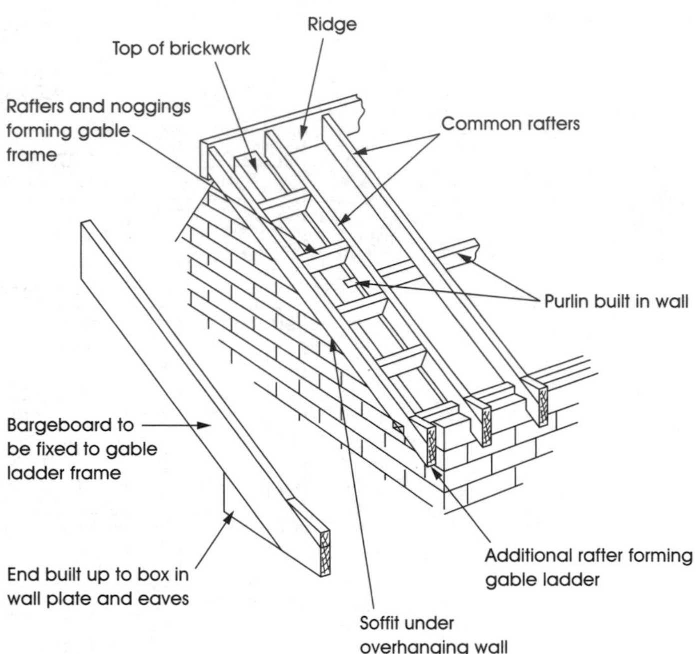

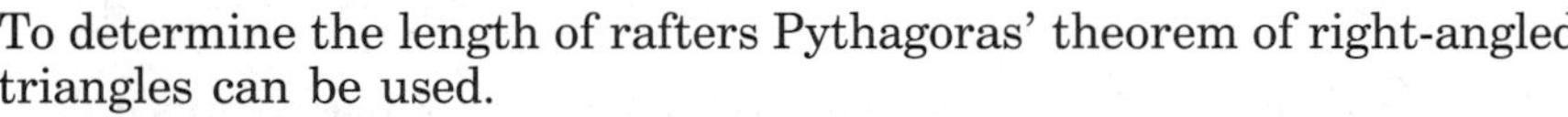

To determine the length of rafters Pythagoras' theorem of right-angled triangles can be used.

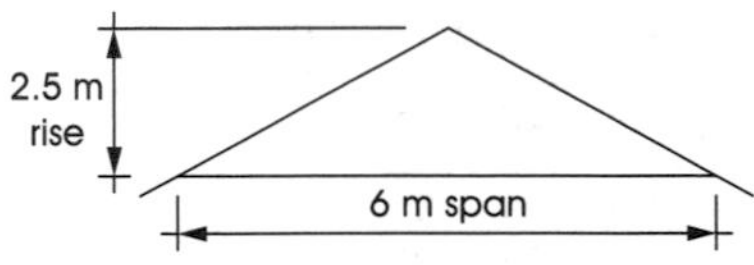

Determine the length of rafter required for a roof of 2.5 m rise and 6 m span

A = Half span = 3 m
B = Rise = 2.5 m
C = Length of rafter = ?

$C^2 = B^2 + A^2$
$= (2.5 \times 2.5) + (3 \times 3)$
$= 6.25 + 9$

$C = \sqrt{15.25}$
$= 3.905$ m

An allowance must be added to this length for the eaves overhang and the cutting. Say 0.5 m and 10%.

Total length of rafter = 3.905 + 0.5 + 10%
= 4.405 × 1.1
= 4.845 m

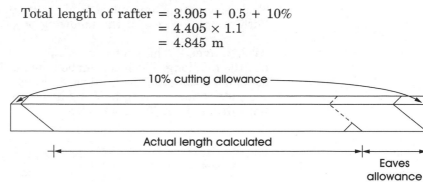

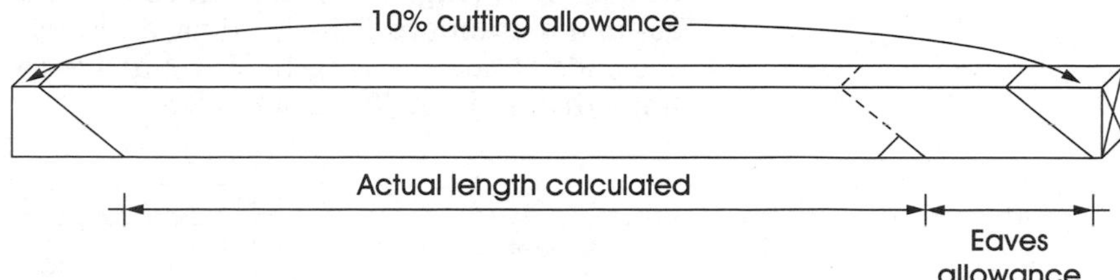

Fascia, barge and soffits

Calculating the lengths of material required for fascia boards, barge boards and soffits is often a simple matter of measuring, allowing a certain amount extra for jointing, and adding lengths together to determine total metres run.

A hipped-end roof requires two 4.4 m lengths and two 7.2 m lengths of ex 25 mm × 150 mm PAR softwood for its fascia boards.

Metres run required = (4.4 × 2) + (7.2 × 2)
= 8.8 + 14.4
= 23 m

It is standard practice to allow a certain amount extra for cutting and jointing. This is often 10%.

Total metres run required = 23.2 × 1.1
= 25.52 m

The length of timber for barge boards may require calculation using Pythagoras' theorem for right-angled triangles.

Sheet material – Where sheet material is used for fascias and soffits the amount which can be cut from a full sheet often needs calculating. This entails dividing the width of the sheet by the width of the fascia or soffit, then using the resulting whole number to multiply by the sheet's length, to give the total metres run.

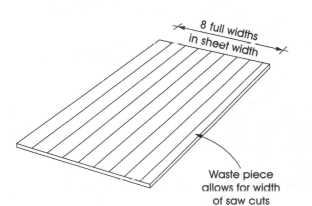

8 full widths in sheet width

Waste piece allows for width of saw cuts

Example

Determine the total metres run of 150 mm wide soffit board that may be cut from a 1220 mm × 2440 mm sheet.

Number of lengths = 1220 ÷ 150
= 8.133, say 8

Total metres run = 8 × 2440
= 19.520 m

Trim

To determine the amount of trim required for any particular task is a fairly simple process, if the following procedures are used.

Architraves – The jambs or legs in most situations can be taken to be 2100 mm long. The head can be taken to be 1000 mm. These lengths assume a standard full-size door and include an allowance for mitring the ends. Thus the length of architrave required for one face of a door lining/frame is 5200 mm or 5.2 m.

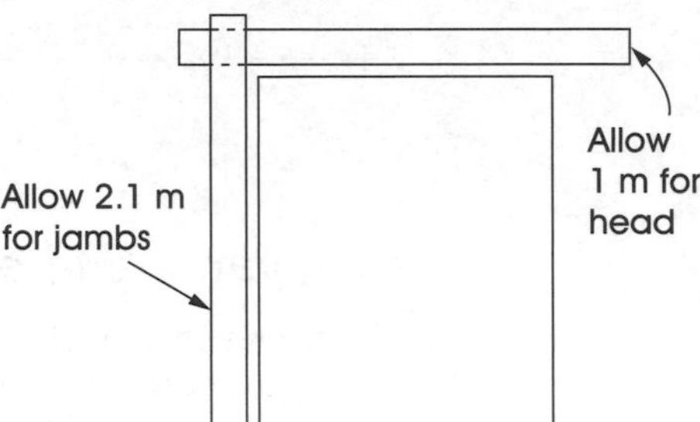

Multiply this figure by the number of architrave sets to be fixed. This will determine the total metres run required, say 8 sets, both sides of four doors:

$5.2 \times 8 = 41.6$ m

Skirtings – and other horizontal trim can be estimated from the perimeter. This is found by adding up the lengths of the walls in the area. The widths of any doorways and other openings are taken away to give the actual metres run required.

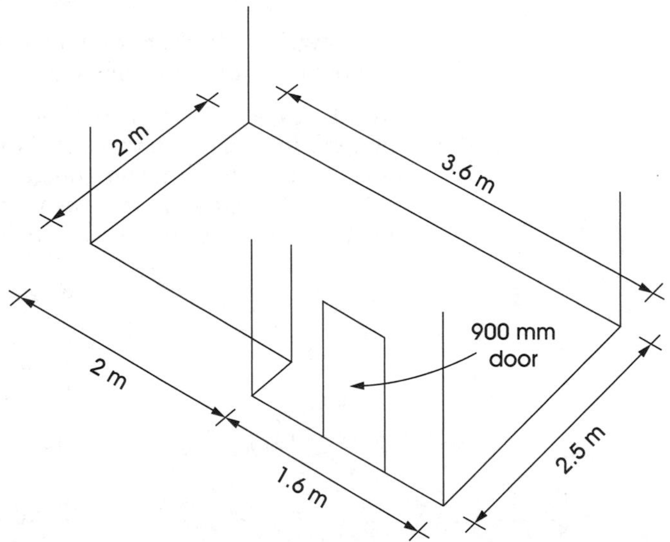

Determine the total length of timber required for the room.

Perimeter = 2 + 3.6 + 2.5 + 1.6 + 0.5 + 2
= 12.2 m

Total metres run required = 12.2 – 0.9 (door opening)
= 11.3 m

An allowance of 10% for cutting and waste is normally included in any estimate for horizontal moulding.

Determine the total metres run of skirting required for the run shown including an allowance of 10% for cutting and waste.

Total metres run required = 11.3 + 1.13

= 12.43 m, say 12.5 m

Brickwork and mortar

There are 60 bricks per m^2 in half brick thick walls and 120 bricks per m^2 in one brick thick walls. An additional percentage of 5% is normally allowed for cutting and damaged bricks.

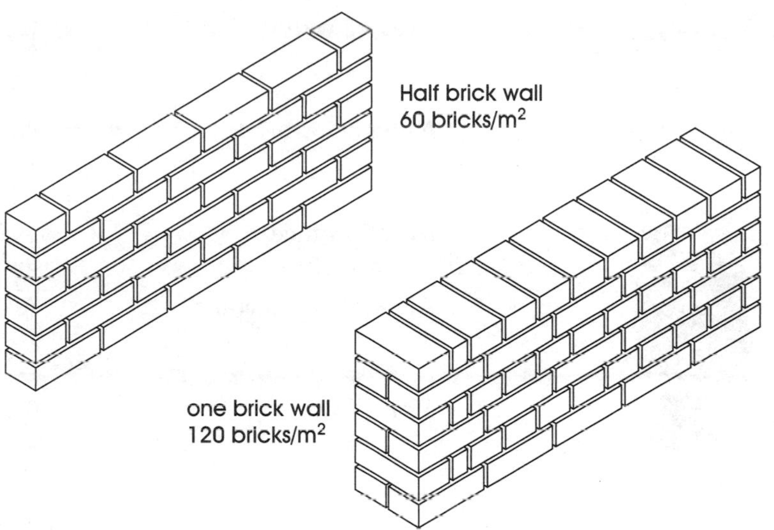

Approximately 1 kg of mortar is required to lay 1 brick. This figure can be used for small areas of new brickwork and making good. A 25 kg bag of mortar mix is sufficient to lay up to 25 bricks, depending on the thickness of joints. One 25 kg bag of cement and six 25 kg bags of sand, is sufficient to lay up to 175 bricks using a 1:6 cement-sand ratio.

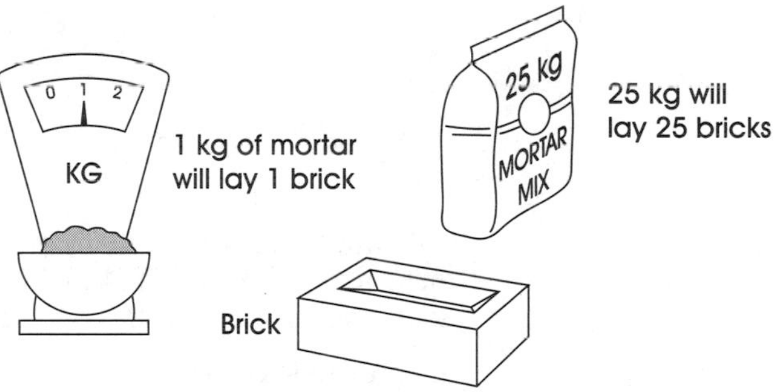

For larger areas of brickwork the amount of mortar can be assessed as 0.03 m^3 per square metre of brickwork or 0.5 m^3 per 1000 bricks for ½-brick walls. Just over double the amount per square metre for one brick walls, 0.07 m^3, or 0.58 m^3 per 1000 bricks. An additional percentage of 10% is normally allowed for wastage.

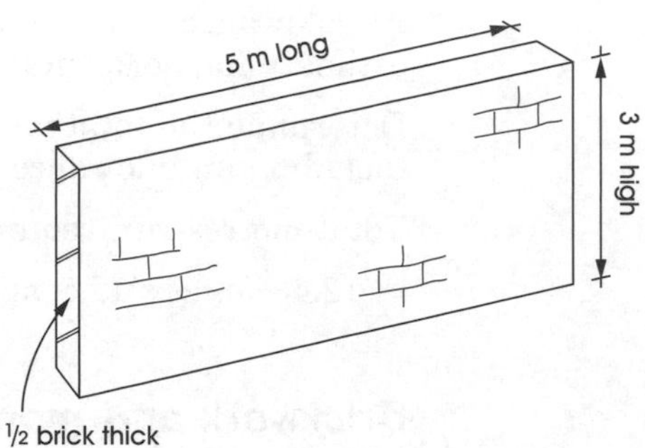

Calculate the number of bricks and amount of mortar required for a half-brick wall 3 m high and 5 m long, allow 5% extra for cutting and 10% for mortar.

Area of wall	=	Height × Length
	=	3 × 5
	=	15 m^2
Number of bricks required	=	Area × Number of bricks per m^2
	=	15 × 60
	=	900 bricks
5% allowance	=	900 × 1.05
	=	945 bricks
Amount of mortar	=	Area × Mortar/m^2
	=	15 × 0.03
	=	0.45 m^3
10% allowance	=	0.45 × 1.1
	=	0.495 m^3, say 0.5 m^3

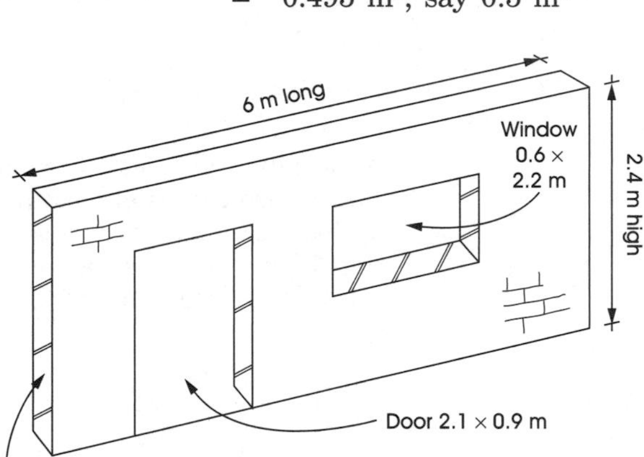

Calculate the number of bricks and the amount of mortar required to form a ½-brick wall 2.4 m high and 6 m long containing a 2.1 × 0.9 m door opening and a 0.6 m × 2.4 m window opening.

Area of brickwork	=	Total area – Area of door and window
Total area	=	2.4 × 6
	=	14.4 m^2
Area of door	=	2.1 × 0.9
	=	1.89 m^2
Area of window	=	0.6 × 2.2
	=	1.32 m^2

Therefore:

Area of brickwork = 14.4 – 1.89 – 1.32
= 11.19 m^2

No. of bricks = Area × No. of bricks/m^2
= 11.19 × 60
= 671.4, say 672

5% allowance = 672 × 1.05
= 705.6, say 706

Amount of mortar = Area × Mortar/m^2
= 11.19 × 0.03
= 0.336 m^3

10% allowance = 0.336 × 1.1
= 0.369 m^3

Where brickwork returns around corners as in a building the wall's centre line length is used to calculate the area.

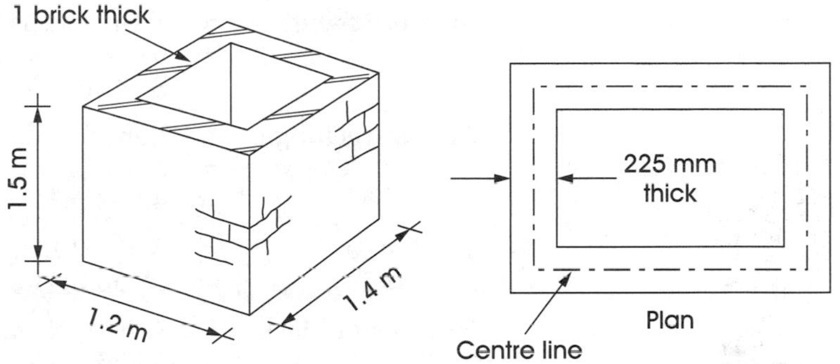

Calculate the number of bricks and amount of mortar to build the one-brick thick inspection chamber.

Length of centre line
= (2 × Length) + (2 × Width) – (4 × Wall thickness)
= 2.8 + 2.2 – 0.9
= 4.1 m

Area = Length of centre line × Height
= 4.1 × 1.5
= 6.15 m^2

Number of bricks = Area × Number of bricks per m^2
= 6.15 × 120
= 738

5% allowance = 738 × 1.05
= 774.9, say 775 bricks

Amount of mortar = Area × Mortar per m^2
= 6.15 × 0.07
= 0.4305 m^3

10% allowance = 0.4305 × 1.1
= 0.4735, say 0.474 m^3

Blockwork and mortar – There are approximately ten 450 mm × 215 mm blocks per square metre. An additional allowance of 5% may be added for cutting and damaged blocks. Each m^2 of 100 mm thick blocks can be assessed as requiring 0.01 m^3 of mortar. An allowance of 10% is normally added for wastage.

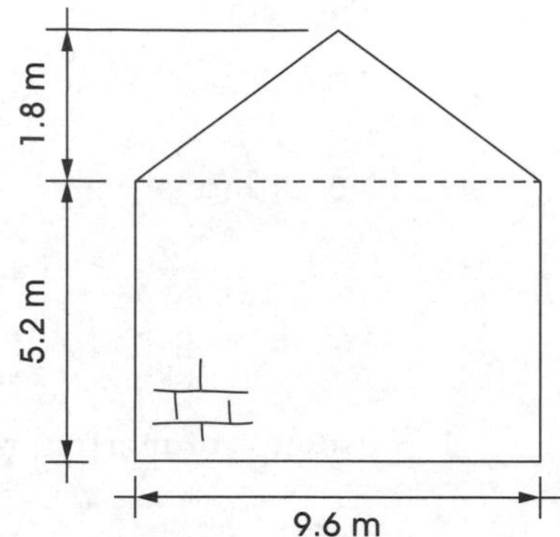

Calculate the number of 450 mm × 215 mm blocks and mortar required for the gable end illustrated. Allow 5% extra on the blocks and 10% extra on the mortar.

Area of blockwork	=	Triangular area + Rectangular area
Area of triangle	=	Base × Height ÷ 2
	=	9.6 × 1.8 ÷ 2
	=	8.64 m^2
Area of rectangle	=	Length × Height
	=	9.6 × 5.2
	=	49.92 m^2
Total area	=	8.6 + 49.92
	=	58.56 m^2
Number of blocks	=	Area × Number of blocks per m^2
	=	58.56 × 10
	=	585.6 blocks
5% allowance	=	585.6 × 1.05
	=	614.8, say 615 blocks
Amount of mortar	=	Area × Mortar per m^2
	=	58.56 × 0.01
	=	0.586 m^3
10% allowance	=	0.586 × 1.1
	=	0.644 m^3

Blockwork walls with returns and openings are calculated in the same ways as those in brickwork using the centre line length when working out the area.

Dry materials

Mortar – The density of mortar is approximately 2300 kg/m^3. Different mixes are specified for different situations.

A 1:6 mix contains 7 parts:

- 1 part cement
- 6 parts sand (fine aggregate)

A 1:1:8 mix contains 10 parts:

- 1 part cement
- 1 part lime
- 8 parts sand

To determine dry materials for a given quantity of mortar, multiply volume by density 2300 kg/m^3 for total mass and divide by number of parts. This gives the amount of cement; multiply cement by required number of parts to give other quantities.

For 0.5 m^3 of 1:6 mortar:

Mass of mortar = 0.5×2300
= 1150 kg

Number of parts in mix = 7

Amount of cement = $1150 \div 7$
= 164.286 kg, say seven 25 kg bags

Amount of sand = 164.286×6
= 985.716 kg,
say 1 tonne or 40×25 kg bags

Mixed on site concrete – It is necessary to determine the amount of cement, fine and coarse aggregates for ordering. The density of compacted concrete is 2400 kg/m^3.

A mix of 1:3:6 contains 10 parts.

- 1 part cement
- 3 parts fine aggregate
- 6 parts coarse aggregate

To determine dry materials for a given quantity of concrete multiply volume by density which gives total mass and then divide by the number of parts. This gives the amount of cement; multiply cement by 3 to give fine aggregate, then multiply cement by 6 to give coarse aggregate.

For 2.5 m^3 of 1:3:6 concrete:

Mass of concrete = 2.5×2400
= 6000 kg

Number of parts = 10

Amount of cement = $6000 \div 10$
= 600 kg or 24×25 kg bags

Amount of fine aggregate = 600×3
= 1800 kg,
say almost two 1 tonne bags
or 72×25 kg bags

Amount of coarse aggregate = 600×6
= 3600 kg, say almost four 1 tonne
bags or 144×25 kg bags

Tiles and paving slabs – for a particular area these can be calculated using the following method:

- Determine the total area to be covered in m^2 (excluding any openings)
- Determine the total area of a tile or slab in m^2
- Divide the area to be covered by the area of the tile or slab
- Add percentage for cutting and waste (typically 5 to 10%)

To determine the number of 1500 mm square tiles required for a kitchen floor.

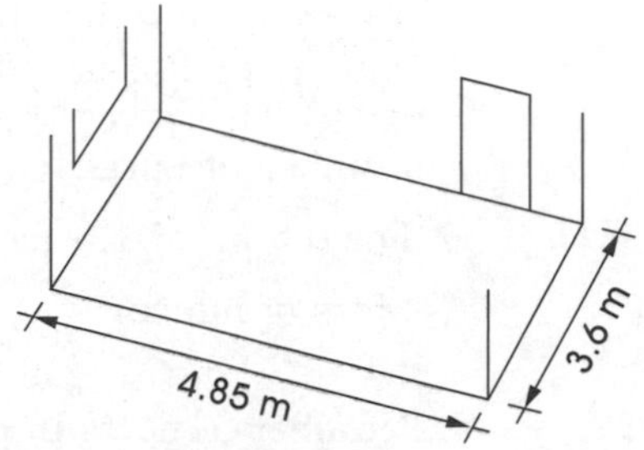

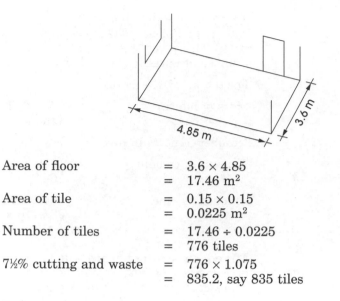

Area of floor	=	3.6 × 4.85
	=	17.46 m^2
Area of tile	=	0.15 × 0.15
	=	0.0225 m^2
Number of tiles	=	17.46 ÷ 0.0225
	=	776 tiles
7½% cutting and waste	=	776 × 1.075
	=	835.2, say 835 tiles

Paint – for a particular area can be calculated using the following method:

- Determine the total area to be covered in m^2 (excluding any openings)
- Divide the total area by the recommended covering capacity (coverage) of the paint to be used. This will give the number of litres required for one coat.
- Multiply litres for one coat by number of coats, to give total litres required.

Determine the amount of paint required to paint the wall of a factory with 2 coats of emulsion. Typical manufacturer's coverage figures are:

- Primers and undercoats cover 12–14 m^2 per litre per coat
- Gloss or satin top coats cover 14–16 m^2 per litre per coat
- Emulsions cover 10–12 m^2 per litre per coat

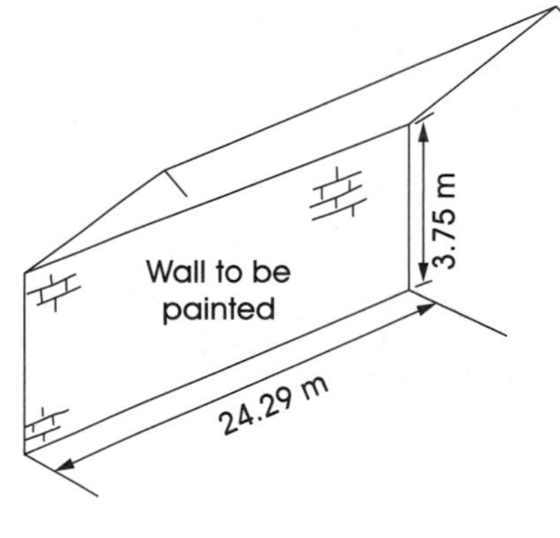

Use the minimum coverage from the table for a brick wall.

Area of wall	=	24.29 × 3.75
	=	91.0875 m^2
Amount of paint for 1 coat	=	91.0875 ÷ 10 (coverage)
	=	9.10875 litres
Total amount of paint	=	9.10875 × 2 (coats)
	=	18.2175 litres,
		say 20 litres or 4 × 5 litre tins.

Costing materials

This can be carried out once the required quantities of material have been calculated. It is a simple matter of finding out prices and multiplying these by the number of items required.

Suppose you were asked to find the total price of four 2440 × 1220 sheets of 18 mm MDF. From the typical extract of the supplier's price list these are £10.97 each including VAT.

Total price	=	Price per sheet × Number of sheets
	=	£10.97 × 4
	=	£43.88

Sheet Materials			**Price per item**	
Product	Size	Product code	£ inc. VAT	£ exc. VAT
Blockboard BB/CC 18 mm	2440 × 1220 mm	5022652518455	20.32	17.29
PLYWOOD				
Hardwood WBP BB/CC 4 mm	2440 × 1220 mm	5022652504588	7.00	5.96
Hardwood WBP BB/CC 6 mm	2440 × 1220 mm	5022652506605	8.50	7.23
Hardwood WBP BB/CC 9 mm	2440 × 1220 mm	5022652519629	11.55	9.83
OSB				
OSB2 11 mm	2440 × 1220 mm	5022652560386	7.41	6.31
OSB3 18 mm	2440 × 1220 mm	5014957148673	15.98	13.60
MDF				
MDF 6 mm	2440 × 1220 mm	5022652550035	6.57	5.60
MDF 12 mm	2440 × 1220 mm	5022652512231	8.48	7.22
MDF 18 mm	2440 × 1220 mm	5022652518267	10.97	9.34
HARDBOARD				
Standard Hardboard 3.2 mm	2440 × 1220 mm	5022652503314	2.60	2.21
White Faced Hardboard 3.2 mm	2440 × 1220 mm	5022652503338	5.17	4.40
CHIPBOARD				
Flooring Grade 18 mm P4	2400 × 600 mm	5014957105706	3.98	3.39
Flooring Grade 22 mm P4	2400 × 600 mm	5014957105720	5.58	4.75
Flooring Grade 18 mm P5	2400 × 600 mm MR	5014957054691	5.31	4.52
Flooring Grade 22 mm P5	2400 × 600 mm MR	5014957088320	7.98	6.79
Standard Grade 12 mm	2440 × 1220 mm	5014957054677	4.95	4.21

Learning task

The following materials are required for a small building extension.

- 12 off 2.7 m long 50 × 200 joists
- 8 off 2.7 m long 50 × 100 sawn softwood
- 12 off 50 mm joist hangers
- 11 off 600 × 2440 T&G chipboard flooring
- 1750 off Redland red drag faced bricks
- 450 off 100 mm blocks
- 31 bags of cement
- 3 tonne of fine aggregate (building sand)
- 2 tonne of fine aggregate (sharp sand)
- 4 tonne of coarse aggregate (gravel 20 mm)

On phoning your material supplier, the prices shown on the telephone message pad were obtained. Use this information to determine the following:

Telephone Message

Date *26 OCT* Time *9.45*

Message for *JAMES*

Message from (Name) *BBS SUPPLIES*

(Address) *LAKESIDE INDUSTRIAL PARK, NOTTINGHAM*

(Telephone) *01159 464368*

Message *All plus VAT at 17.5%*

50 × 200 Joists	*£2.35 per m*
50 × 100 Sawn	*£0.85 per m*
Joist Hangers	*£1.69 each*
600 × 2400 Chipboard Flooring	*£3.98 each*
Redland Brick	*£0.70 each*
100 mm Blocks	*£1.62 each*
Cement	*£4.99 per bag*
Building Sand	*£1.06 per 25 kg*
Sharp Sand	*£1.14 per 25 kg*
Gravel	*£24 per tonne*

Message taken by *CHRIS*

(a) Total cost at list price

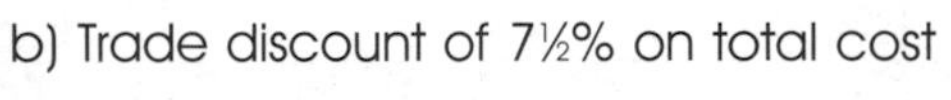

b) Trade discount of 7½% on total cost

(c) VAT at 17½% on discounted total cost

(d) Actual cost payable including VAT

WELL, HOW MANY DID YOU GET?

Questions for you

43. Calculate the number of eight-hour days required for a four person gang of bricklayers to build 145 m^2 of brickwork. (Use the rate of 0.3 m^2 per bricklayer per hour)

44. $A = B \times C$ find the value of B if $A = 4.5\ m^2$ and $C = 1.2\ m$

45. The following materials are required for a refurbishing contract:

Softwood
Sawn softwood at £158.50 per m^3

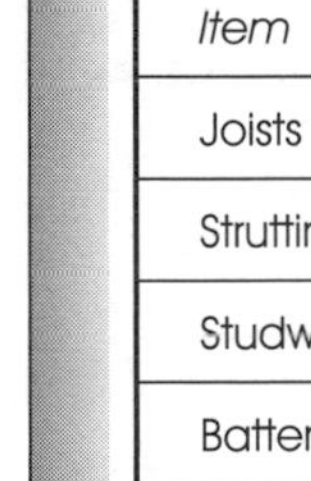

Item	*Number*	*Size*
Joists	16	50 × 225 × 3600
Strutting	10	50 × 50 × 4200
Studwork	84	50 × 100 × 2400
Battening	50	25 × 50 × 4800

Flooring

18 mm flooring grade chipboard at £49 per 10 m^2; 30 sheets 600 × 2400 mm

Calculate the total cost including an allowance of 10 per cent for cutting and wastage and 17½ per cent for VAT.

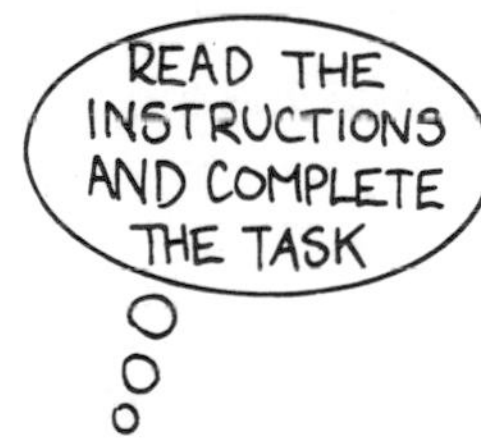
READ THE INSTRUCTIONS AND COMPLETE THE TASK

46. What radius is required to set out a centre for a segmental arch having a rise of 550 mm and a span of 4.500 m?

47. What is the diameter of a circular rostrum if its perimeter measures 12 m?

48. A rectangular room 4.200 m × 6 m is to be floored using hardwood boarding costing £9.55 per square metre.

(a) Allowing 12½ per cent for cutting wastage, how many square metres are required?

(b) What would be the total cost including 17½ per cent VAT?

49. A builders' merchant sells cement in three sizes. Which is the best value?

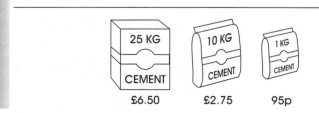

25 KG CEMENT £6.50

10 KG CEMENT £2.75

1 KG CEMENT 95p

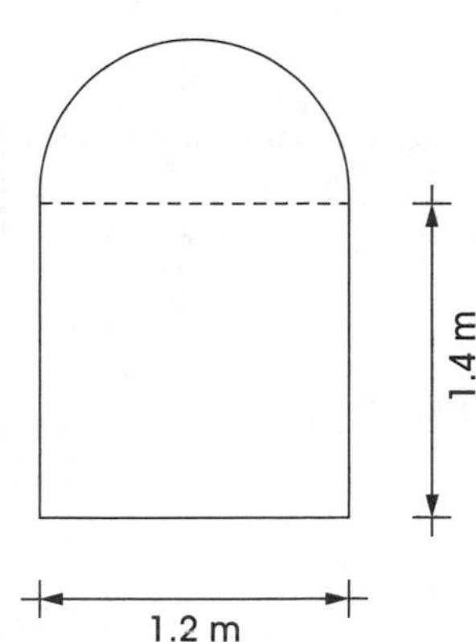

50. A window is in the shape of a rectangle and a semicircle. Find the area of the window.

51. Find the angles, and give the reason:

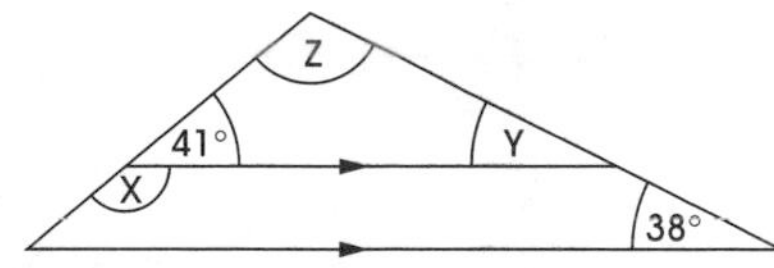

(a) Angle X

(b) Angle Y

(c) Angle Z

52. Details of a one brick thick garden wall 15.750 m long are illustrated. Calculate:

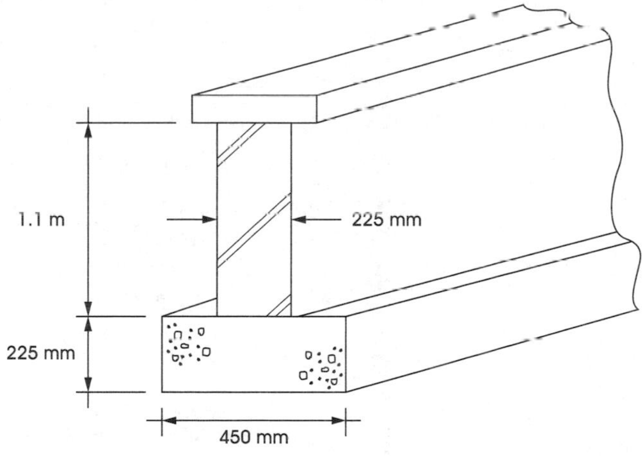

(a) The number of bricks required to build the wall.

(b) The amount of cement and fine aggregate required to build the wall using a 1:6 mix.

(c) The total volume of concrete for the foundation.

(d) The number of coping stones if each covers a length of 600 mm.

53. Use the following time sheet to determine how much Ivor Carpenter has earned for a week. Take the rate of pay for weekdays as £9.25 for the first 8 hours each day and time and a half after that. Saturdays is also paid at time and a half and Sundays at double time.

BBS Recruitment Solutions: WEEKLY TIME SHEET | Name: **I. CARPENTER** | Works No. **26**

Day	Date	Start Time	Lunch	Finish Time	Total basic hours	Total overtime hours
Monday	**15/3**	**7.30**	**½ hr**	**6.00**		
Tuesday	**16/3**	**7.30**	**½ hr**	**5.30**		
Wednesday	**17/3**	**8.00**	**½ hr**	**6.15**		
Thursday	**18/3**	**8.00**	**½ hr**	**6.00**		
Friday	**19/3**	**7.30**	**½ hr**	**5.45**		
Saturday	**20/3**	**6.00**		**12.00**		
Sunday	**21/3**	**6.00**		**12.00**		
Signature: I Carpenter						

54. What would be the net pay in Question 11 if a total of 26% deductions were taken from the gross pay?

55. Determine for the room shown:

(a) the number of 900 × 1800 mm sheets of plasterboard required to cover the ceiling

(b) the total length of skirting required allowing 10% for cutting and waste

(c) the total length of timber required to cover the floor if the boards have a covering width of 95 mm

(d) the amount of paint required for two coats of emulsion to the walls and ceilings if one litre covers 10 m^2 , making a reduction for the door and the window.

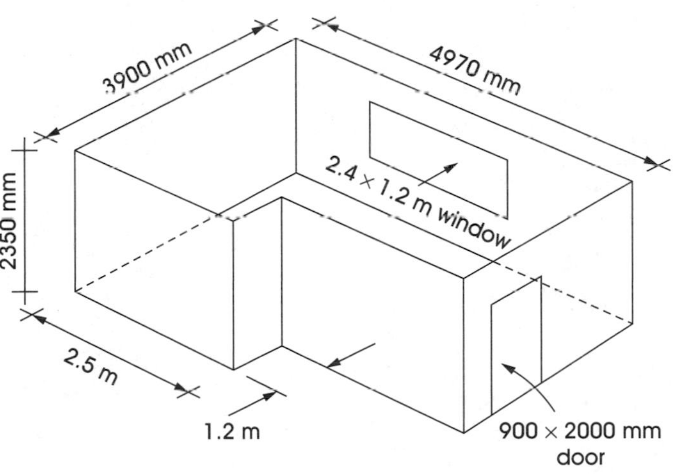

56. A cavity wall consists of a half-brick thick outer skin and a 100 mm thick blockwork inner skin. Calculate the number of bricks, blocks and mortar required for the cavity wall, 9 m long and 4.8 m high.

57. A 1:6 mortar mix is used for question 44. Determine the amount of cement and fine aggregate required.

58. A wall area 3.6 m × 2.4 m containing a window 1.2 m × 900 mm is to be tiled using 100 mm square tiles. Calculate the number of tiles required including an allowance of 5% for cutting and waste.

WELL, HOW DID YOU DO?
WORK THROUGH THE SECTION AGAIN IF YOU HAD ANY PROBLEMS

5 Scaffolding

Using scaffolding

Scaffolding is a temporary structure which is used in order to carry out certain building operations at height. It must, as applicable, provide a safe means of access to heights and a safe working platform.

Scaffolding should only be erected, altered and dismantled by a trained scaffolder. Craft operatives have to work on scaffolding 'safely'. It is therefore essential that you have an understanding of scaffolding principles, types, materials and statutory regulations.

Scaffolding and working platforms are covered by the Construction (Health, Safety and Welfare) Regulations which are recommended for further reading.

Tubular scaffolding

All materials and methods must conform to the regulations.

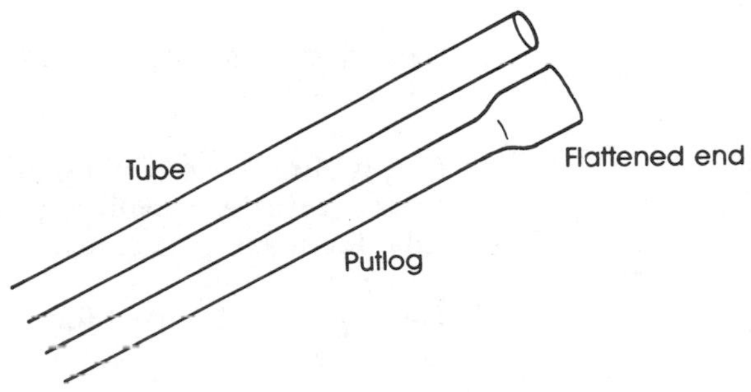

Scaffold tubes

These may be either tubular steel or tubular aluminium. Both types have an outside diameter of 48 mm. However they should not be used in the same scaffold, as aluminium tubes deflect more than the tubes made from steel under the same loading conditions.

Putlog – a type of short tube with flattened end, to bear in a brick joint; used for putlog scaffolds.

Scaffold fittings

These may be manufactured from either steel or aluminium. They are both normally suitable for use with both types of tube, unless the manufacturer or supplier states otherwise.

Double coupler – a one-piece coupler which connects two scaffold tubes together at right angles.

Universal coupler – connects two scaffold tubes together at right angles or parallel to each other.

Swivel coupler – a one-piece coupler which connects two scaffold tubes together at any angle.

Putlog coupler – a non-loadbearing one-piece coupler which connects putlogs to ledgers.

Joint pin – an internal fitting which expands and grips against the wall of the tube, used for joining two vertical scaffold tubes end to end.

Sleeve coupler – an external fitting used for joining two horizontal or bracing scaffold tubes end to end.

Base plate – a steel plate 150 mm square with an integral spigot, used for distributing loads from standards and has fixing holes for nailing to sole plates.

Adjustable base plate – a base plate for use on uneven ground. It incorporates a robust screw thread which enables adjustment for levelling.

Reveal pin – inserted into the end of a short piece of scaffold tube. When adjusted it forms a rigid fixing member in a window reveal or other opening.

Putlog end – attached over the end of a scaffold tube to convert it into a putlog.

Guard board clip – connects a guard board or toe board to a scaffold standard.

Gin wheel – used for raising and lowering equipment from a scaffold. Incorporates a swivel ring at the top which completely encircles the tube for maximum safety.

Castor wheel – used for mobile towers, must be fitted with a locking device or brake.

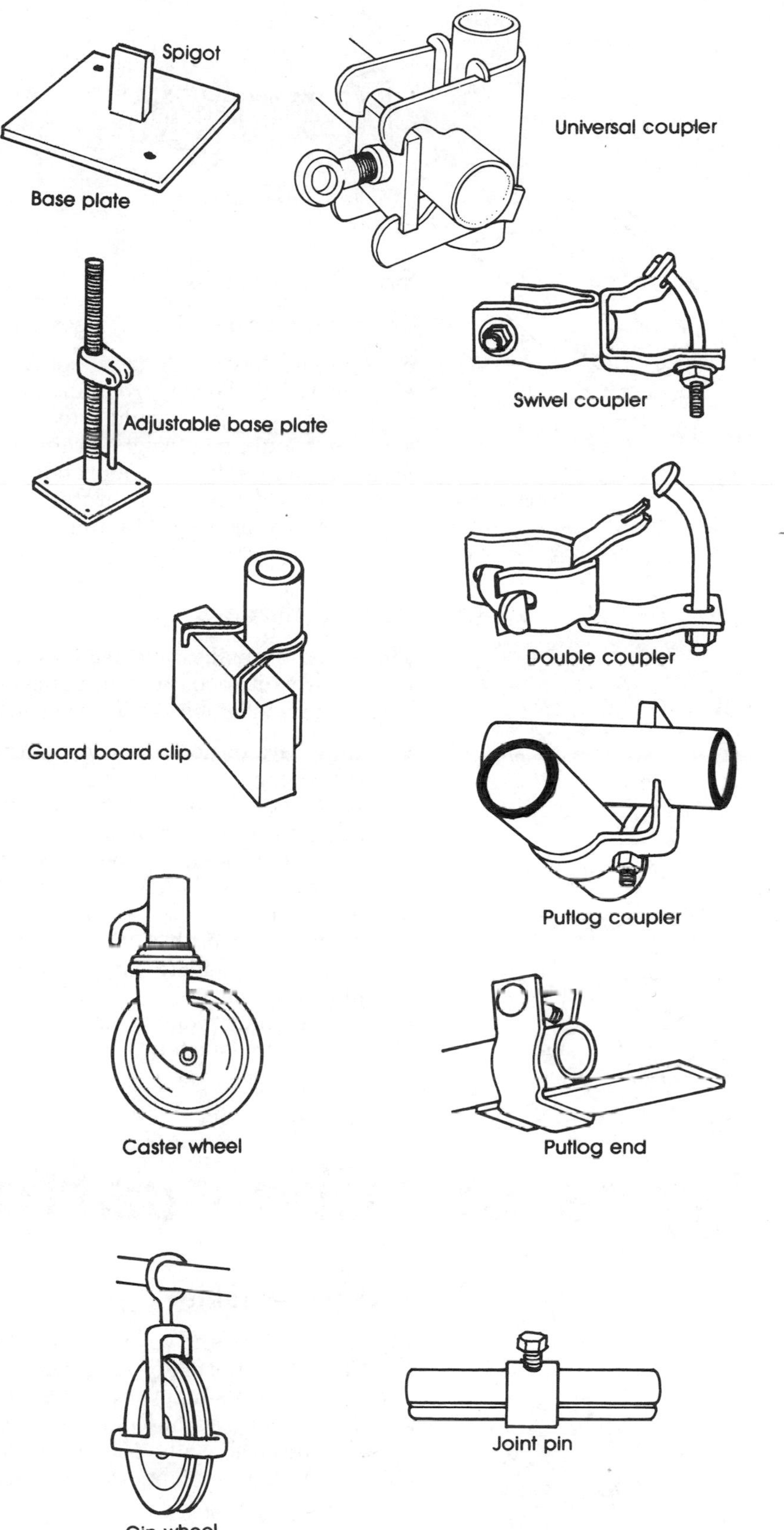
Spigot
Base plate
Universal coupler
Swivel coupler
Adjustable base plate
Double coupler
Guard board clip
Putlog coupler
Caster wheel
Putlog end
Joint pin
Gin wheel

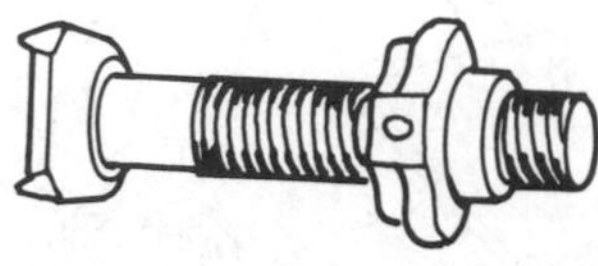
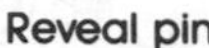

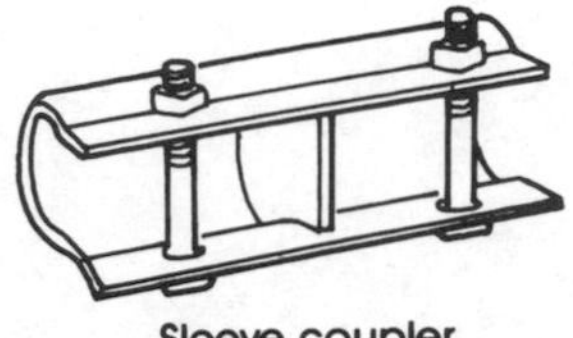

Scaffold boards

The main considerations for scaffold boards are that:

- they are best made from spruce, fir, redwood or whitewood
- they should be free from any defects, such as splits, checks, shakes or damage which could affect their strength
- they should be straight grained, not twisted or warped
- to prevent splitting their ends should be bound with galvanised or sheradised (corrosion protection) hoop iron, which extends along each edge at least 150 mm.

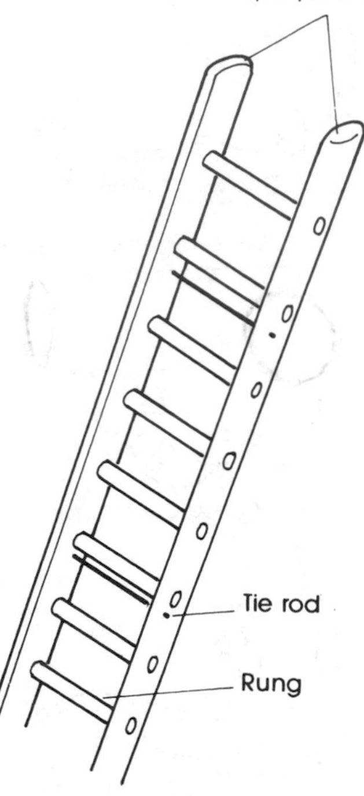

Access ladders

These are normally pole ladders. Stiles are made from one piece of European whitewood which has been cut down the middle. This is to ensure the ladder has an even strength and flexibility.

- Rungs are either round or rectangular and made from oak, birch or hickory.
- Rungs at the top and bottom of the ladder must be at least 100 mm from the ends of the stiles.
- Steel tie rods should be fitted at intervals of not more than nine rungs apart and under the second rung from each end of the ladder.
- Steel reinforcing wire may be incorporated in the edge of the stile for extra strength.
- Before use, ladders can be protected with a coat of clear, exterior quality varnish.
- **Never** paint ladders as paint may conceal potentially dangerous damage or defects.

Types of tube scaffolding

Putlog scaffolds

These are often known as either bricklayers' scaffold or single scaffold. They are normally used when constructing new brick buildings and consist of a single row of vertical standards which are connected together by horizontal ledgers. Putlogs are coupled to the ledgers and are built into the wall as the brickwork proceeds. This type of scaffold obtains most of its support and stability from the building.

Types of tube scaffolding

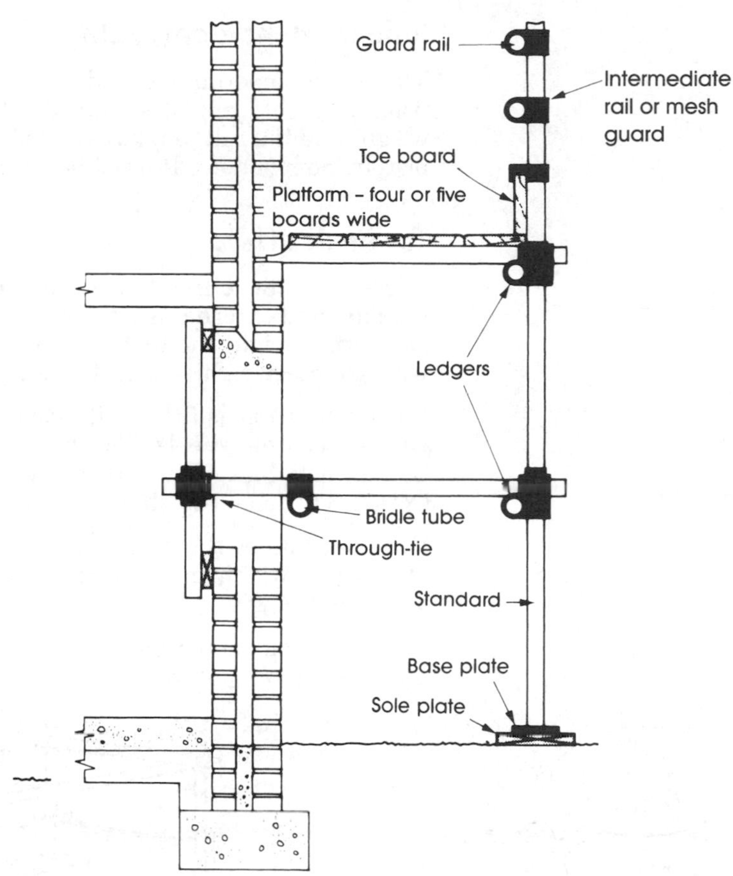

Putlog scaffold

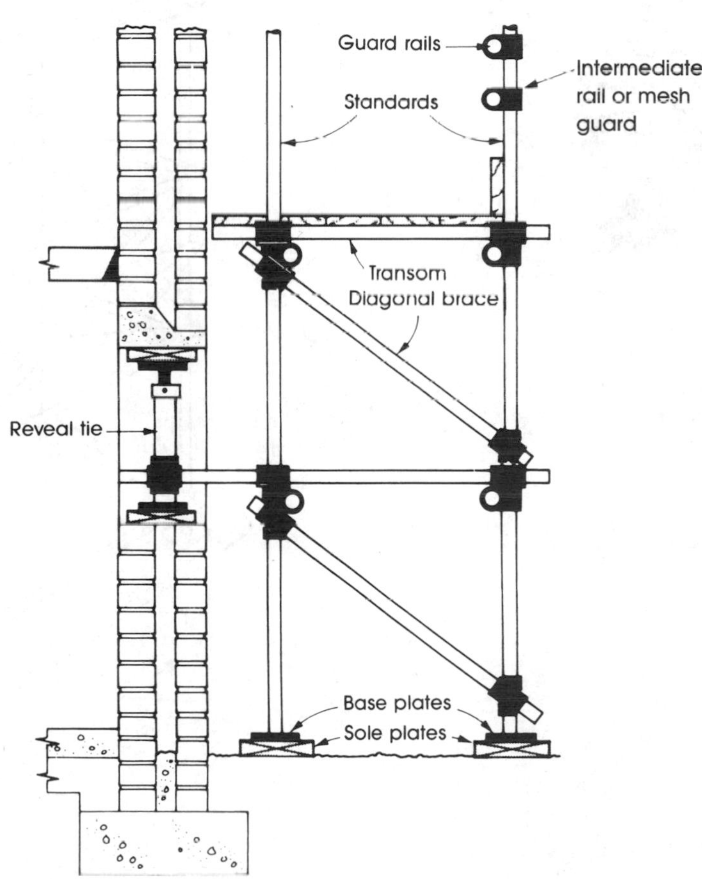

Independent scaffold

Independent scaffolds

These are sometimes called double scaffolds as they are constructed using a double row of standards. This type of scaffold carries its own weight and the full weight of all loads imposed upon it, but it is not completely independent. It must be suitably tied to the building for stability.

Tower scaffold

These may be either static or mobile. They are suitable for both internal and external use for work up to about 6 m in height. For work above this height the tower should be tied into the building or be fitted with counterweights to stabilise it.

When the tower is fitted with castors, these should incorporate brakes which lock the wheels. The height is limited according to the size of the tower's base. For internal use the maximum height should not exceed $3\frac{1}{2}$ times the shorter base dimension.

For external use the height is restricted to three times this dimension. Outriggers which increase the base dimension can be used to permit additional height.

Tower scaffolds must only be used on firm, level ground. Mobile towers must never be moved while people or equipment are on them.

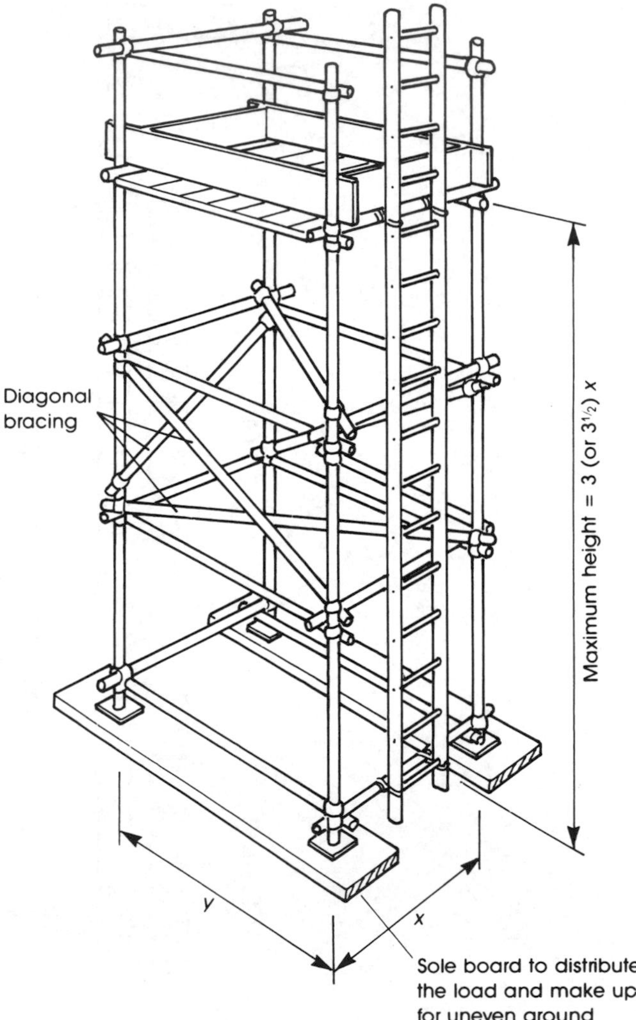

Static tower

Learning task

Identify the following scaffold fittings:

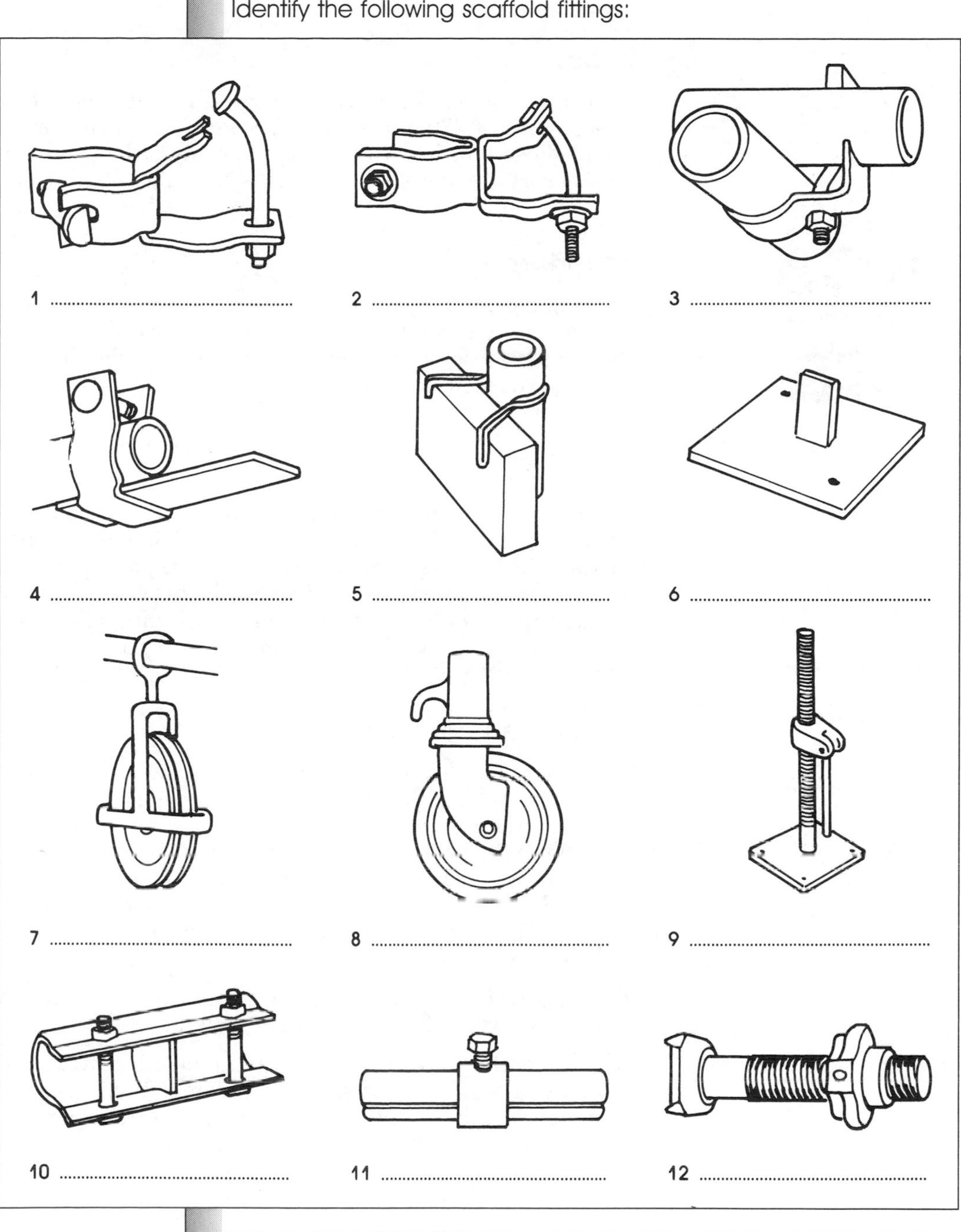

Scaffold components

Base

A good foundation for a scaffold is essential, it should be made using at least 35 mm × 220 mm sole plates laid on a firm, level base of well-rammed earth or hardcore. Base plates must be fixed to the sole plates under every standard. Bricks, blocks or timber offcuts must not be used.

Standards

These must be erected plumb or lean slightly towards the building, spaced close enough together to provide adequate support, joints staggered and positioned as near as possible to a ledger (joints should never occur in adjacent standards on the same lift).

Ledgers

These must be horizontally level and connected on the inside of the standards with right angle couplers. The spacing between ledgers varies with the height of the lift. A spacing of 1.2 m to 1.5 m is found to be the most convenient height for bricklayers to build a wall before moving up to the next lift (minimises bending and stretching). Joints in ledgers should be staggered and never occur in the same bay.

Putlogs and transoms

These are coupled to the ledgers at centres of about 1.2 m apart. Maximum distance between centres depends on the board thickness used: 1.5 m for 38 mm boards and 2.6 m for 50 mm boards. Each standard should have a putlog or transom as close as possible to it and in any case not more than 300 mm away from it. Double putlogs are required where the scaffold boards are butted together. These must be placed so that no board overhangs more than four times its thickness or less than 50 mm.

The flattened end of a putlog must be pushed right into the brick joint giving a bearing of approximately 75 mm.

Where a putlog is required opposite an opening in the building, a short tube called a bridle tube should be clamped with right angle couplers to the underside of the putlogs adjacent to the opening. The intermediate putlog can then be fixed to the ledger and bridle using right angle couplers.

Tying in to a building

REFER TO
P. 207 FOR DETAILS
OF THROUGH AND
REVEAL TIES

Putlog scaffold

It is most important that this is tied in correctly as putlogs can easily work loose in green (newly laid) brickwork. All ties should be of the through type, one tie for each 32 m^2 of scaffold area (sheeted scaffolds will require more ties due to increased wind resistance: one tie per 25 m^2). All couplers used for tying should be right angle couplers and the ties should be next to, or as close as possible to, the intersection of a standard and ledger.

Until the ties become effective, temporary rakers should be fixed to each alternate standard, to stabilise the scaffold.

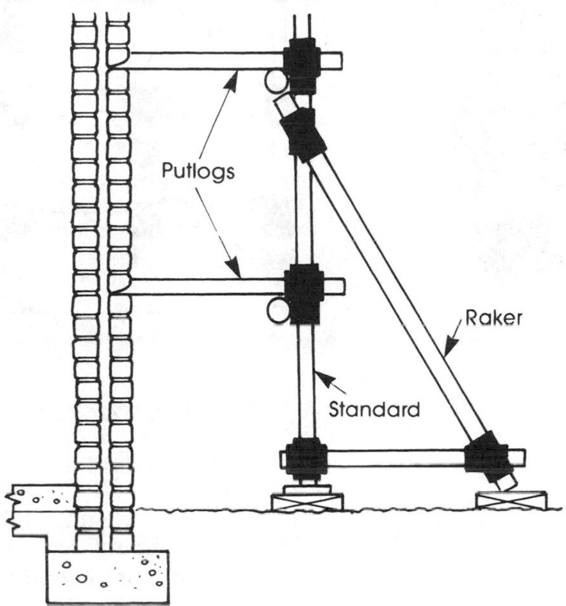
Putlogs
Raker
Standard

Independent scaffold

Tying is essential in order to prevent the scaffold moving away from or into the building. One tie is required for each 32 m^2 of scaffold area. Again, sheeted scaffolds will require more ties due to increased wind resistance. Ties can either be reveal ties, through ties, box ties or ties provided by casting or drilled in anchorages.

Again, until the ties become effective, temporary rakers should be fixed to each alternate standard, to stabilise the scaffold.

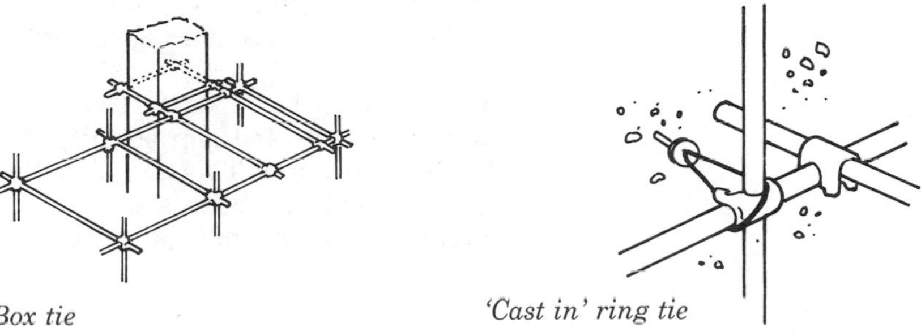

Box tie

'Cast in' ring tie

Bracing – can be one of two types:

Longitudinal or zig-zag bracing is fixed to the standards at an angle of approximately 45 degrees every 30 m or less. This is intended to prevent sideways movement of the scaffold.

Diagonal bracing is fixed diagonally to each alternate pair of standards at right angles to the building. They may be fixed either parallel to each other or in a zig-zag pattern. Whichever method is used, the bracing should continue the full height of the scaffold. This tends to hold the scaffold into the building.

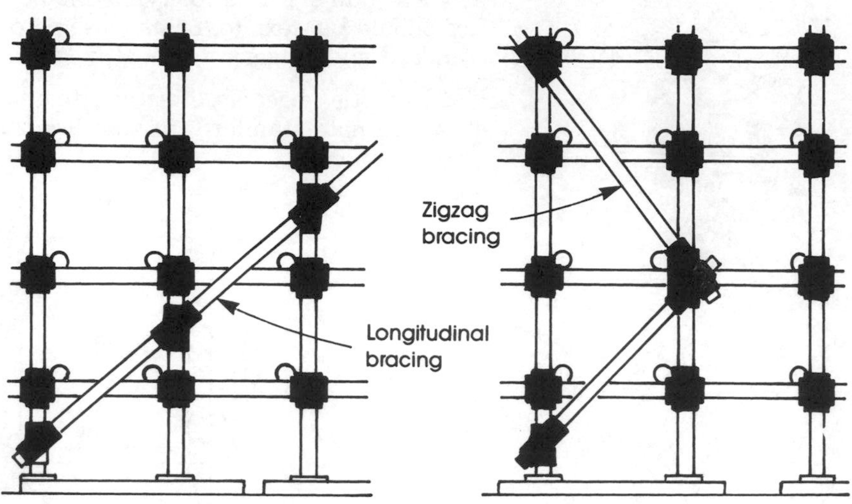

Working platform

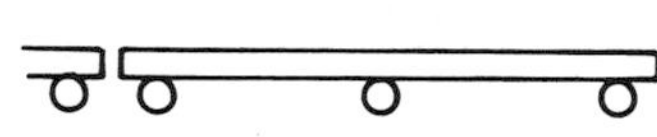

This should be four or five 38 mm × 22 mm scaffold boards in width. Each board should normally have at least three supports in order to prevent undue sagging. Where there is a danger of high winds the scaffold boards should be clipped down to putlogs or transoms. A clear passage of 440 mm (640 mm for barrows) must be maintained when loading out with materials.

Guard rails

These must be fitted to all working platforms where it is possible for a person or materials to fall 2 m or more. Fixed to the inside of the standards along the outside edge and the ends of the working platform, they are also required on the inside of the scaffold in the following circumstances:

- where the gap between the scaffold and the inside of an existing building exceeds 300 mm
- where the scaffold rises above a building
- where recesses occur in the building.

Guard rails must be at least 910 mm above the working platform. The maximum unprotected gap permitted between the guard rail and the toe board is 470 mm. This can be achieved by the use of intermediate rails, additional toe boards or mesh guards. Mesh guards are also recommended where materials are stacked on a scaffold.

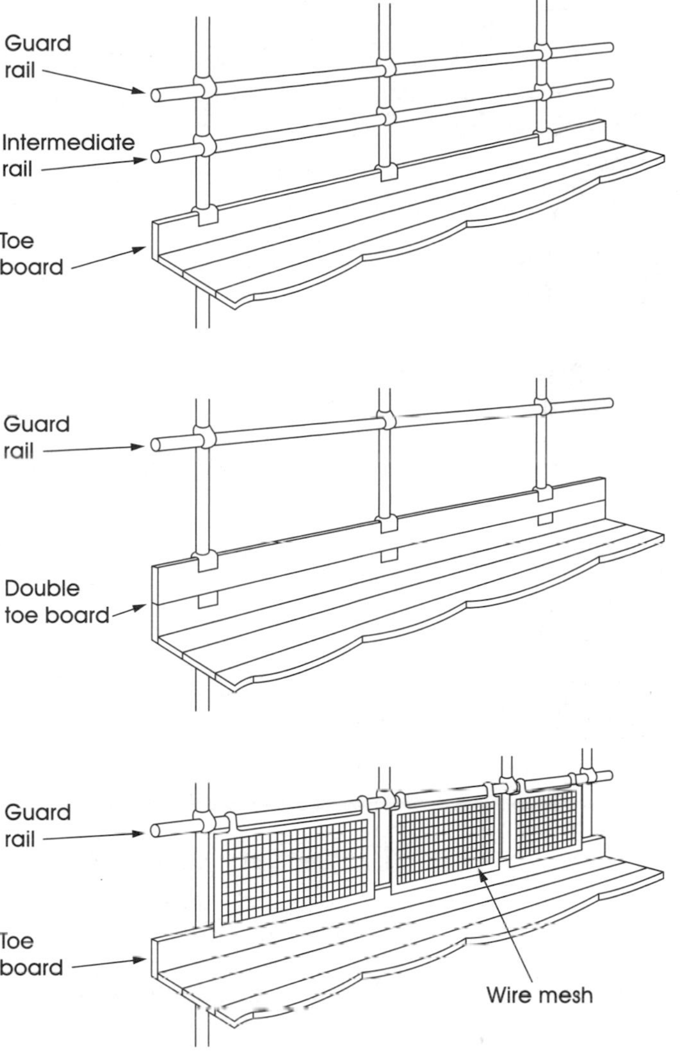

Toe boards – accompany the guard rail and must rise at least 150 mm above the working platform. They prevent items being kicked or falling off.

Ladder access

The ladder should be set at a working angle or 75 degrees. This is a slope of four units vertical to one unit horizontal. The stiles of the ladder should stand on a firm base and be securely held at the top and bottom to prevent sideways or outward movement. The top of the ladder must rise sufficiently above the landing point, unless other adequate handhold is provided. Where ladders are required to rise more than 9 m, a properly guarded intermediate landing stage must be provided.

Ladders should be boarded over to prevent access after working hours.

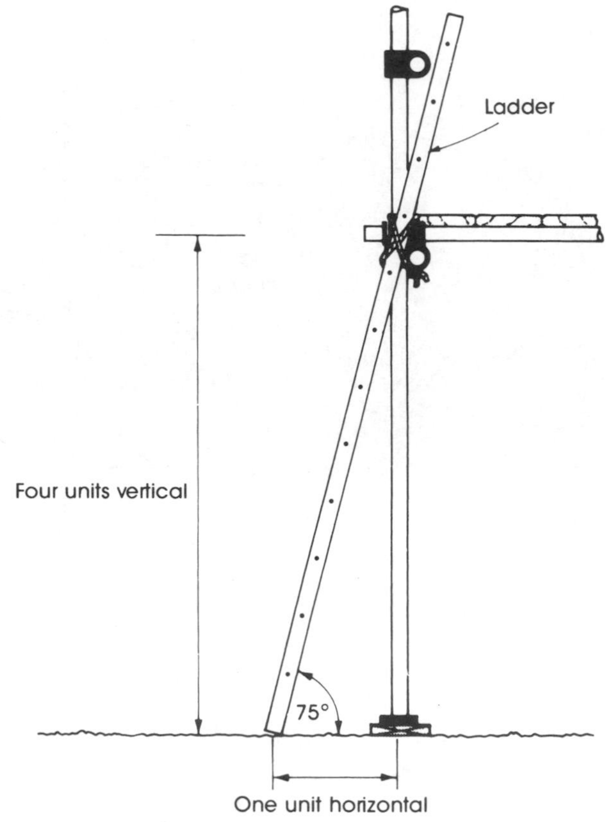

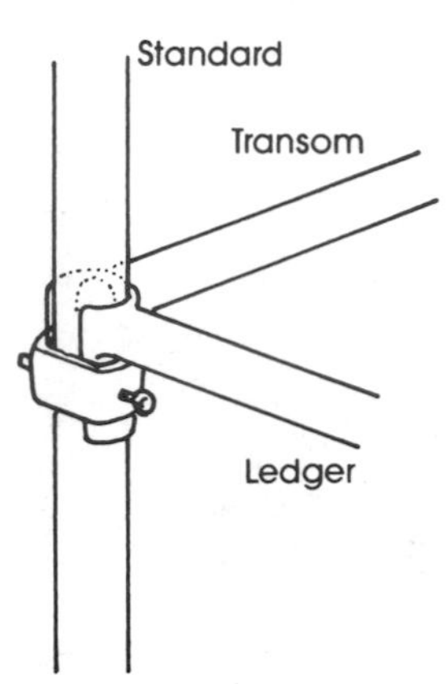

'Interlock' system

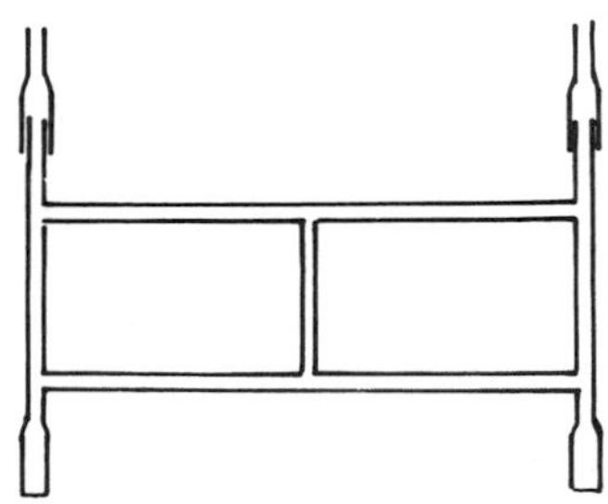

'H' frame system

Proprietary scaffolds

These are prefabricated and require very simple erection procedures on site. They are either an independent or tower type, consisting of a range of interlocking or no-bolt jointed components. These are made to individual manufacturers' 'systems' and parts cannot normally be interchanged. Reference should be made to manufacturers' publications for this type of scaffold.

Protection of the public

Nets, steel wire and brick guards should be used where scaffolds are erected adjacent to public access areas. Scaffolds on pavements should have night-time lighting and their tube ends and threads should be protected with a soft material to prevent a contact injury.

Inspection

Each scaffold must be inspected by an experienced, competent person. This is normally the responsibility of the site safety supervisor and the inspection should be carried out as follows:

- after erection, before the scaffold is used and at least once every seven days
- after cold weather, heavy rainfall or high winds.

Each inspection must be recorded in a scaffold register which is kept on site. The safety supervisor should be looking at all of the different points to ensure that the scaffold complies with the regulations.

Learning task

Carry out a safety inspection of a scaffold on your site or training establishment using this form:

Construction (Health, Safety and Welfare) Regulations 1996

INSPECTION REPORTS: NOTES

Place of work requiring inspection	Timing of frequency of inspection					
	Before being used for the first time.	After substantial addition, dismantling or alteration.	After any event likely to have affected its strength or stability.	At regular intervals not exceeding 7 days.	Before work at the start of every shift.	After accidental fall of rock, earth or any material.
Any working platform or part thereof or any personal suspension equipment.	✓	✓	✓	✓		
Excavations which are supported in pursuit of paragraphs (1), (2) or (3) of regulation 12.			✓		✓	✓
Cofferdams and caissons.			✓		✓	

NOTES

General

1. The inspection report should be completed before the end of the relevant period.
2. The person who prepares the report should, within 24 hours, provide either the report or a copy to the person on whose behalf the inspection was carried out.
3. The report should be kept on site until work is complete. It should then be retained for three months at the office of the person for whom the inspection was carried out.

Working platforms only

1. An inspection is only required where a person is liable to fall more than 2 metres from a place of work.
2. Any employer or any other person who controls the activities of persons using a scaffold shall ensure that it is stable and of sound construction and that the relevant safeguards are in place before his employees or persons under his control first use the scaffold.
3. No report is required following the inspection of any mobile tower scaffold which remains in the same place for less than 7 days.
4. Where an inspection of a working platform or part thereof or any personal suspension equipment is carried out.
   i. before it is taken into use for the first time; or
   ii. after any substantial addition, dismantling or other alterations;

   not more than one report is required for any 24 hour period.

Excavations only

1. The duties to inspect and prepare a report apply only to any excavation which needs to be supported to prevent any person being trapped or buried by an accidental collapse, or dislodgement of material from its sides, roof or area adjacent to it. Although an excavation must be inspected at the start of every shift, only one report of such inspections is required every 7 days. Reports must be completed for all inspections carried out during this period for other purposes, e.g. after accidental fall material.

Checklist of typical scaffolding faults

Footings	Standards	Ledgers	Bracing	Putlogs and transoms	Couplings	Bridles	Ties	Boarding	Guard-rails and toe-boards	Ladders
Soft and uneven	Not plumb	Not level	Some missing	Wrongly spaced	Wrong fitting	Wrong spacing	Some missing	Bad boards	Wrong height	Damaged
No base plates	Jointed at same height	Joints in same bay	Loose	Loose	Loose	Wrong couplings	Loose	Trap boards	Loose	Insufficient length
No safe plates	Wrong spacing	Loose	Wrong fittings	Wrongly supported	Damaged	No check couplers	Not enough	Incomplete	Some missing	Not bad
Undermined	Damaged	Damaged	–	–	No check couplers	–	–	Insufficient supports	–	–

READ THE INSTRUCTIONS AND COMPLETE THE TASK

Other working platforms

Hop-up working platforms

These are purpose-made by craft operatives, normally from softwood boarding or plywood. They should be at least 400 mm wide, about 500 mm high and have two steps. The top should be at least 500 mm square when used as a working platform.

Single hop-ups provide an isolated working platform, whereas two hop-ups spanned by boards enable a much greater area to be covered.

Hop-ups are intended to provide a working platform for reasonably low surfaces, up to about 2.4 m, without too much stretching.

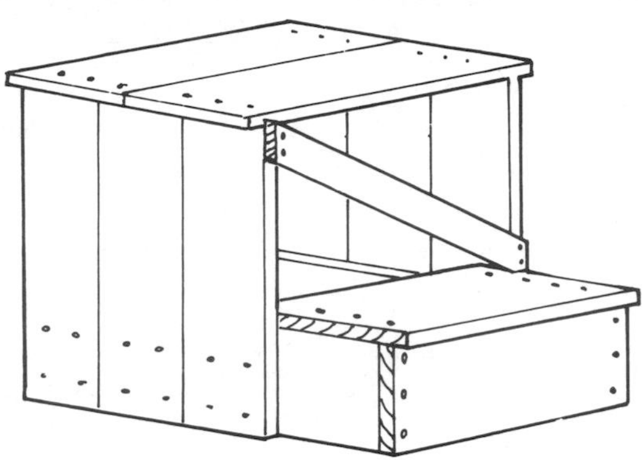

Split-head type working platforms

These utilise tripod-type adjustable metal split heads in conjunction with timber joists and scaffold boards. A series of holes and pin or screw jacks adjusts the working platform's height. These must be cleaned after use to ensure ease of working.

Four split heads are required to support the smallest working platform, eight or more split heads can be combined to board out larger areas.

These structures are intended to provide a working platform for work on ceilings.

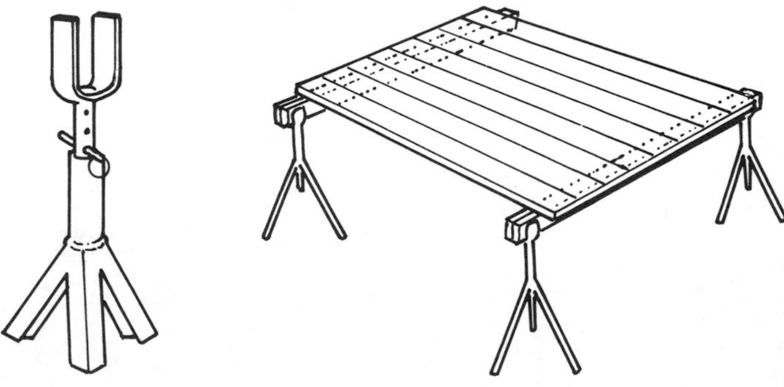

Staging boards

These are proprietary items used in conjunction with hop-ups, split heads, steps or trestles, to enable longer spans between supports. They

are manufactured in timber with metal tie rods and reinforcing wires and are available in a variety of lengths from 1.8 m to about 7 m and at a standard width of 450 mm and are suitable for supporting up to three persons. Consult the manufacturers' information for specific details.

Staging boards must not be used if they have any broken, damaged, repaired or missing parts. In addition they should not be painted, as this may hide defects. On finding defects the item must be taken out of use immediately, labelled as defective with 'DO NOT USE', and reported as soon as possible to your chargehand/foreman.

In use, staging boards must overhang their support by at least 50 mm but not by more than four times their thickness.

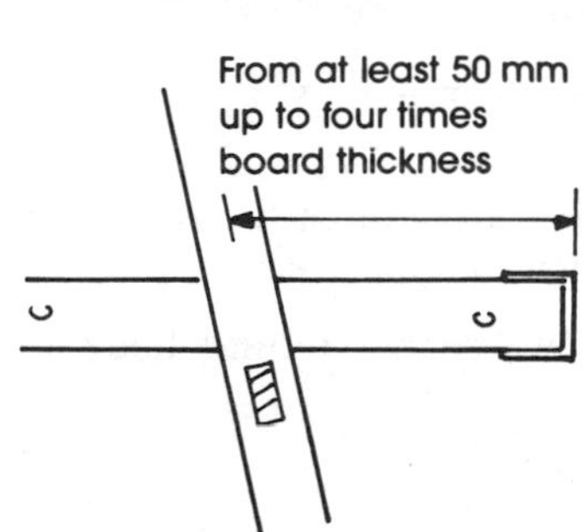

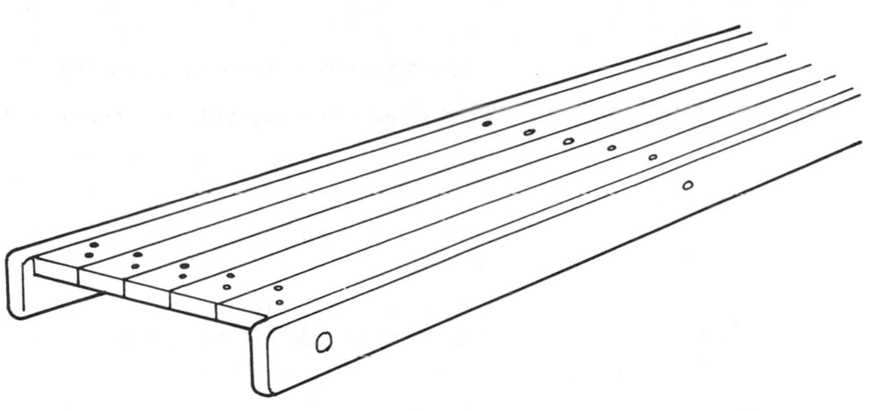

Ladders

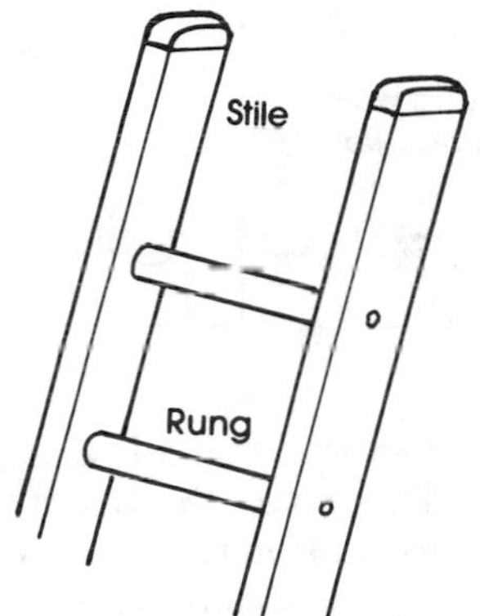

These are used to enable access to higher level working platforms or to provide a very short-term position for light work (painting) at heights. They may be either timber or aluminium and they consist of two long stiles which support rungs at about 250 mm centres.

Single-section ladders, termed standing ladders, can be obtained up to around 7 m in length.

Multi-section ladders, termed extension ladders (double or triple according to the number of sections), are fitted with latching hooks to the bottom of the extension section(s) and guide brackets to the top of the lower section(s). The latching hooks locate over the rung of the section below when extended and the guide brackets keep the sections together. Lengths of extension ladders vary from about 3 to 7 m when closed, extending up to approximately 19 m.

The extending section may be rope-operated to facilitate erection.

Aluminium ladders are often fitted with ladder feet. These are patent non-slip devices such as serrated rubber blocks or suction pads.

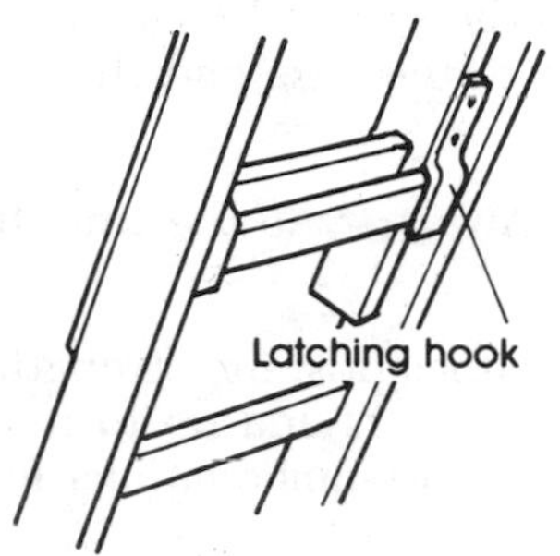

Aluminium ladders are often preferred to timber ladders. They are lighter, stronger, rot-proof and in addition will not warp, twist or burn, but do not use them near overhead electric cables.

Timber pole ladders are used mainly for scaffold access.

Timber standing ladders are made from Douglas fir, redwood, whitewood or hemlock. Rungs are round or rectangular. Steel tie rods should be fitted at intervals of not more than nine rungs apart and also under the second rung from each end of the ladder. Stiles may be reinforced by wires housed and clipped to their undersides.

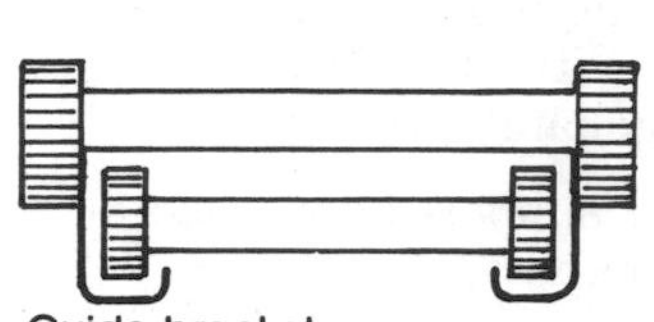

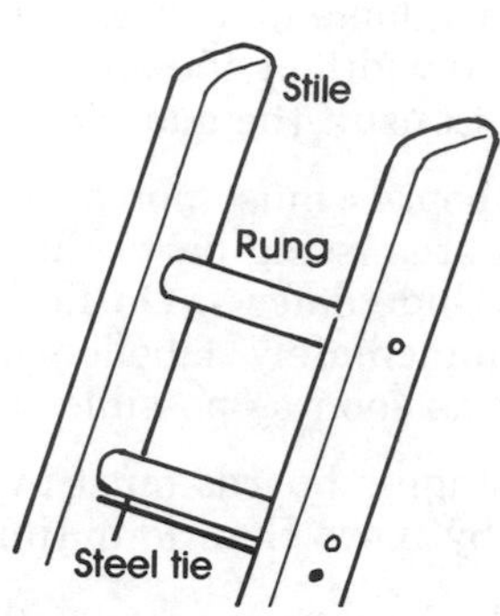

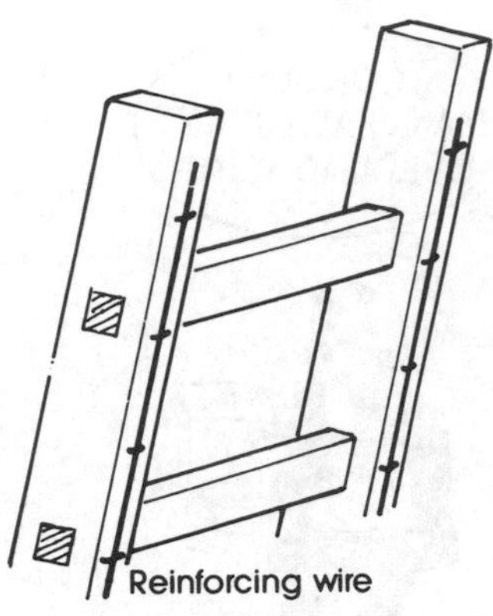

Moving and erecting ladders

This should be done with the sections closed.

Small ladders may, over short distances, be carried by one person almost vertically over the shoulder.

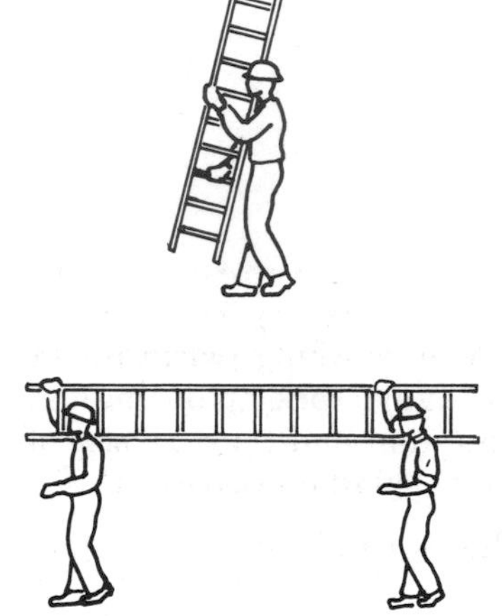

For greater distances and longer ladders, two people must be used, one towards each end with the ladder at shoulder height.

Extensions are raised one at a time ensuring each latching hook is located in position before progressing.

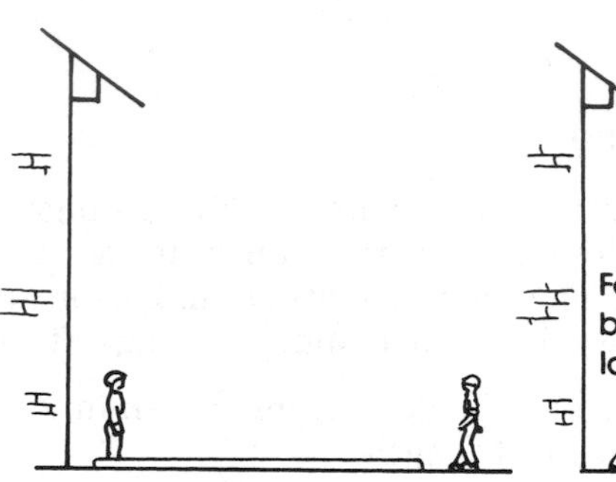

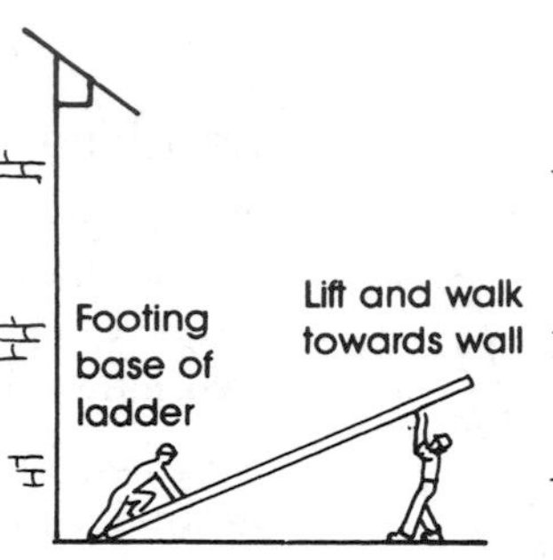

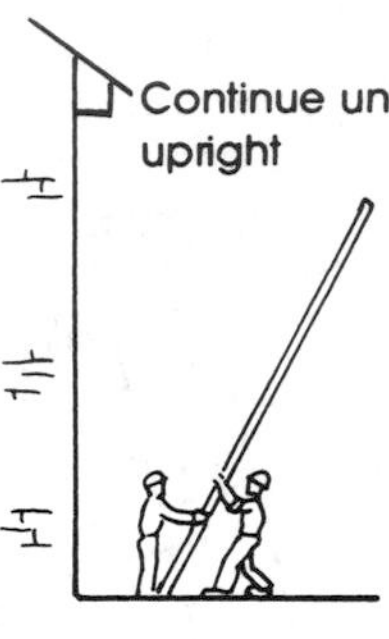

When erected the working angle should be 75 degrees, which is a slope of four vertical units to one horizontal unit. This gives the person a comfortable angle at which to ascend, descend and work, vertically, with arms extended. In addition it minimises the possibility of the ladder slipping outwards from the base.

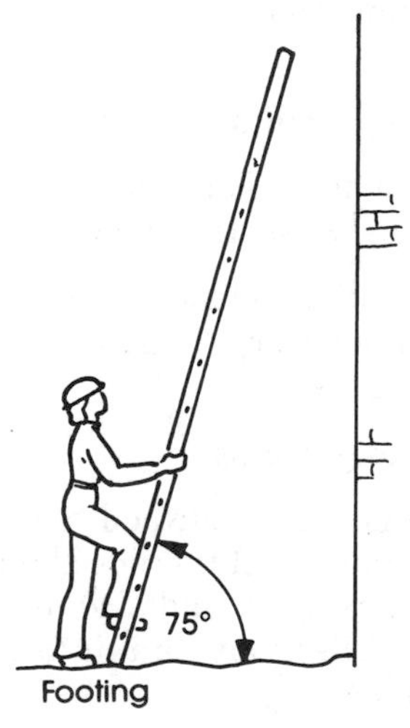

Ladders must be of sufficient length for the work in hand, sited on a firm, level base. They must be securely fixed by tying at the top or where this is not possible either by the stake and guy rope method or by having someone foot the ladder. (The footer must pay attention to the work in hand at all times and not 'watch the scenery'.) This is to prevent the ladder from slipping outwards from the base and the top sliding sideways.

Never over-reach when working on a ladder. Always take the time to stop and move it.

Sections of extension ladders must overlap sufficiently for strength. This is normally at least two rungs for short ladders and up to four rungs for longer ones. Consult manufacturers' recommendations for specific details.

Ladders should be lowered at the end of the working day and locked away. Where this is not practical, the lower rungs should be securely covered with a tied scaffold board to prevent public access. Always store flat to prevent twisting.

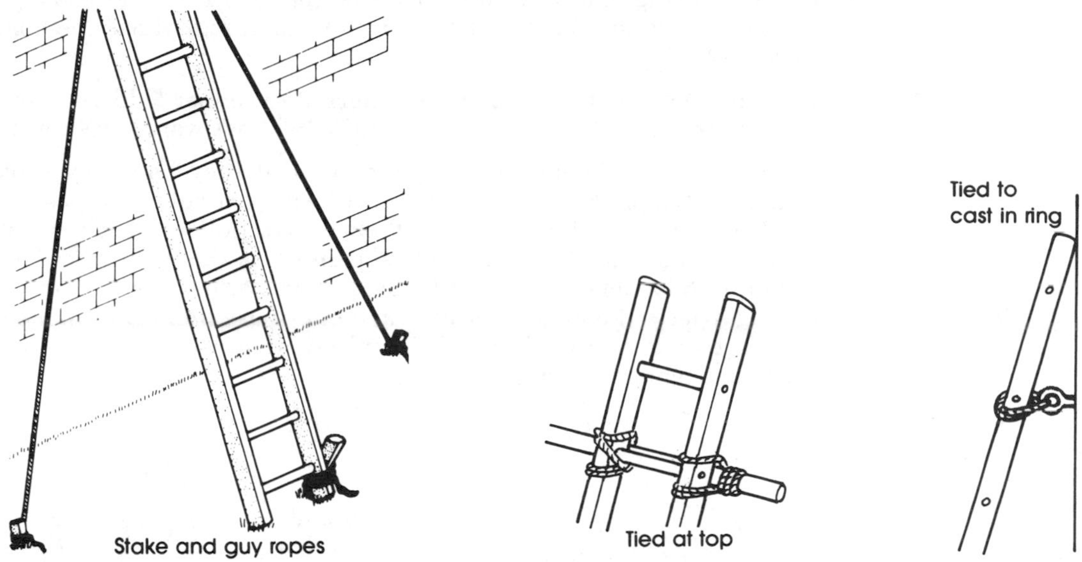

Stepladders

These are used mainly internally on firm flat surfaces and must be fully opened. They provide a working platform where the work area is just out of reach. Like ladders, they may be timber or aluminium, consisting of stiles supporting flat treads at about 250 mm centres. A back frame hinged at the top and secured towards the bottom with a cord or locking bar ensures the correct working angle and checks the opening to prevent collapse.

Stepladders are available in a range of tread lengths, commonly from five to fourteen.

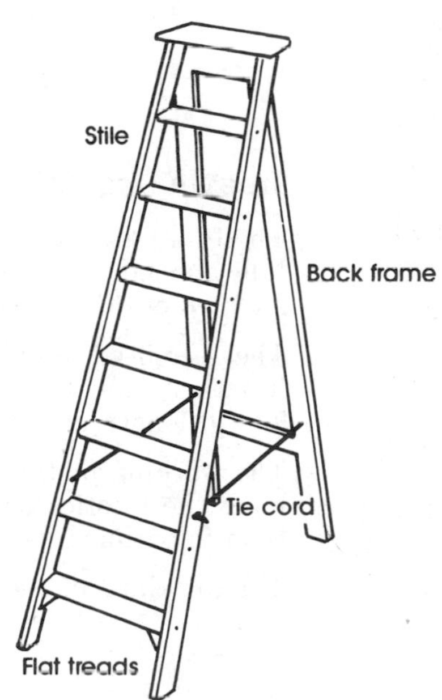

They must be used at right angles to the workface whenever possible (this reduces risk of sideways overturning). The user's knees should be below the top step.

They may be used in pairs in conjunction with boarding to provide a longer working platform, in which case the board must overhang the treads by at least 50 mm but by not more than four times the board's thickness.

Aluminium stepladders are often preferred to timber ladders; they are lighter, stronger, rot-proof and in addition will not warp, twist or burn.

Ladders and stepladders must not be used if they have any broken, damaged, repaired or missing parts. Check particularly ropes, guide brackets, latching hooks, locking bars and pulley wheels. Timber ladders and stepladders must not be painted as this may hide defects. Aluminium ladders should not be used near overhead power lines. On finding defects the item must be taken out of use immediately, labelled as defective with 'DO NOT USE', and reported as soon as possible to your changehand/foreman.

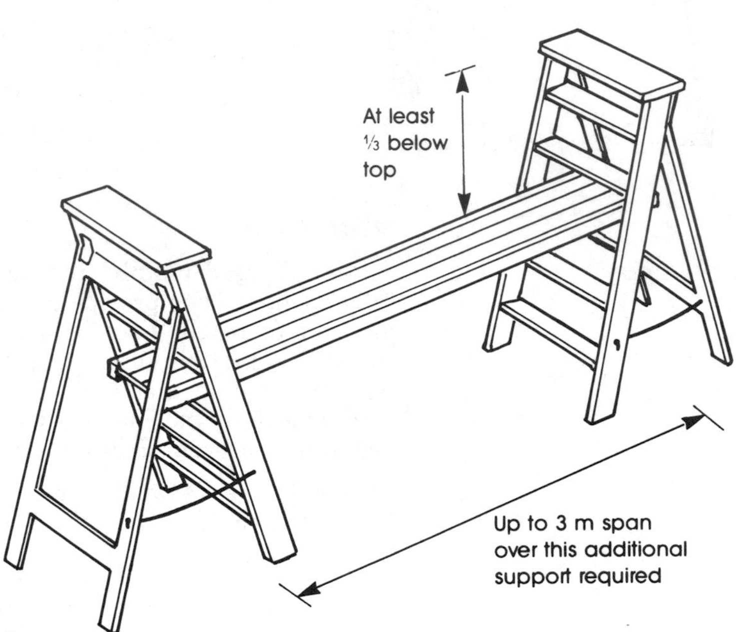

Trestle scaffolds

These are used mainly internally on firm, flat surfaces and must be fully opened. They provide a working platform for light work (painting etc.), where the working platform is up to 4.5 m in height.

They are used in pairs in conjunction with boarding and are available in either timber or aluminium and in a variety of sizes up to about 7 m in length and a standard width of 450 mm.

The working platform must be at least 430 mm wide and must overhang the trestles at either end by at least 50 mm, but by not more than four times its thickness.

The top third of each trestle must be above the working platform.

Working platforms over 2 m high will require a separate pair of steps or ladder for access and those over 3.6 m should be securely tied to the building or other suitable anchorage for stability.

It is recommended that work is carried out in the sitting position when the platform is above about 1.8 m.

Folding trestle platforms do not require guardrails or toe boards but fixed trestle platforms do, where a person may fall more than 2 m.

Trestles must not be used if they have any broken, damaged, repaired or missing parts. Timber trestles must not be painted as this may hide defects. Aluminium trestles should not be used near overhead power lines. On finding defects the item must be taken out of use immediately, labelled as defective with 'DO NOT USE', and reported as soon as possible to your chargehand/foreman.

Working safety

All scaffolding and working platforms must be clean, in good order and checked by a competent person before use and periodically after erection.

For your own safety –

- Carry out a check yourself prior to use.
- Do not use if unsafe; consult chargehand/supervisor for approval.
- Report any defects to your chargehand/foreman.
- Do not remove any part from a scaffold or working platform. *You* may be responsible for its total collapse.
- Do not use a working platform in adverse weather conditions, high winds, heavy rain, snow or ice, etc.
- Do not block the working platform. Ensure a free passage for other people.
- Clear up your 'mess' as you go and also before leaving the working platform.
- Do not push your mess over the edge. Ensure it is properly lowered along with tools and materials.

Questions for you

1. State the reason for **not** painting timber scaffold equipment.

2. State the purpose of:
(a) a toe board (b) a guardrail

3. Name two parts of a ladder or stepladder.

4. Ladders should be fixed to the scaffold at a working angle of:
(a) 60 degrees
(b) 45 degrees
(c) 90 degrees
(d) 75 degrees

a	b	c	d

5. Why does a putlog have a flattened end?

6. State the purpose of longitudinal bracing.

7. State the procedure to be adopted on finding a pair of stepladders with a defective tread.

8. State **TWO** methods which may be used to secure ladders.

9. Explain why materials stacked on a scaffold should be kept to a minimum.

TRY AND ANSWER THESE

COMPLETE THE WORD SQUARE

WORD-SQUARE SEARCH

Hidden in the word-square are the following 20 words associated with '*Scaffolding*'. You may find the words written forwards, backwards, up, down or diagonally.

Scaffold	Ledgers
Putlog	Bracing
Ladder	Transom
Trestle	Toe board
Hop up	Couplers
Split heads	Sole plates
Regulations	Tower
Independent	Platform
Board	Proprietary
Standards	Guard rail

Draw a ring around the words, or line in using a highlight pen thus:

EXAMPLE

EXAMPLE

P	L	A	T	F	O	R	M	S	S	D	L	O	F	F	A	C	S
R	U	S	P	L	I	T	H	E	A	D	S	R	R	E	W	O	T
O	R	T	S	L	L	M	F	D	N	D	H	O	P	U	P	P	A
P	I	C	L	A	D	D	E	R	L	S	P	M	B	M	E	O	N
R	A	D	D	O	Y	A	N	E	T	E	C	O	O	U	H	R	D
I	U	H	L	N	G	O	T	G	G	T	C	S	L	N	E	T	A
E	A	N	O	E	P	A	K	U	O	A	D	N	A	I	L	T	R
T	L	W	V	P	E	S	S	L	G	L	S	A	H	C	M	N	D
A	I	T	O	S	U	S	Y	A	I	P	O	R	T	A	E	E	S
R	A	C	O	C	C	P	R	T	R	E	S	T	L	E	T	D	N
Y	R	O	R	I	H	O	G	I	A	L	H	O	C	I	S	N	I
U	D	N	E	L	E	G	T	O	S	O	A	E	A	O	T	E	W
R	R	S	P	V	N	D	A	N	C	S	D	B	N	N	P	P	A
O	A	T	V	I	U	S	T	S	M	E	R	O	I	C	C	E	R
L	U	R	C	O	U	P	L	E	R	S	I	A	T	I	O	D	D
K	G	A	T	I	E	L	E	T	T	E	R	R	O	D	C	N	V
F	R	C	T	A	S	B	M	L	A	D	E	D	E	R	S	I	E
B	S	P	B	O	A	R	D	C	A	T	S	R	E	G	D	E	L

WELL, HOW MANY DID YOU GET?

6 Materials

On-site provision for storage of materials

Site storage provision

A building site can be seen as a temporary factory, a workshop and materials store from which a contractor will construct a building. An important consideration when planning the layout of this temporary factory is the storage of materials.

Their positioning should be planned in relation to where on site they are to be used, whilst at the same time providing protection and security.

Large valuable items

Frames, pipes and drainage fittings, etc. should be stored in a lockable, fully fenced compound.

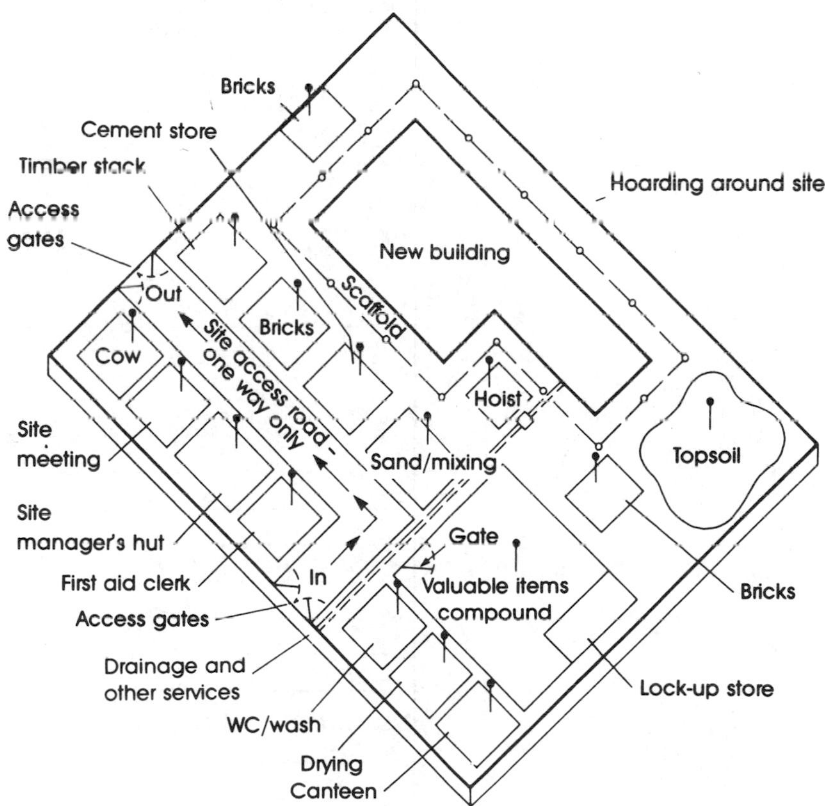

Smaller valuable items

Carpenters' ironmongery, fixings; plumbers' copper pipe, fittings, appliances; electricians' wire, fittings; and paint, should be kept secure in one or more lockable site huts, depending on the size of the site. Like items should be stored adjacent to each other on shelving or in a bin system, as appropriate.

Each shelf of bin should be clearly marked with its contents, and each item entered on a tally card.

If materials are returned to stock they should be put back in the correct place and the tally card amended.

Heavy items should be stored at low level.

New deliveries should be put at the back of existing stock. This ensures stock is used in rotation and does not deteriorate due to an exceeded shelf-life, making it useless.

Issues of stores should be undertaken by a storeperson or supervisor against an authorised requisition. Each issue should be recorded on the tally card. On some sites the tally card system of recording the issue of materials may have been superseded by the use of a computerised system.

BBS SUPPLIERS TALLY CARD

Description of materials: WHITE GLOSS Ref No: BSB 14/2

Size or No: 5 LITRE

Date	Order No.	Amount Inwards	Amount Outwards	Signature	Balance
15·3·02	B14	50		PB	50
18·3·02			10	PB	40
27·3·02			12	PB	28
19·4·02	B75	30		PB	58
20·4·02			16	PB	42

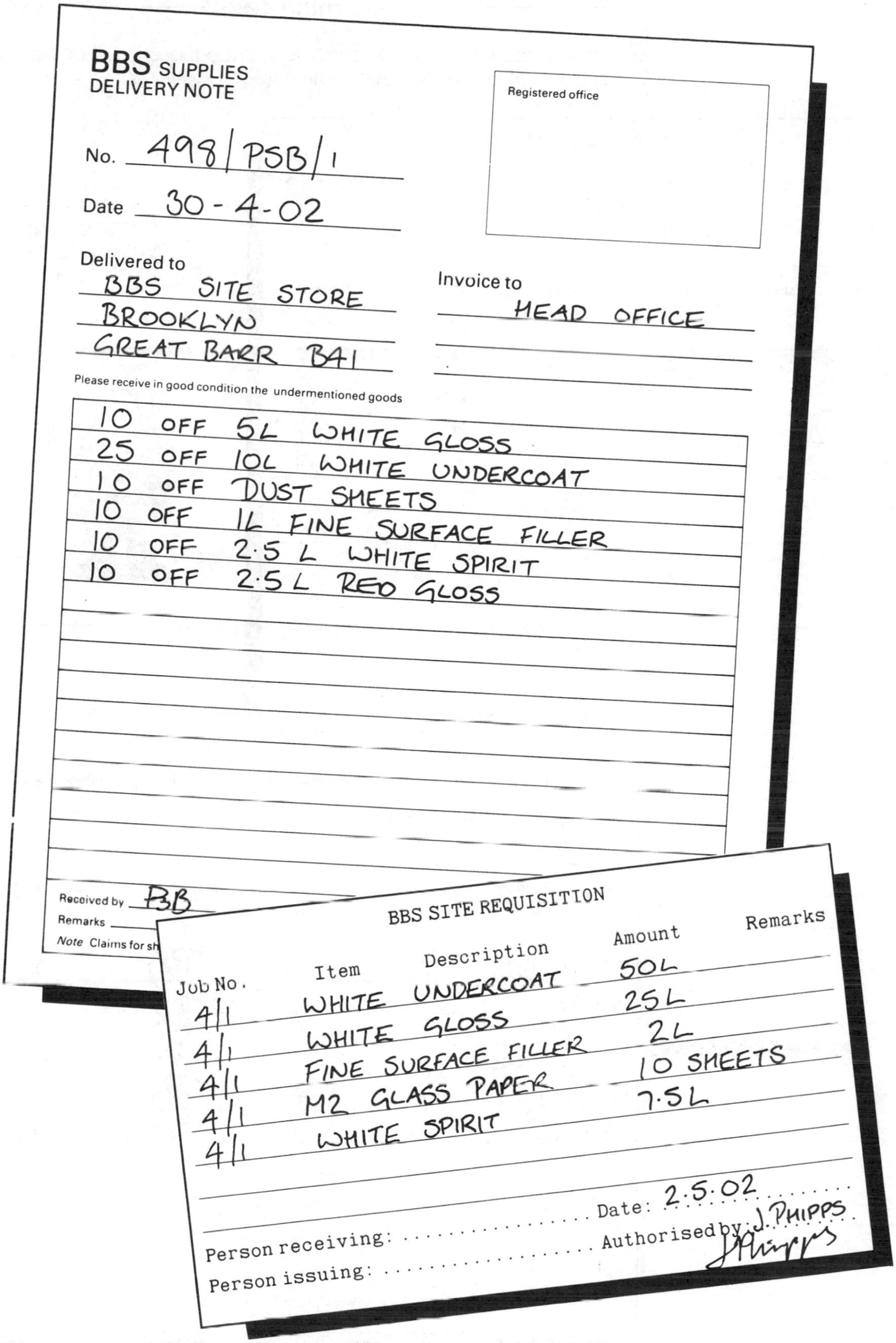

BBS SUPPLIES
DELIVERY NOTE

Registered office

No. 498/PSB/1

Date 30-4-02

Delivered to
BBS SITE STORE
BROOKLYN
GREAT BARR B41

Invoice to
HEAD OFFICE

Please receive in good condition the undermentioned goods

10 OFF 5L WHITE GLOSS
25 OFF 10L WHITE UNDERCOAT
10 OFF DUST SHEETS
10 OFF 1L FINE SURFACE FILLER
10 OFF 2.5 L WHITE SPIRIT
10 OFF 2.5 L RED GLOSS

Received by BB
Remarks
Note Claims for sh

BBS SITE REQUISITION

Job No.	Item	Description	Amount	Remarks
4/1	WHITE	UNDERCOAT	50L	
4/1	WHITE	GLOSS	25L	
4/1	FINE	SURFACE FILLER	2L	
4/1	M2	GLASS PAPER	10 SHEETS	
4/1	WHITE	SPIRIT	7.5L	

Date: 2.5.02
Person receiving: ....
Authorised by: J. PHIPPS
Person issuing: ....

Learning task

Update the following tally cards to include these latest deliveries and the site requisition information. (See page 227.)

BBS SUPPLIERS TALLY CARD

Description of materials: WHITE UNDERCOAT   Ref No: BSB 15/3

Size or No: 10 LITRE

Date	Order No.	Amount Inwards	Amount Outwards	Signature	Balance
8.3.02	B11	65		BB	65
15.3.02			10	BB	55
16.3.02			1	BB	54
18.3.02			15	BB	39
27.4.02			12	BB	27

READ THE INSTRUCTIONS AND COMPLETE THE TASK

BBS SUPPLIERS TALLY CARD

Description of materials: WHITE SPIRIT   Ref No: BSB 15/50

Size or No: 2.5 LITRE

Date	Order No.	Amount Inwards	Amount Outwards	Signature	Balance
8.3.02	B12	10		BB	10
18.3.02			2	BB	8
27.3.02			3	BB	5
30.3.02	B54	10		BB	15
20.4.02			4	BB	11
28.4.02			3	BB	8

Storage requirements

Different materials have different storage requirements. Points to bear in mind are:

Delivery dates – phased deliveries of material should be considered in line with the planned construction programme. This will prevent unnecessarily long periods of site storage and unnecessarily large storage areas being needed.

Physical size, weight and delivery method – will determine what plant (crane or fork-lift truck, etc.) and labour is required for off-loading and stacking.

Protection – many materials are destroyed by extremes of temperature, absorption of moisture, or exposure to sunlight, etc.

Stores should be maintained as far as possible at an even temperature of about 15°C.

High temperatures cause adhesives, paints, varnishes, putties and mastics, etc. to dry out and harden.

Flammable liquids such as white spirit, thinner, paraffin, petrol, some paints and varnishes, some timber preservatives and some formwork release agents must be stored in a cool, dry, lockable place. Fumes from such liquids present a fire hazard and can have an overpowering effect if inhaled. Stores of this type should always have two or more fire exists and be equipped with suitable fire extinguishers in case of fire. (See 'Health and Safety'.)

Water-based materials, such as emulsion paints and formwork release agents, may be ruined if allowed to freeze.

Non-durable materials such as timber, cement and plaster require weather protection to prevent moisture absorption.

Boxed or canned dry materials, such as powder adhesives, wallpaper paste, fillers, detergent powders and sugar soap quickly become useless if exposed to any form of dampness. Dampness will also rust metal containers which may result in leakage and contamination of the contents.

Where materials are stored in a building under construction, ensure the building:

- has dried out after so-called 'wet trades', such as brickwork and plastering, are finished
- is fully glazed and preferably heated
- is well ventilated – this is essential to prevent the build-up of high humidity (warm moist air).

Transit – Non-durable materials should be delivered in closed or tarpaulin-covered lorries. This will protect them from both wet weather and moisture absorption from damp or humid atmospheres.

Handling – Careless or unnecessary repeated handling will result in increased costs through damaged materials and even personal injury.

Security – Many building materials are 'desirable' items – they will 'grow legs' and walk away if site security lapses.

Safety – Finally, take care of your personal hygiene. This is just as important as any physical protection measure. Certain building materials, e.g. cement, admixtures and release agents, can have an irritant effect on skin contact; they are poisonous if swallowed and can result in narcosis if their vapour or powder is inhaled. By taking proper precautions these harmful effects can be avoided. Follow manufacturers' instructions; avoid inhaling spray mists, fumes and powders; wear disposable gloves or a barrier cream; thoroughly wash hands before eating, drinking, smoking, and after work. In case of accidental inhalation, swallowing or contact with skin, eyes, etc. medical advice should be sought immediately.

Caution, risk of fire

Caution, toxic hazard

Caution, corrosive substance

Take note of material labelling and manufacturers' instructions

TRY AND ANSWER THESE

Questions for you

1. State the purpose of storing materials on site.

2. Briefly describe **THREE** main points to be considered when determining on-site storage requirements.

Storage of bulk durable building materials

Bricks

These are walling unit components having a standard size, including a 10 mm mortar allowance, of 225 mm × 112.5 mm × 75 mm.

Bricks are normally made from either calcium silicate or clay. Clay bricks are usually pressed, cut or moulded and then fired in a kiln at very high temperatures. Their density, strength, colour and surface texture will depend on the variety of clay used and the firing temperature. Calcium silicate bricks are pressed into shape and steamed at high temperature. Pigments may be added during the manufacturing process to achieve a range of colours.

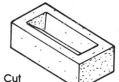

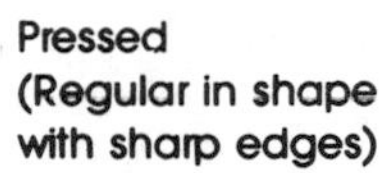

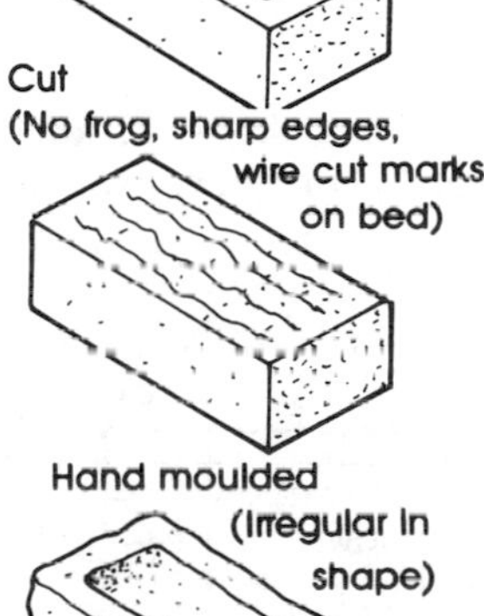

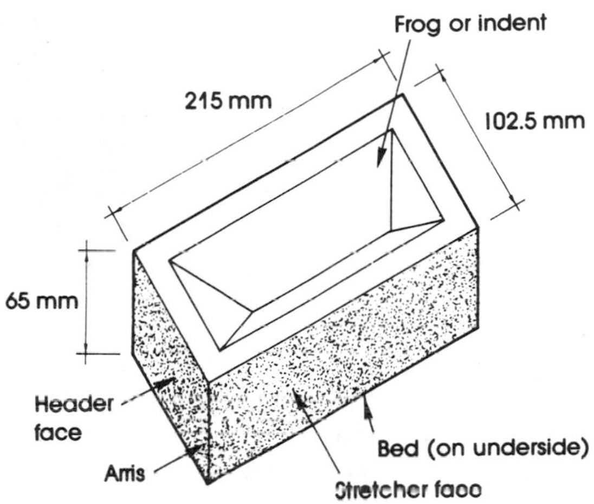

The three main types of bricks are as follows:

Common or Fletton bricks are basic bricks used in the main for internal or covered (rendered or cladded) external work, although sand-faced Flettons are available for use as cheap facing bricks.

Facing bricks are made from selected clays and are chosen for their attractive appearance rather than any other performance characteristic.

Engineering bricks have a very high density and strength and do not absorb moisture; they are used in both highly loaded and damp conditions such as inspection chambers, basements and other substructure work.

Bricks may be supplied loose or banded in unit loads, shrink-wrapped in plastic and sometimes on timber pallets.

Loose bricks should be off-loaded manually, never tipped; they should be stacked on edge in rows, on level, well-drained ground. Do not stack too high: up to a maximum of 1.8 m.

Careless handling, can chip the faces and arrises (corners), and also lead to fractures, making the bricks useless for both face and hidden work. Poor stacking creates an untidy workplace and unsafe conditions for those working or passing through the area.

Banded loads of up to 500 are off-loaded mechanically using either the lorry-mounted device, a fork-lift truck or a crane. Bricks stacked on polythene or timber pallets will be protected from the absorption of sulphates and other contaminants which could later mar the finished brickwork.

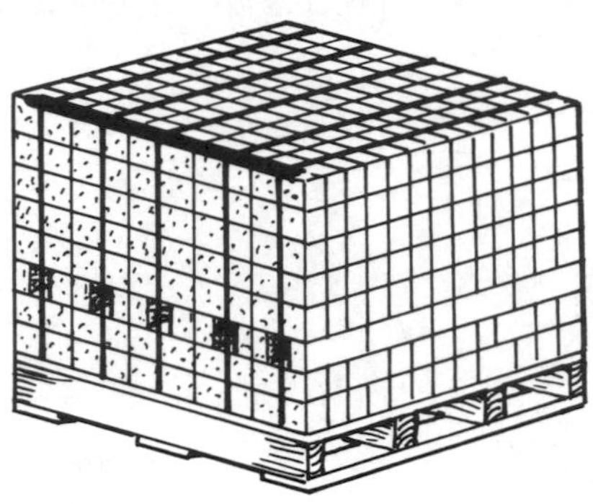

To protect bricks against rain, frost and atmospheric pollution, all stacks should be covered with a tarpaulin or polythene sheets weighted at the bottom.

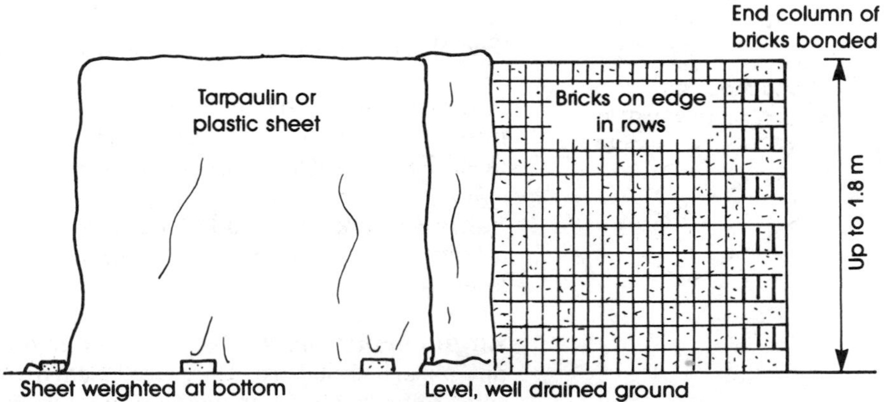

Blocks

These are walling units which are larger than bricks, normally made either from concrete or natural stone.

Concrete blocks can be either dense or lightweight; dense blocks are often made hollow to lighten them, lightweight blocks can use either a lightweight aggregate or a fine aggregate that is aerated to form air bubbles. Concrete blocks are most often used for internal partition walls of the inner leaf of cavity walls. When used externally, they are normally either rendered (covered with a thin layer of cement mortar) or covered in cladding (tiles, slates or timber), to provide a waterproof construction. The main advantage of blocks over bricks is their increased speed of laying and also the good thermal insulation qualities of the aerated variety.

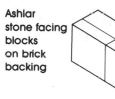

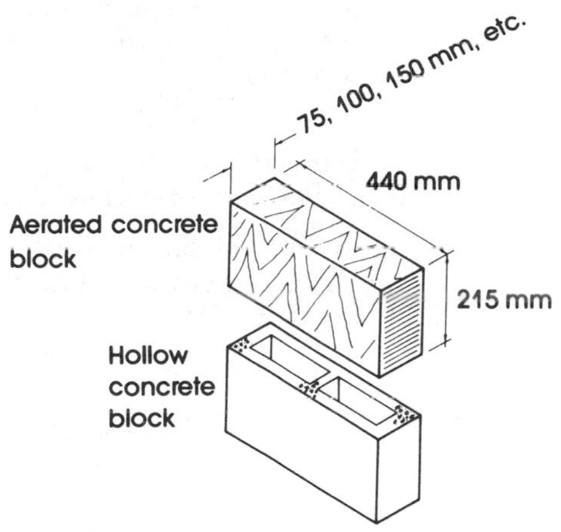

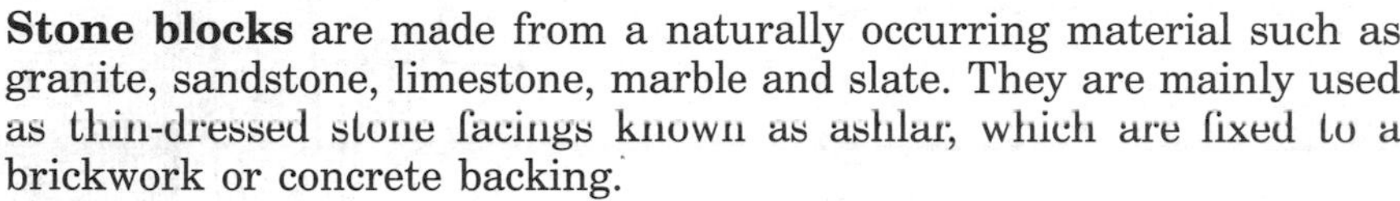

Stone blocks are made from a naturally occurring material such as granite, sandstone, limestone, marble and slate. They are mainly used as thin-dressed stone facings known as ashlar, which are fixed to a brickwork or concrete backing.

Blocks may be either supplied loose, banded in unit loads, shrink-wrapped in plastic packs and sometimes on timber pallets.

Loose blocks should be off-loaded manually, never tipped; they should be stacked on edge in rows or columns, on level, well drained ground. Do not stack too high: six to eight courses maximum.

Banded or palleted loads are off-loaded mechanically using either the lorry-mounted device, a fork-lift truck or a crane.

To protect blocks against rain, frost and atmospheric pollution, all stacks should be covered with a tarpaulin or polythene sheets weighted at the bottom.

Stone blocks are often stored in straw, or other similar soft packing to protect arrises (corners) from impact damage.

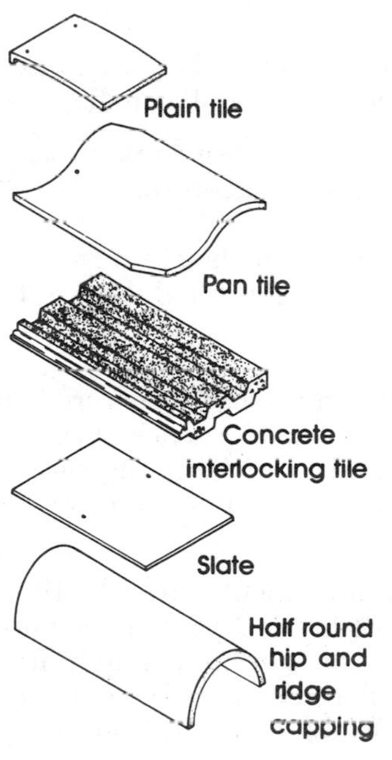

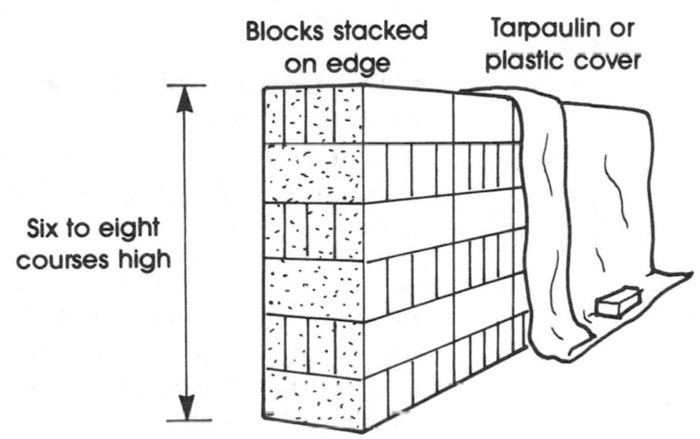

Roof tiles

These are normally either a terracotta clay product, a natural, metamorphic stone slate, or a cast concrete product.

Roof tiles may be either supplied loose, in banded packs, in shrink-wrapped plastic packs or in unit loads on timber pallets.

Loose tiles should be off-loaded manually, never tipped; they should be stacked on edge in rows, on level, well-drained ground. Do not stack too high: four to six rows maximum; and taper the stack towards the top. End tiles in each course should be laid flat to prevent toppling. Ridge and hip cappings should be stored on end.

Banded, packed or palleted loads are off-loaded mechanically using either the lorry-mounted device, a fork-lift truck or a crane.

To protect tiles against rain, frost and atmospheric pollution, all stacks should be covered with a tarpaulin or polythene sheets weighted at the bottom.

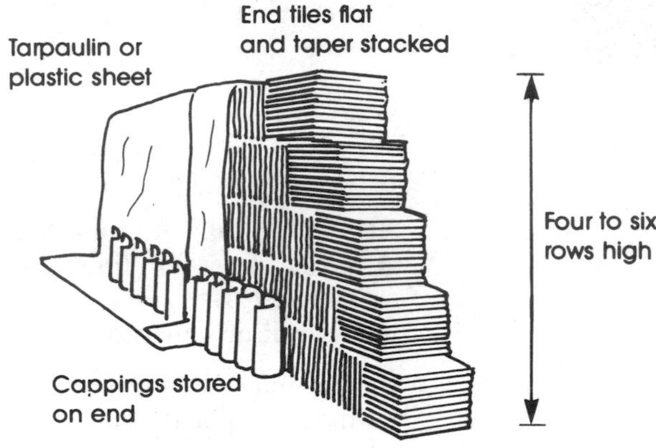

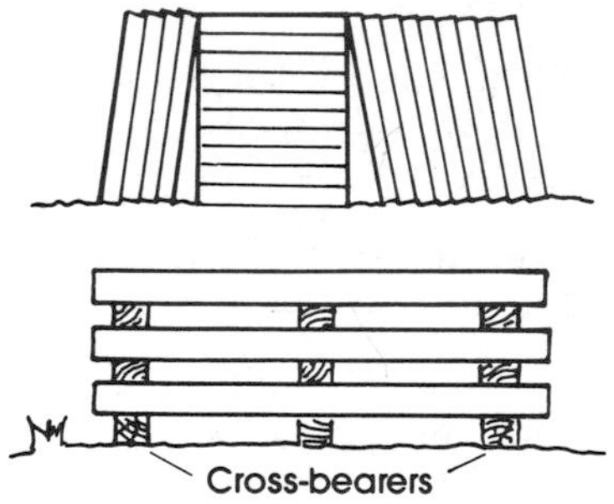

Concrete units (paving slabs, kerbs and lintels)

These are normally pre-cast in factory conditions and transported to site.

Concrete units may be either supplied loose singly, in banded packs, in shrink-wrapped plastic packs or in unit loads on timber pallets.

Loose concrete units should be off-loaded manually, never tipped. Items of equipment can be used to lift single units, e.g. kerb lifter.

Paving slabs should be stacked on edge in single height rows, on level, well-drained ground. Intermediate stacks of slabs laid flat can be introduced to prevent toppling.

Kerbs and lintels should be stacked flat on timber cross-bearers laid on level ground. Cross-bearers should be laid between each layer, to provide a level surface, to spread the load and to prevent the risk of distortion and chipping damage. Do not stack too high: four to six layers maximum.

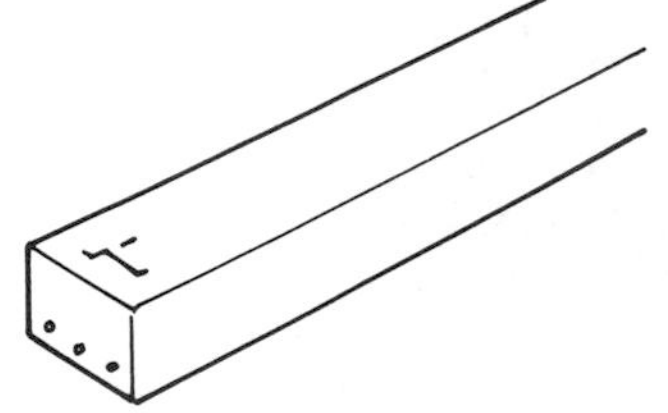

Lintels are designed to contain steel reinforcement towards their bottom edge, to resist tensile forces. It is most important that they are moved in the plane of intended use, otherwise they may simply fold in two. Where the steel bars or wires cannot be seen on the end, the top edge is often marked with a 'T' or 'TOP' for identification.

Banded, packed or palleted loads are off-loaded mechanically using either the lorry-mounted device, a fork-lift truck or a crane.

To protect concrete units against rain, frost and atmospheric pollution, they may be covered with a tarpaulin or polythene sheets weighted at the bottom.

Drainage pipes and fittings

These may be glazed or unglazed clay products, cast iron or UPVC (a rigid or unplasticised polyvinylchloride).

Drainage pipes and fittings may be either supplied loose singly, in banded packs, in shrink-wrapped plastic packs or in unit loads on timber pallets.

Loose pipes and fittings should be off-loaded manually, never tipped. Pipes should be stacked horizontally in rows and wedged or chocked to prevent rolling, on level, well-drained ground. Do not stack too high: 1.5 m maximum; and taper the stack towards the top. Spigot and socket pipes should be stacked on timber cross-bearers, alternate rows should be reversed to allow sockets to project beyond spigots. Gullies and other fittings should be stacked upside down and supported so that they remain level.

Banded, packed or palleted loads are off-loaded mechanically using either the lorry-mounted device, a fork-lift truck or a crane.

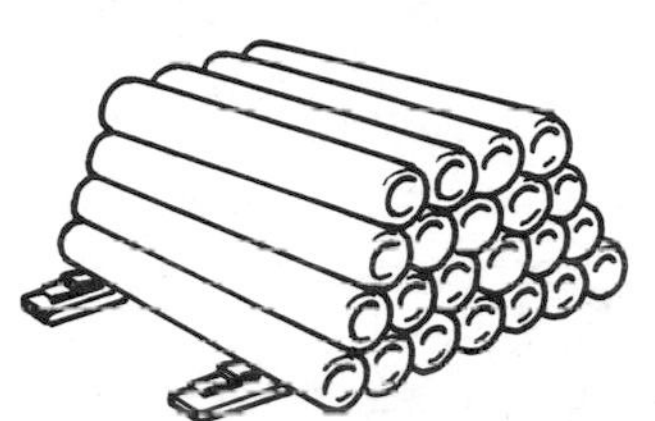

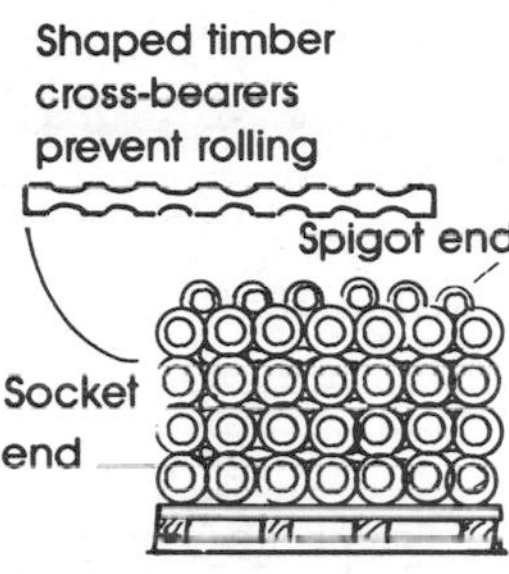

Aggregates

These are sands, gravel and crushed rock which are added to cement as a filler material to produce concrete and mortar.

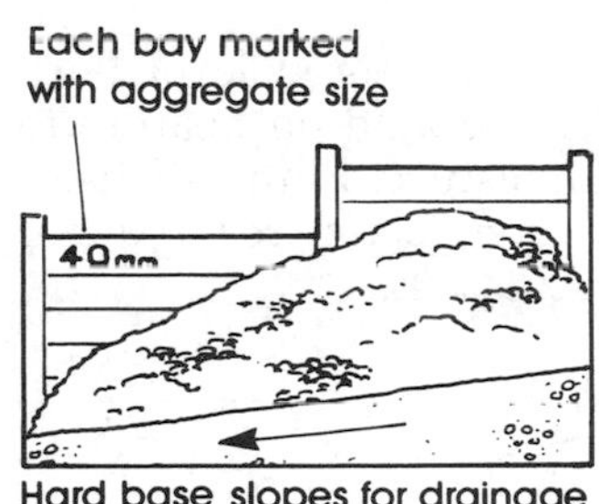

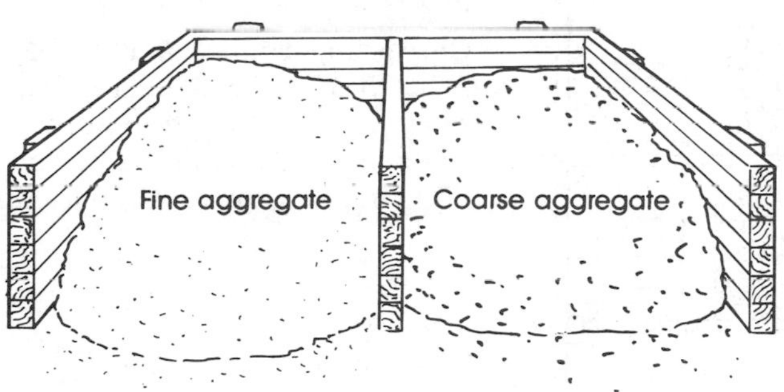

Normally supplied in bulk by tipper lorries; small amounts are available bagged. Each size should be stored separately adjacent to the mixer. Stockpiles should ideally be on a hard, concrete base, laid so that water will drain away, and separated into bays by division walls.

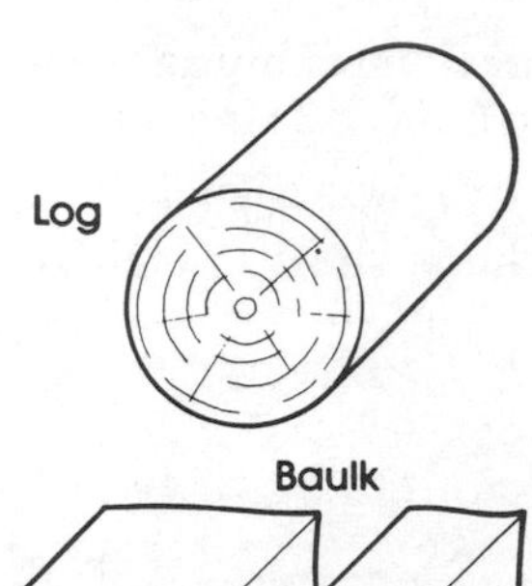

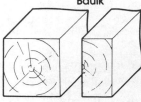

Stock piles should be sited away from trees to prevent leaf contamination and kept free from general site and canteen rubbish.

Tarpaulins or plastic covers can be used to protect stock piles from leaves, rubbish and rainwater.

In severe winter conditions the use of insulating blankets is to be recommended, to provide protection from frost and snow.

Timber

This is sawn or planed wood in its natural state – softwoods from coniferous trees and hardwoods from broadleaf trees. In general, softwoods are less decorative and tend to be used for structural work, painted joinery and trim. Hardwoods are more often used for decorative work, polished joinery and trim. Timbers readily absorb and lose moisture to achieve a balance with their surroundings. However this causes the timber to expand and shrink, which can cause it to distort, split and crack. In addition damp and wet timber is highly susceptible to fungal decay.

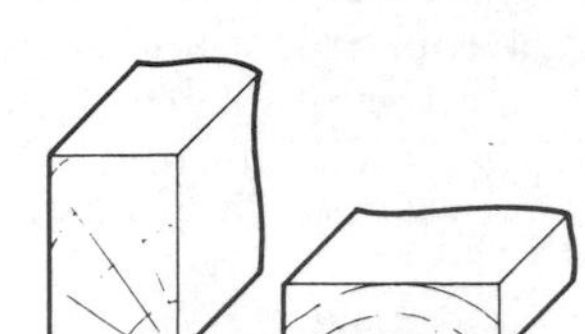

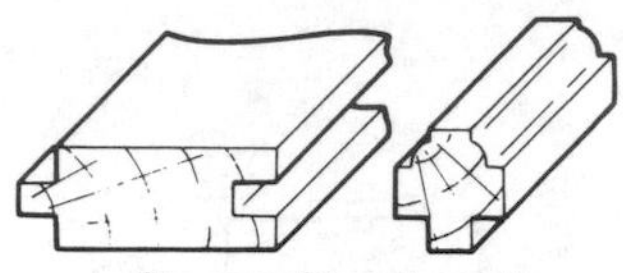

Timber may be either supplied loose in individual lengths, in banded packs, or in shrink-wrapped plastic packs.

Individual lengths should be off-loaded manually; long lengths and large sections may require a person at each end. Banded or wrapped loads are best off-loaded mechanically using either the lorry-mounted device, a fork-lift truck or a crane.

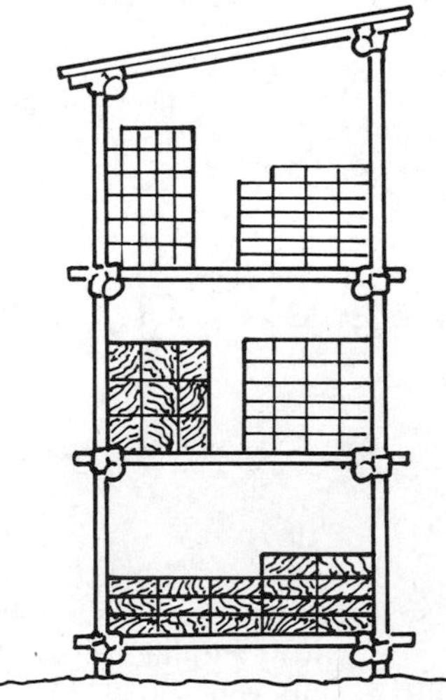

Timber supplied in shrink-wrapped plastic packs should be stored in them until required for use. Care must be taken not to damage the plastic.

Carcassing timber and external joinery should be stored level horizontally to prevent distortion, and clear of the ground on bearers to prevent distortion, and absorption of ground moisture. Piling sticks or cross-bearers should be placed between each layer, at centres of about 600 mm, to provide support and allow air circulation.

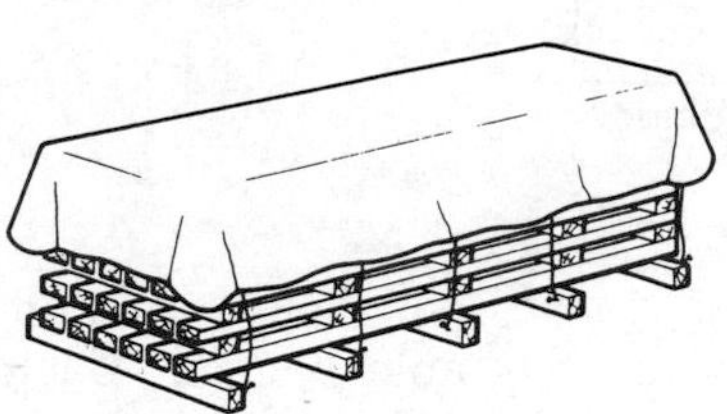

To protect timber against rain, frost, direct sunlight and atmospheric pollution, all stacks should be covered with a tarpaulin or polythene sheets, weighted or tied at the bottom. Care must be taken to allow free air circulation through the stack, to reduce the risk of fungal attack and condensation problems.

Internal trim and other planed sections may be stored horizontally in open-ended covered racks. Priming or sealing of trim and planed sections should be carried out on receipt, if it has not been done prior to delivery.

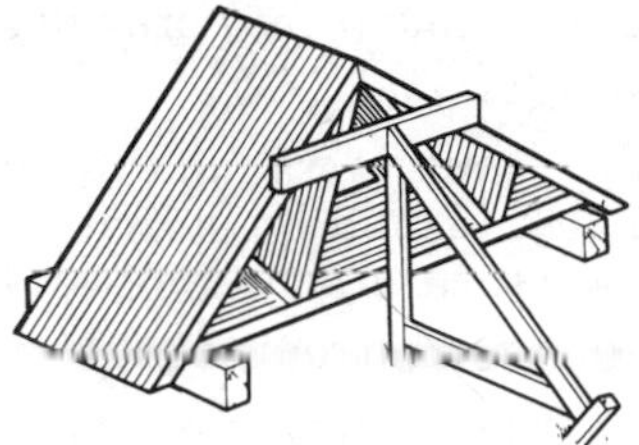

Trussed rafters are either supplied singly or in banded sets. They should be racked upright against a firm support, on eaves bearers. Alternatively, trussed rafters may be stored horizontally on close-spaced bearers, to give level support and prevent deformation. Stacks should be covered with a tarpaulin or polythene sheets, weighted or tied at the bottom. Care must be taken to allow free air circulation through the stack, to reduce the risk of fungal attack, condensation or connector corrosion problems.

Storage of hazardous products

Paint

This is a thin decorative and/or protective coating which is applied in a liquid or plastic form and later dries out or hardens to a solid film covering a surface. Paints consist of a film former, known as the vehicle; a thinner or solvent (water, white spirit or methylated spirit, etc.) to make the coating liquid enough; and a pigment suspended in the vehicle to provide covering power and colour. Paint schemes require either the application by brush, spray or roller, of one or more coats of the same material (varnish, emulsion and solvent paints) or a build-up of different successive coats, each having their own functions (primer, undercoat and finishing coat).

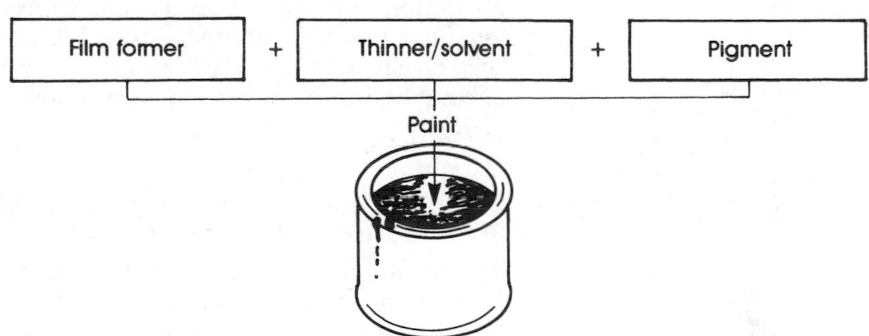

Varnish – a paint without a pigment, used for clear finishing.

Emulsion – a water-thinned paint for use on walls and ceilings.

Solvent paint – based on rubber, bitumen or coal tar and used for protecting metals and water-proofing concrete, etc.

Primer – may form a protective coat against moisture and corrosion, or act as a barrier between dissimilar materials. Also provides a good surface for bonding subsequent coats.

Undercoat – a paint used on primed surfaces to give it a uniform body and colour on which a finishing coat can be successfully applied.

Finishing coat – seals the surface, gives the final colour and provides the desired surface finish (flat, eggshell, gloss).

Paints are mainly supplied in 1, 2.5 and 5 litre containers, and more rarely in bulk or trade 25 litre containers. These should be stored on shelves, in a secure store, at an even temperature. Each shelf and container should be marked with its contents.

Large containers should be placed on lower shelves to avoid necessary lifting.

Highly flammable liquids

White spirit, cellulose, most special paint thinners and some formwork release agents give off a vapour that can ignite at room temperatures. The use of these liquids is controlled by the Highly Flammable Liquids and Liquefied Petroleum Gases Regulations. Up to 50 litres may be stored in a normal store, but because of the increased fire risk, larger quantities must be stored in special fire-resistant stores, normally placed at least 4 m away from buildings, materials, work places and boundary fences.

All storerooms used for hazardous products should be no smoking areas; signs stating 'NO SMOKING – FLAMMABLE' and 'HIGHLY FLAMMABLE SUBSTANCES' must be prominently displayed.

Suitable fire extinguishers must be available to deal with the potential hazard. See 'Health and Safety'.

Rags used to mop up a spillage must **never** be left in the store. Rolled up dirty rags start to generate heat and can eventually burst into flames (spontaneous combustion). This creates an ignition source for the entire store contents.

Hazardous products may be used directly from their container or a smaller amount decanted into a 'kettle' for convenience. The surplus should be returned to the main container after use. **Never** store any substance in an unlabelled container.

Decanting – Consult the manufacturer's instructions on the container prior to decanting as certain substances **must not** be used or stored in another container for reasons of safety (it parts them from their use/safety instructions and renders the contents unknown).

To decant from container to kettle: dust the top of container; remove the lid with an opener; thoroughly stir the contents with a mixing knife to achieve an even consistency; pour the required quantity into the kettle, from the side opposite the manufacturer's instructions; use a brush to mop up the surplus on the rim or side of the container; scrape the brush on the edge of the kettle to remove surplus; firmly replace the container lid to prevent vapours escaping and dust and dirt getting in.

Decanting of hazardous products should be carried out in well-ventilated conditions.

Gases

Liquefied petroleum gas (LPG) sold commercially as propane and butane is supplied in pressurized metal cylinders ranging from 4 kg to 50 kg in size.

Cylinders should be stored upright in a well-ventilated fire-resisting storeroom, or in a secure compound away from any heat source. A minimum of two exits are normally required for an LPG store.

A sign stating 'NO SMOKING, HIGHLY FLAMMABLE SUBSTANCES' should be displayed. Empty and full cylinders should be stored separately and empty ones clearly marked as such.

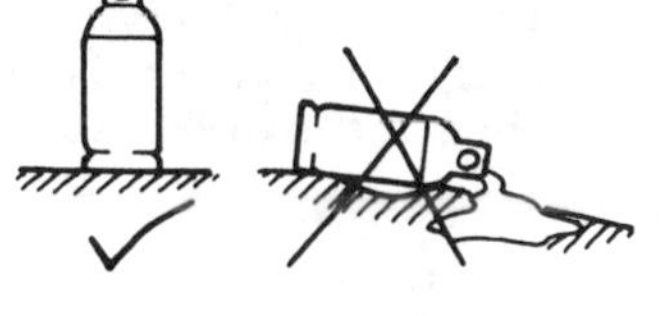

Never use or store cylinders on their sides, nor near to or in excavations, in confined spaces or in other areas with restricted ventilation.

All valves should be fully closed. LPG is heavier than air and it will collect at low levels where an ignition source (discarded match, sparking power tool, dirty rags) could ignite it.

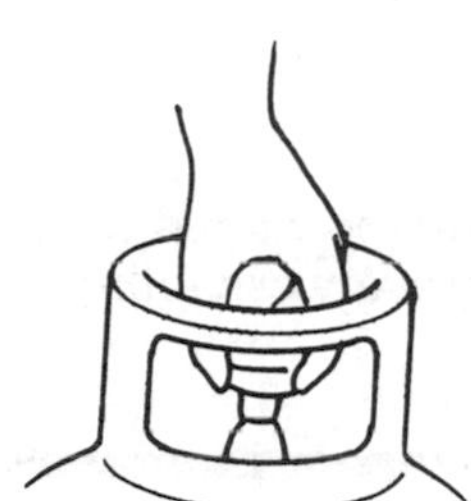

Storage of fragile or perishable materials

Bagged materials

Cement – manufactured from chalk or limestone and clay which are ground into a powder, mixed together and fired in a kiln causing a chemical reaction. On leaving the kiln, the resulting material is ground to a fine powder. Hence a popular site term for cement is 'dust'. When water is added to the cement, another reaction takes place causing it to gradually stiffen, harden and develop strength.

Ordinary Portland cement (OPC) – when hardened its appearance resembles Portland stone.

Rapid hardening Portland cement (RHPC) – for cold weather use.

Sulphate-resisting Portland cement (SRPC) – for use underground in high sulphate conditions.

White or coloured Portland cement – made using white china clay; pigments are added for coloured cements.

High alumina cement (HAC) – uses bauxite (aluminium oxide) instead of clay. It develops very early strength which is much higher than OPC, although in the long term it has been found unstable and thus is now rarely favoured for structural work.

Cement is used in all forms of *in situ* and pre-cast concrete products, cement mortar, cement screeds and rendering.

Plaster – applied on internal walls and ceilings to provide a jointless, smooth, easily decorated surface. External plastering is normally called rendering. Plaster is a mixture that hardens after application; it is based on a binder (gypsum, cement or lime) and water with or without the addition of aggregates. Depending on the background (surface being plastered) plastering schemes may require the application of either one coat, or undercoats to build up a level surface followed by a finishing coat.

Gypsum plaster – for internal use, different grades of gypsum plaster are used according to the surface and coat. For undercoats, browning is generally used for brick and blockwork or bonding for concrete; for finishing coats, finish is used on an undercoat or board finish for plasterboard.

Cement-sand plaster – used for external rendering, internal undercoats and water-resisting finishing coats.

Lime-sand plaster – used for both undercoats and (rarely) finishing coats, although lime can be added to other plasters to improve their workability.

Lime – ground, powdered white limestone added to plaster and mortar mixes to improve workability.

Bagged material storage – Bagged materials may be supplied lose in individual bags or in unit loads shrink-wrapped in plastic on timber pallets.

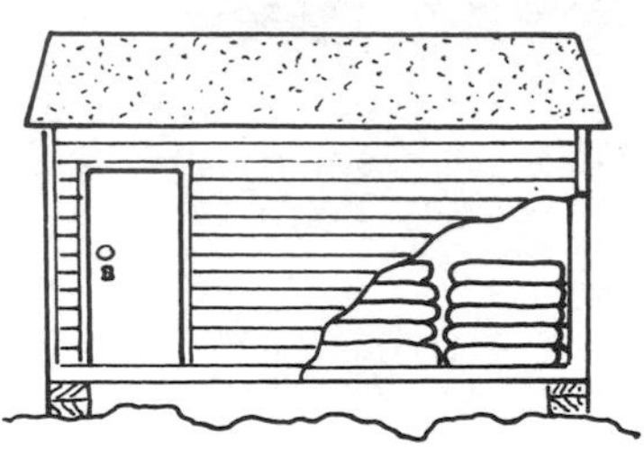

Individual bags should be off-loaded manually, with the bag over the shoulder as the preferred method. Palleted loads are best off-loaded mechanically using either the lorry-mounted device, a fork-lift truck or a crane.

Bags supplied in shrink-wrapped loads are best stored in these until required for use. Care must be taken not to damage the plastic.

Bags should be stored in ventilated, waterproof sheds, on a sound dry floor, with different products having their own shed to avoid confusion.

Bags should be stored clear of the walls and no more than eight to ten bags high. This is to prevent bags becoming damp through a defect in the outside wall, causing the contents to set in the bag. It also reduces the risk of compaction ('warehouse setting') of the lower bags due to the excessive weight of the bags above.

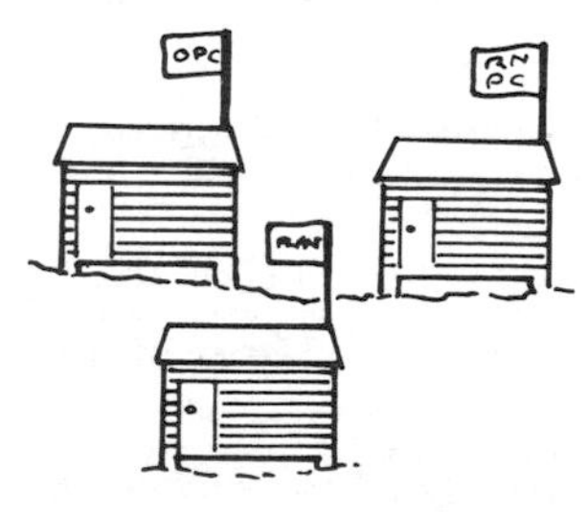

Bags should be used in the same order as they were delivered, known as 'first in, first out' (FIFO). This is to minimise the storage time and prevent the bag contents becoming stale or 'air setting'.

Where small numbers of bags are stored and a shed is not available, they may have to be stored in the open. Stack no more than six to eight bags high on timber pallets and cover with tarpaulins or polythene sheets weighted or tied at ground level.

Sheet materials

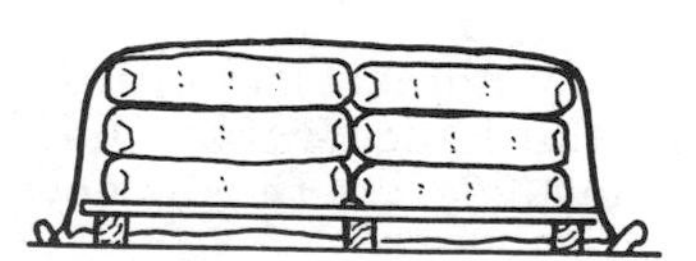

Supplied either as individual sheets, taped face to face in pairs (plasterboard), in banded bundles or unit loads on timber pallets; they may also be supplied in shrink-wrapped plastic packs.

Plywoods – usually consist of an odd number of thin layers glued together with their grains alternating, for strength and stability. Used for flooring, formwork, panelling, sheathing and cabinet construction.

Laminated boards – consist of strips of timber which are glued together, sandwiched between two plywood veneers. They are used for panelling, doors and cabinet construction.

Particle boards – either chipboard (small chips and flakes) or waferboard (large flakes or wafers); both are manufactured using wood chips and/or flakes impregnated with an adhesive. They are used for flooring, furniture and cabinet construction.

Fibres boards – made from pulped wood, mixed with an adhesive and pressed forming hardboard, medium board, medium density fibre board (MDF) and insulation board. They are used for floor, wall, ceiling and formwork linings, insulation, display boards, furniture and cabinet construction.

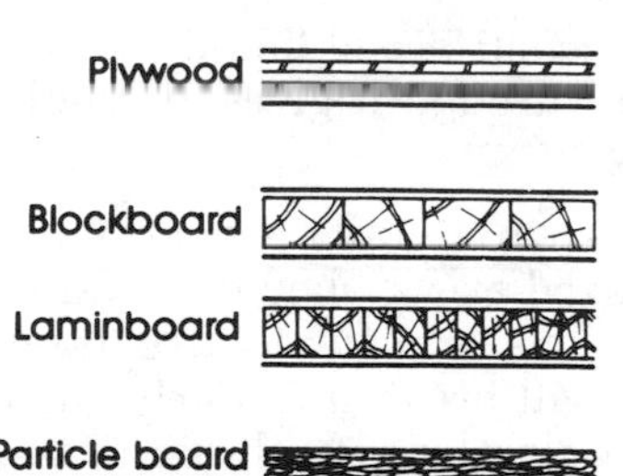

Woodwool slabs – made from wood shavings coated with a cement slurry; used for roof decks and as a permanent formwork lining.

Plastic laminate – made from layers of paper impregnated with an adhesive; used for worktops and other horizontal and vertical surfaces requiring decorative hygienic and hard-wearing properties.

Plasterboard – comprises a gypsum plaster core sandwiched between sheets of heavy paper; used for wall and ceiling linings.

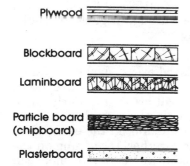

Individual sheets should be off-loaded manually. To avoid damage they should be carried on edge, which may require a person at each end. Banded bundles or palleted loads are best off-loaded mechanically using either the lorry-mounted device, a fork-lift truck or a crane.

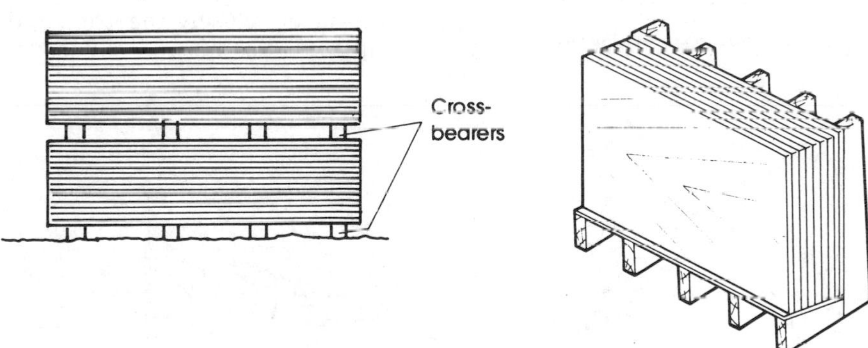

Sheets supplied in shrink-wrapped plastic packs should be stored in them until required for use. Care must be taken not to damage the plastic.

All sheet materials should preferably be stored in a warm dry place; ideally stacked flat on timber cross-bearers, spaced close enough together to prevent sagging. Alternatively, where space is limited, sheet material can be stored on edge in a purpose-made rack, which allows the sheets to rest against the back board in a true plane.

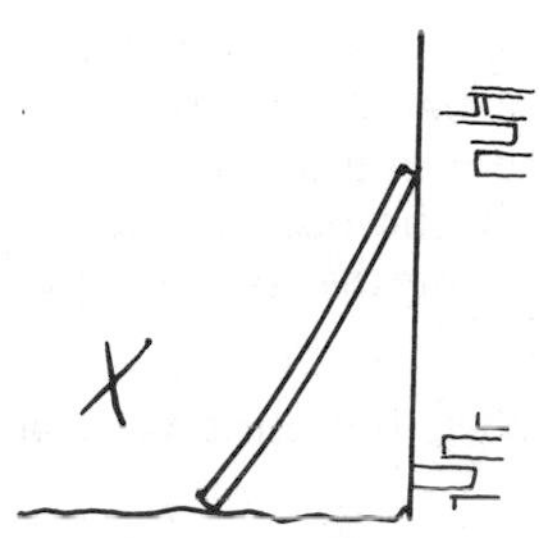

Leaning sheets against walls on edge or end is not recommended as they will take on a bow which is difficult to reverse.

Veneered or other finished surfaced sheets should be stored good face to good face, to minimise the risk of surface scratching.

Glass – a mixture of sand, soda, ash, limestone and dolomite that is heated in a furnace to produce molten glass. On cooling the molten mixture becomes hard and clear. Glass is supplied individually in single sheets or in banded timber packs. Sealed units and cut sizes may be supplied in shrink-wrapped plastic packs for specific purposes.

Drawn glass – molten glass is drawn up between rollers in a continuous flow, cooled in water towers and cut into sheets. Patterned rollers may be introduced to create rough cast and patterned glass. Wire can be incorporated in the glass during the drawing to form wired glass, used for fire-resistant purposes. The surfaces of drawn glass are not perfectly flat, so when you look through your view is distorted. Thus large sheets of glass for shop fronts, etc. have to be ground and polished perfectly flat to give undistorted vision. This type of glass is known as *polished plate*.

Float glass – molten glass is floated on to the surface of liquid tin, and subsequently allowed to cool. When looked through it gives an undistorted view without the need for polishing.

Safety glazing – 'at risk' areas of glazing such as fully glazed doors, patio doors, side panels and other large glazed areas, should contain a safety glazing material. Toughened safety glass is up to five times stronger than standard glass. If broken it will break into fairly small pieces with dulled edges. Laminated safety glass is a sandwich of two or more sheets of glass interlaid with a plastic film. In the event of an impact the plastic holds the sheets of glass together. Depending on the number of layers, impacts from hammer blows and even gun shots can be resisted.

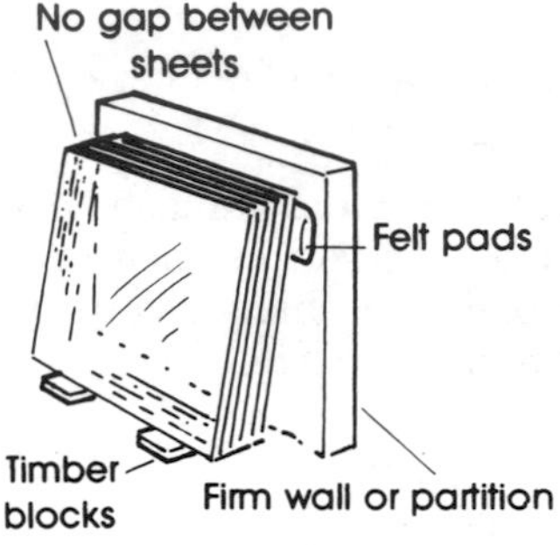

Storing glass – Glass should be stored in dry, wind-free conditions. Never store glass flat as it will distort and break. Sheets should be stood on one long edge almost upright at an angle of about 85 degrees. Timber and/or felt blocks should be used to prevent the glass coming into contact with rough surfaces which can result in scratches or chips. Always stack sheets closely together and never leave spaces, as again this can lead to distortion and subsequent breakage.

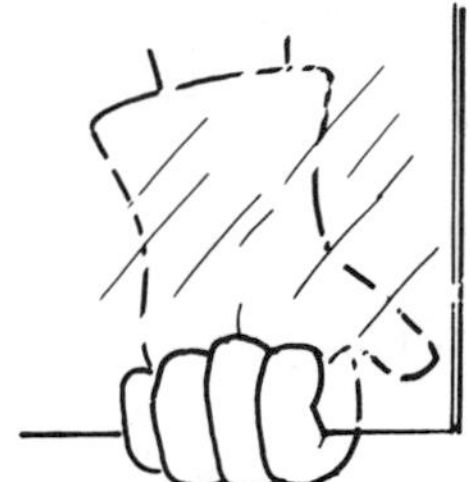

Handling glass – When handling glass use laps to protect your palms. In addition gauntlets may be worn to protect your lower arms and wrists.

Glass should be held firmly, but not too tightly as it may break. To balance the glass correctly: small panes should be carried under one arm and steadied on the front edge by the other hand; larger panes are held towards your body with one hand under the bottom edge and the other steadying the front; large sheets will require two people to handle them, walking in step, one on each side.

To prevent large panes and sheets 'whipping' when handling it is safer to carry two or more at a time.

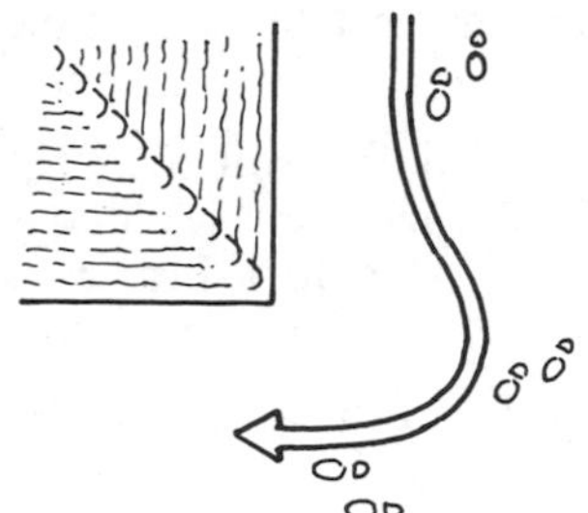

Ensure your route is clear of obstructions, keep about a metre out from the wall or building and take a wide path at corners. Do not stop or step back suddenly, since serious injury can be caused by a collision with another person.

If the pane of glass you are carrying breaks or slips, step clear and let it fall freely. Never attempt to catch it.

Vitreous chinaware

Sanitary appliances, such as WC (water closet) pans, WWP (waste water preventer) cisterns, wash basins and shower trays. They are supplied in various ways: as individual items, shrink-wrapped in plastic or with corners taped for protection; in unit loads shrink-wrapped on timber pallets, or increasingly, in bathroom sets including bath, shrink-wrapped on a timber pallet.

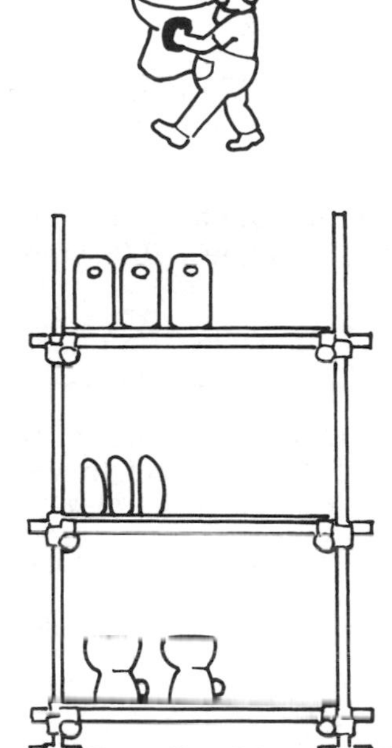

Vitreous chinaware – a ceramic material consisting of a mixture of sand and clay which has been shaped, dried and kiln fired, to produce a smooth, hard, glassy surface material. Different colours are achieved by coating items with a prepared glaze solution before firing.

Individual items should be off-loaded manually, with the item firmly gripped with both arms. Palleted loads are best off-loaded mechanically using either the lorry-mounted device, a fork-lift truck or a crane.

Items are supplied in shrink-wrapped plastic loads and are best stored in them until required for use.

Individual items should be nested together on timber bearers, or alternatively stored in a racking system.

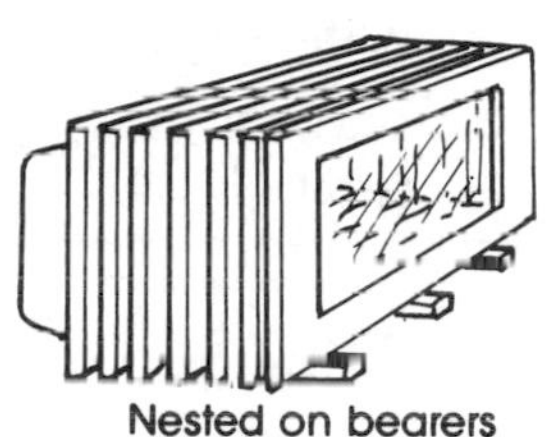
Nested on bearers

Storage of miscellaneous materials

Rolled materials

Bitumen – either occurs naturally or distilled from petroleum, used for roofing felt and damp-proof courses (DPCs).

Metal – mainly non-ferrous (not containing iron) such as copper, lead and zinc. Used for roof coverings, DPCs and flashing.

Plastic – polythene, a thermoplastic (is softened by heat or solvent), used for DPCs and damp-proof membranes (DPMs).

Supplied as individual rolls, in banded bundles or unit loads shrink-wrapped on timber pallets.

Individual rolls should be off-loaded manually, with the roll over the shoulder as the preferred method. Palleted loads are best off-loaded mechanically using either the lorry-mounted device, a fork-lift truck or a crane.

Rolls are supplied in shrink-wrapped plastic loads and are best stored in them until required for use.

All rolls should be stacked vertically on end, on a level, dry surface. Alternatively they may be stored in a racking system. However, again they should be vertical to prevent them rolling off and to reduce the risk of compression damage (e.g. the layers of bitumen rolls adhere together under pressure) due to excessive loads (this occurs if rolls are stacked horizontally on top of each other).

Ironmongery

Carpenters' locks, bolts, handles, screws and nails, etc. are 'desirable' items which are most likely to 'walk' unless stored securely under the control of a storeperson.

They are supplied either as individual items, plastic bubble-packed on cards or in boxed sets, by amount, e.g. 10 pairs of hinges, 200 screws or 25 kg of nails, etc.

Large ironmongery deliveries may be supplied in mixed unit loads shrink-wrapped on timber pallets.

Individual items should be off-loaded manually and transferred immediately to the store. Palleted loads are best off-loaded mechanically using either the lorry-mounted device, a fork-lift truck or a crane.

Large or heavy items should be stored on lower shelves to avoid unnecessary lifting.

Joinery

Doors, frames and units are supplied as individual items, in banded bundles, boxed, flat-packed or as ready assembled units and in unit loads on timber pallets. They may also be supplied in shrink-wrapped plastic packs for protection.

Individual joinery items should be off-loaded manually. To avoid damage and stress they should be carried on edge or in their plane of use. This may require a person at each end. Banded bundles or palleted loads are best off-loaded mechanically using either the lorry-mounted device, a fork-lift truck or a crane.

Items supplied in shrink-wrapped packs should be stored in them until required for use. Care must be taken not to damage the plastic.

All joinery items should preferably be stored in a warm dry place; ideally stacked flat to prevent twisting and on timber cross-bearers, spaced close enough together to prevent sagging. Also see 'Timber' on page 236.

Leaning items against walls on edge or end is not recommended, as they will take on a bow, rendering them unusable.

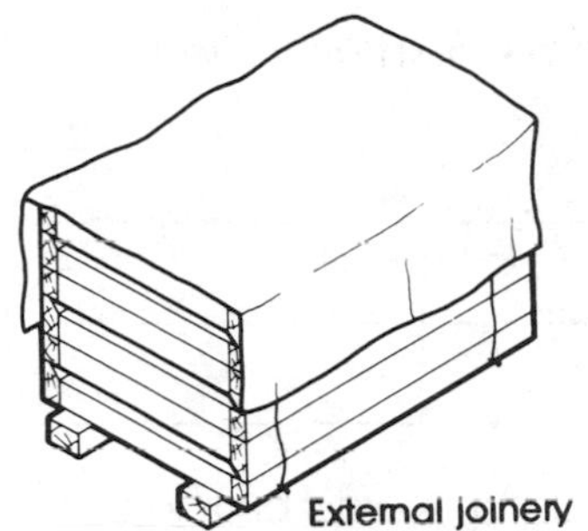

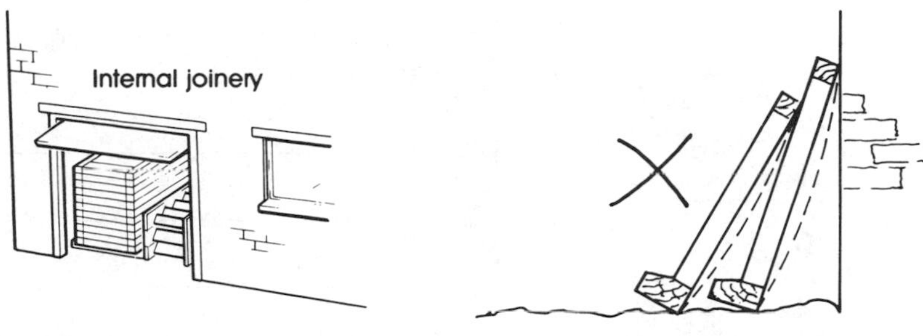

Questions for you

3. Name a bulk building material and state the reason why it should be stored clear of the ground.

4. State the purpose of covering stored building materials.

5. Explain why piling sticks or cross-bearers are used when stacking timber.

6. State why it is not good practice to store liquids in unmarked containers.

7. Explain the term 'first in, first out' when applied to the use of bagged materials.

TRY AND ANSWER THESE

8. State the reasons why rolled materials are normally stored vertically on end.

______

______

______

9. Describe how cut panes of glass should be stored prior to use.

______

______

______

10. Explain why flat storage is recommended for doors, frames and sheet materials.

______

______

______

11. Name an item of safety clothing used when handling glass.

______

12. Identify the materials from the following descriptions:

(a) a walling unit component having a standard format size including a 10 mm mortar allowance of 225 mm × 112.5 mm × 75 mm.

(b) a paint used to form a protective coat against moisture and corrosion, or act as a barrier between dissimilar materials.

(c) A sheet material that is formed by being floated on to the surface of liquid tin.

(a) ______

(b) ______

(c) ______

WELL, HOW DID YOU DO?
WORK THROUGH THE SECTION AGAIN IF YOU HAD ANY PROBLEMS

COMPLETE THE WORD SQUARE

WORD-SQUARE SEARCH

Hidden in the word-square are the following 20 words associated with '*Materials*'. You may find the words written forwards, backwards, up, down or diagonally.

Handling	Aggregate
Security	Bearers
Bricks	Emulsion
Timber	Paint
Protection	Flammable
Tarpaulin	Hazardous
Mortar	Plaster
Banded	Vitreous
Tile	Ironmongery
Lintel	Plasterboard

Draw a ring around the words, or line in using a highlight pen thus:

EXAMPLE

EXAMPLE

H	A	N	D	L	I	N	G	S	V	Y	T	I	R	U	C	E	S
R	U	S	P	I	I	T	B	E	I	P	I	R	R	E	W	O	T
O	R	T	S	N	L	M	R	D	T	L	M	O	R	T	A	R	A
P	I	C	L	T	D	D	I	R	R	A	B	M	B	M	E	O	N
R	A	D	D	E	Y	A	C	E	E	S	E	O	O	U	H	R	D
I	U	H	L	L	G	O	K	G	O	T	R	S	L	N	E	T	A
E	A	N	O	E	P	A	S	U	U	E	R	N	T	N	I	A	P
H	A	Z	A	R	D	O	U	S	S	R	D	A	H	C	M	N	D
A	G	G	R	E	G	A	T	E	I	B	O	R	T	A	E	E	F
R	A	C	O	C	C	D	R	M	R	O	S	T	L	N	T	D	L
P	L	A	S	T	E	R	G	U	A	A	H	O	C	I	S	N	A
U	D	N	E	D	E	G	T	L	S	R	A	E	A	L	T	E	M
R	R	S	N	V	N	D	A	S	C	D	D	B	N	U	P	P	M
O	A	A	V	I	U	S	T	I	M	E	E	O	I	A	C	E	A
L	B	R	C	O	U	P	L	O	R	S	L	A	T	P	O	D	B
K	Y	R	E	G	N	O	M	N	O	R	I	R	O	R	C	N	L
F	R	C	T	A	S	B	M	L	A	D	T	D	E	A	S	I	E
B	E	A	R	E	R	S	D	N	O	I	T	C	E	T	O	R	P

WELL, HOW MANY DID YOU GET?

REMEMBER.
THINK
SAFETY
THINK
SAFETY
DON'T BECOME
A STATISTIC!
ACCIDENTS
ARE CAUSED!

WELL THAT'S
IT. SEE YOU IN
THE NEXT
PACKAGE